Semi-Lagrangian Approximation Schemes for Linear and Hamilton–Jacobi Equations

Semi-Lagrangian Approximation Schemes for Linear and Hamilton–Jacobi Equations

Maurizio Falcone
Sapienza – Università di Roma
Rome, Italy

Roberto Ferretti
Università di Roma Tre
Rome, Italy

siam.
Society for Industrial and Applied Mathematics
Philadelphia

10 9 8 7 6 5 4 3 2 1

Library of Congress Cataloging-in-Publication Data

Falcone, Maurizio, author.
Semi-Lagrangian approximation schemes for linear and Hamilton–Jacobi equations / Maurizio Falcone, Sapienza – Università di Roma, Rome, Italy, Roberto Ferretti, Università di Roma Tre, Rome, Italy.
pages cm. -- (Applied mathematics)
Includes bibliographical references and index.
ISBN 978-1-611973-04-4
1. Hamilton–Jacobi equations. 2. Approximation theory. 3. Differential equations, Partial–Numerical solutions. I. Ferretti, Roberto, author. II. Title.
QA377.F35 2013
515'.353--dc23

2013033334

To our daughters

Cecilia and Flavia,

who will never read this book.

Contents

Preface

This book is for people interested in the numerical solution of PDEs, with an emphasis on high-order schemes for hyperbolic problems. In fact, evolutive linear and nonlinear first-order PDEs appear in a variety of applications ranging from geophysical and atmospheric problems to control theory, Fluid Dynamics, and image processing. Although every application has its own features, a usual requirement for the numerical solution is to keep accuracy and stability without overly restrictive assumptions on the time step, which usually imply a high computational complexity.

We will describe and analyze basic and recent developments of the Semi-Lagrangian technique for the approximation of first-order PDEs with a special focus on Hamilton–Jacobi equations. The main purpose is to obtain methods which are unconditionally stable with respect to the choice of the time step. The basic idea dates back to the method proposed by Courant, Isaacson, and Rees in 1952 [CIR52] for first-order systems of linear equations and has gone through a number of improvements and extensions since. In particular, the main developments have been proposed within the Numerical Weather Prediction community [SC91], where this class of schemes has become of common use.

As for the advection equation, the driving force of this method is the method of characteristics which accounts for the flow of information in the model equation. At the numerical level, the Semi-Lagrangian approximation mimics the method of characteristics looking for the foot of the characteristic curve passing through every node and following this curve for a single time step.

In order to derive a numerical method from this general idea, several ingredients should be put together, mainly a technique for ODEs to track characteristics and a reconstruction technique to recover pointwise values of the numerical solution. In this framework, the order of space and time discretizations is typically reversed with respect to more conventional schemes—the first discretization is done in time, the second in space—although the final result is clearly a fully discrete scheme. This reversal is the key to getting weaker stability requirements.

This book has two main purposes. The first is to present Semi-Lagrangian approximation schemes, reviewing their construction and theory on model equations which range from the simple advection equation to Hamilton–Jacobi equations. In the basic situation of first-order schemes, we will also make a comparison with more standard approximation techniques, namely finite difference schemes of upwind and centered types. The reader will find information on the construction of the schemes together with a detailed theoretical analysis, at least for the more popular choices of their elementary building blocks. We have included a number of practical algorithmic recipes, and we believe that after reading this book it should be possible to program and apply the methods described here from scratch.

The second goal of the book is to present some applications and show the Semi-Lagrangian approximations at work on a variety of nonlinear problems of applicative interest. We also aim at showing that this technique is well suited for the approximation of problems with nonsmooth solutions, like weak solutions in the viscosity sense in the case of Hamilton–Jacobi equations.

The book is intended to be accessible to a reader with a basic knowledge of linear PDEs and standard numerical methods. The analysis of Hamilton–Jacobi equations will, of course, require specific analytical tools, which have been included in the section devoted to viscosity solutions. The presentation is conceived to be reasonably self-contained, but at the same time it will collect in a unified framework results which are spread over the literature. Clearly, more information on the theory of viscosity solutions can be obtained looking at the specific references, and in particular at the books [Ba98, BCD97, FS93]. Some standard finite difference schemes for Hamilton–Jacobi equations can be found in [Se99, OF03], whereas other schemes based on control techniques are presented in [KuD01].

Although Hamilton–Jacobi equations and their applications are the object of an extensive literature, we will mostly limit our presentation to the Semi-Lagrangian approach and to its relationship with Dynamic Programming. A detailed overview of numerical methods for Hamilton–Jacobi equations goes beyond the scopes of this book, although we hope to develop this topic in greater detail in the future.

Acknowledgments We wish to thank all colleagues and students who contributed to the development of the results reviewed in this book or took part in enlightening discussions on algorithmic and theoretical issues. We are also indebted to L. Bonaventura, E. Carlini, J.-M. Qiu, M. Restelli, G. Rosatti, C.-W. Shu, and G. Tumolo for providing some of the numerical tests presented here. Special thanks go to our Acquisitions Editor, Elizabeth Greenspan, for her endless patience during the writing of the manuscript, and to our families for their constant support.

Maurizio Falcone and Roberto Ferretti

Rome, May 20th, 2013

Notation

a.e.	almost everywhere (with respect to the Lebesgue measure)
$\mathbb{R}^d$	euclidean d-dimensional space
$x \cdot y$	scalar product $\sum_{i=1}^d x_i y_i$ of two vectors $x, y \in \mathbb{R}^d$
$\lvert x\rvert$	euclidean norm of $x \in \mathbb{R}^d$, $\lvert x\rvert = (x \cdot x)^{1/2}$
$B(x_0, r)$	open ball of center x_0 and radius r in $\mathbb{R}^d$, $\{x : \lvert x - x_0\rvert < r\}$
$\overline{B}(x_0, r)$	closed ball of center x_0 and radius r in $\mathbb{R}^d$, $\{x : \lvert x - x_0\rvert \le r\}$
∂E	boundary of the set E
int E	interior of the set E
$\overline{E}$	closure of the set E
co E	convex hull of the set E
$a \vee b$	$\min\{a, b\}$ for $a, b \in \mathbb{R}$
$a \wedge b$	$\max\{a, b\}$ for $a, b \in \mathbb{R}$
$Du(x)$	gradient of a function $u : \mathbb{R}^d \to \mathbb{R}$ at the point x
$d(x, C)$	distance from the point x to the closed set C
$\Delta t, \Delta x$	discretization steps, in compact form $\Delta = (\Delta x, \Delta t)$
sgn(x)	sign of x: $\operatorname{sgn}(x) := \begin{cases} +1 & \text{if } x > 0, \\ 0 & \text{if } x = 0, \\ -1 & \text{if } x < 0. \end{cases}$
i	imaginary unit
$\mathbf{1}_\Omega(x)$	characteristic function of the set Ω: $\mathbf{1}_\Omega(x) := \begin{cases} 1 & \text{if } x \in \Omega, \\ 0 & \text{if } x \notin \Omega. \end{cases}$
δ_{ij}	Kronecker's symbol: $\delta_{ij} := \begin{cases} 1 & \text{if } i = j, \\ 0 & \text{if } i \neq j. \end{cases}$
$\lfloor x \rfloor$	lower integer part of x: $\lfloor x \rfloor := \max\{m \in \mathbb{Z} : m \le x\}$
$I_k[G](x)$	value at the point x of the interpolate of degree k of the function g based on the vector $G = (g(x_i))$ of grid values of g
$\nu(x)$	exterior normal to a surface at the point x
$y(x, t; s)$	position at time s of the trajectory starting at x at time t
$y_{x,t}(s; \alpha)$	position at time s of the trajectory starting at x at time t and driven by the control α
$y_x(s; \alpha)$	position at time s of the trajectory starting at x at time $t = 0$ and driven by the control α
$TV(w)$	total variation of the univariate function w: $TV(w) = \int_D \lvert w'(x)\rvert\, dx$
supp w	support of the function w
$\omega(\delta)$	a modulus of continuity
$BUC(\Omega)$	space of bounded uniformly continuous functions on the open set Ω
$C^k(\Omega)$	space of functions $u : \Omega \to \mathbb{R}$ with continuous kth derivative
$C_0^k(\Omega)$	space of functions with compact support belonging to $C^k(\Omega)$
$L^p(\Omega)$	space of functions $u : \Omega \to \mathbb{R}$ such that $\int_\Omega \lvert u\rvert^p(x)\, dx < +\infty$, $1 \le p < +\infty$
$L^\infty(\Omega)$	space of functions $u : \Omega \to \mathbb{R}$ such that ess sup$\lvert u(x)\rvert < +\infty$

$L^p_{loc}(\Omega)$	space of functions $u \in L^p(K)$ for any compact set $K \subset \Omega$, $1 \le p \le +\infty$
$\|\|u\|\|_p$	norm of u in $L^p(\Omega)$, $1 \le p \le \infty$
$W^{m,p}(\Omega)$	Sobolev space of functions $u : \Omega \to \mathbb{R}$ with the first m derivatives in $L^p(\Omega)$
$W^{m,p}_{loc}(\Omega)$	Sobolev space of functions u with the first m derivatives in $L^p(K)$ for any compact set $K \subset \Omega$
$C([0,T];X)$	space of functions $u : [0,T] \times \mathbb{R} \to \mathbb{R}$ such that $\|\|u(t,\cdot)\|\|_X \in C([0,T])$
$L^\infty([0,T];X)$	space of functions $u : [0,T] \times \mathbb{R} \to \mathbb{R}$ such that $\|\|u(t,\cdot)\|\|_X \in L^\infty([0,T])$
l^p	space of pth power summable vectors or sequences
l^∞	space of bounded vectors or sequences
V, v_j	numerical solution as a vector and as a value at the node x_j
V^n, v^n_j	numerical solution as a vector and as a value at the node (x_j, t_n)
$V^{(k)}$, $v^{(k)}_j$	numerical solution at the iteration k as a vector and as a value at the node x_j
$\|\|V\|\|_p$	norm in l^p, renormalized if referred to a numerical solution: $\|\|V\|\|_\alpha := \begin{cases} \left(\Delta x_1 \cdots \Delta x_d \sum_j \|v_j\|^\alpha\right)^{1/\alpha} & \text{if } \alpha < \infty, \\ \max_j \|v_j\| & \text{if } \alpha = \infty. \end{cases}$
B^t	transpose of a matrix B
$\mathrm{tr}(B)$	trace of a matrix B
$\rho(B)$	spectral radius of a square matrix $B \in \mathbb{R}^{m \times m}$
$\|\|B\|\|_p$	for a (possibly infinite) matrix B, natural operator norm in l^p: $\|\|B\|\|_p := \sup_{V \neq 0} \frac{\|\|BV\|\|_p}{\|\|V\|\|_p}$
$\|\|B\|\|_\infty$	∞-norm of a matrix B: $\|\|B\|\|_\infty = \max_i \sum_j \|b_{ij}\|$
$\|\|B\|\|_1$	1-norm of a matrix B: $\|\|B\|\|_1 = \max_j \sum_i \|b_{ij}\|$
$\|\|B\|\|_2$	2-norm of a matrix B: $\|\|B\|\|_2 = \sqrt{\rho(B^t B)}$
$\underline{w}$, $\overline{w}$	lower and upper semicontinuous envelopes: $\begin{cases} \underline{w}(x) := \liminf_{y \to x} w(y), \\ \overline{w}(x) := \limsup_{y \to x} w(y). \end{cases}$

Chapter 1
Models and motivations

This book is devoted to the study of some class of hyperbolic PDEs and to their approximation of Semi-Lagrangian type. Throughout the book, we will try to keep a mathematically rigorous analysis as much as possible; however, before getting into the most technical details and results, we would like to start by presenting some examples and applications which have motivated much of the research in this area. In fact, the analysis and approximation of first-order PDEs of Hamilton–Jacobi (HJ) type has been driven by their application to fields such as Fluid Dynamics, optimal control and differential games, image processing, and material science just to mention the main ones.

In all the aforementioned fields, the notion of weak solutions has had a crucial impact in giving a sound theoretical framework for both the analytical and the numerical studies. The goal of this chapter is precisely to introduce the reader to this theory via some model problems. We will start from the linear advection equation and then turn to nonlinear models, either the evolutive ones arising in front propagation and optimal control or the stationary ones including, e.g., the eikonal equation of the classical Shape-from-Shading problem in image processing. We will also discuss some difficulties related to the numerical approximation of such models and conclude the chapter by giving some hints on how to use the material presented in the book.

1.1 ▪ Linear advection equation

As a first model, we start by considering the advection equation

$$u_t + f(x,t)\cdot Du = g(x,t), \qquad (x,t)\in\mathbb{R}^d \times (t_0,T). \tag{1.1}$$

Here, $f:\mathbb{R}^d \times (t_0,T)\to\mathbb{R}^d$ is a vector field termed as *drift* and $g:\mathbb{R}^d\times(t_0,T)\to\mathbb{R}$ is the *source* term. We look for a solution $u:\mathbb{R}^d\times(t_0,T)\to\mathbb{R}$ of (1.1) satisfying the initial condition

$$u(x,t_0)=u_0(x), \qquad x\in\mathbb{R}^d. \tag{1.2}$$

This is a classical model describing the transport of a scalar field, e.g., the distribution of a pollutant emitted by one of more sources (represented by g) and transported by a water stream or a wind (represented by f).

This physical interpretation is better clarified through the *method of characteristics* which provides an alternative characterization of the solution. In the simplest case, for

$g(x,t) \equiv 0$ and $f(x,t) \equiv c$ (constant) the solution v is given by the representation formula

$$u(x,t) = u_0(x - c(t - t_0)), \qquad (x,t) \in \mathbb{R}^d \times [t_0, T). \tag{1.3}$$

In fact, assume that a regular solution u exists. Differentiating u with respect to t along a curve of the form $(y(t), t)$ we obtain

$$\frac{du}{dt}(y(t),t) = Du(y(t),t) \cdot \dot{y}(t) + u_t(y(t),t), \tag{1.4}$$

so that (using the fact that u solves the advection equation) the total derivative will identically vanish along the curves which satisfy

$$\dot{y}(t) = c. \tag{1.5}$$

Such curves have the physical meaning of flow lines along which the scalar field u is transported. They are known as *characteristics* and, in this particular case, are straight lines. Then, to assign a value to the solution at a point (x,t) it suffices to follow the unique line passing through (x,t) until it crosses the x-axis at the point $z = x - c(t - t_0)$, z being the *foot* of the characteristic. Since the solution u is constant along this line, we get (1.3). Conversely, it is easy to check that a function of x and t in the form (1.3) satisfies (1.1) for $f \equiv c$ and $g \equiv 0$, provided u_0 is differentiable.

In the case of variable coefficients f and g in (1.1), we have the more general form

$$u(x,t) = u_0(y(x,t;t_0)) + \int_{t_0}^{t} g(y(x,t;s),s)ds, \tag{1.6}$$

where $y(x,t;s)$ is the (backward) solution at time s of the Cauchy problem, called *system of base characteristics*

$$\begin{cases} \dot{y}(x,t;s) = f(y(x,t;s),s), \\ y(x,t;t) = x. \end{cases} \tag{1.7}$$

In this more general formula, which will be proved in Chapter 2, the value $u(x,t)$ depends on the value at the foot of the characteristic curve (which might no longer be a straight line), as well as on the integral of the source term along the characteristic itself. Again, it can be easily shown that a function defined by (1.6)–(1.7) solves (1.1)–(1.2) if u_0 is smooth. The representation formula (1.6)–(1.7) shows that the regularity of the initial condition is preserved along the characteristics if both drift and source terms are smooth enough, so that $u_0 \in C^p(\mathbb{R}^d)$ and $f, g \in C^{p-1}(\mathbb{R}^d \times (t_0, T))$ imply that $u \in C^p(\mathbb{R}^d \times (t_0, T))$. A key point in this respect is that characteristics cannot cross each other, a property which holds only in the linear case.

A typical result of existence and uniqueness of classical solutions to the advection equation in one space dimension is the following.

Theorem 1.1. *Consider* (1.1)–(1.2), *and assume that* $f, g \in C^1(\mathbb{R} \times (t_0, T))$ *and that* $u_0 \in C^1(\mathbb{R} \times (t_0, T))$. *Then, there exists a unique classical solution of* (1.1), *i.e., a function* u *satisfying* (1.1) *at every point* $(x,t) \in \mathbb{R} \times (t_0, T)$.

The proof of this result is based on the method of characteristics, which will be better examined in Chapter 2 (a more general result in an arbitrary dimension can be found in [E98]). A crucial point, besides the regularity of the coefficients and of the initial condition, is that the field of characteristics $(1, f(x,t))$ should be transversal to the x-axis at t_0, so that the solution can be defined by the initial condition on a neighborhood of the x-axis. In practice, this corresponds to the requirement that f should not vanish over an interval.

At a closer look, however, well-posedness results requiring differentiability of the initial condition might not be satisfactory in a number of applications which naturally lead to a nonsmooth definition of u_0. For example, if we try to simulate the advection of a pollutant spread over a precise area ω_p, it would be natural to consider a discontinuous initial condition such as

$$u_0(x) = \mathbf{1}_{\omega_p} = \begin{cases} 1 & \text{if } x \in \omega_p, \\ 0 & \text{else.} \end{cases} \tag{1.8}$$

Although we expect the physically relevant solution to be given again by (1.6)–(1.7), the lack of regularity in the initial condition does not enable the same formal derivation of the representation formula. Hence, it becomes necessary to introduce a notion of a weak solution which could allow us to treat the Cauchy problem in a more general setting. One way to achieve this goal is to look for weak solutions in the sense of distributions.

Definition 1.2. *A function $u : \mathbb{R} \times (t_0, T) \to \mathbb{R}$ in $L^1(\mathbb{R} \times (t_0, T))$ is a weak solution of the Cauchy problem* (1.1)–(1.2) *if and only if for any test function $\phi \in C_0^1(\mathbb{R} \times (t_0, T))$ the following integral condition is satisfied:*

$$\begin{aligned} \int_{t_0}^{T} \int_{\mathbb{R}} u(x,t)\phi_t(x,t) + [u(x,t)f(x,t)]\, \phi_x(x,t)dx\, dt \\ = \int_{t_0}^{T} \int_{\mathbb{R}} g(x,t)\varphi(x,t)\, dx\, dt. \end{aligned} \tag{1.9}$$

In recent years, this strategy of definition for weak solutions has also been extended to include bounded variation solutions. Anyway, we will not pursue this line of work here, and rather focus on the concept of *viscosity solution*, which has proved to be well suited for the case of HJ equations.

1.2 ▪ Nonlinear evolutive problems of HJ type

Among all extensions to nonlinear equations, this book will mostly address the following *HJ equation*:

$$\begin{cases} u_t + H(x, Du) = 0, & (x,t) \in \mathbb{R}^d \times (0, T), \\ u(x, 0) = u_0, & x \in \mathbb{R}^d, \end{cases} \tag{1.10}$$

where $H : \mathbb{R}^d \to \mathbb{R}$ is called the *Hamiltonian* of the equation and is typically assumed to be convex. We soon sketch a couple of applications which fall in this class of problems.

Front propagation via the level set method A special case of the HJ equation arises in modeling a front which evolves in the normal direction with a given velocity $c : \mathbb{R}^d \to \mathbb{R}$. The *level set method* considers the Cauchy problem

$$\begin{cases} u_t + c(x)|Du| = 0, & (x,t) \in \mathbb{R}^d \times (0, T), \\ u(x, 0) = u_0, & x \in \mathbb{R}^d, \end{cases} \tag{1.11}$$

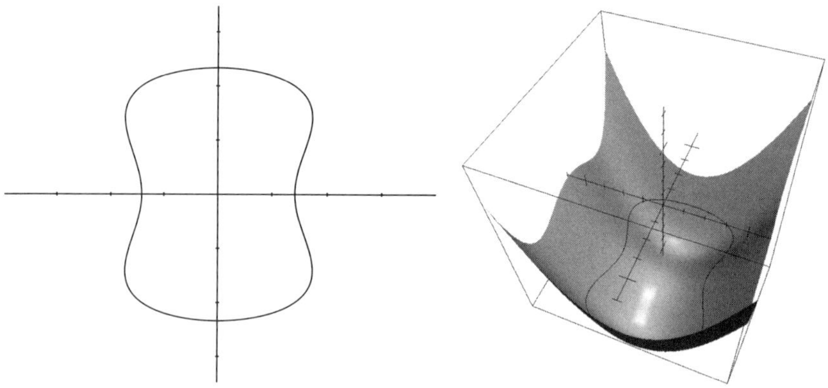

Figure 1.1. *A curve Γ_0 and a possible representation function u_0.*

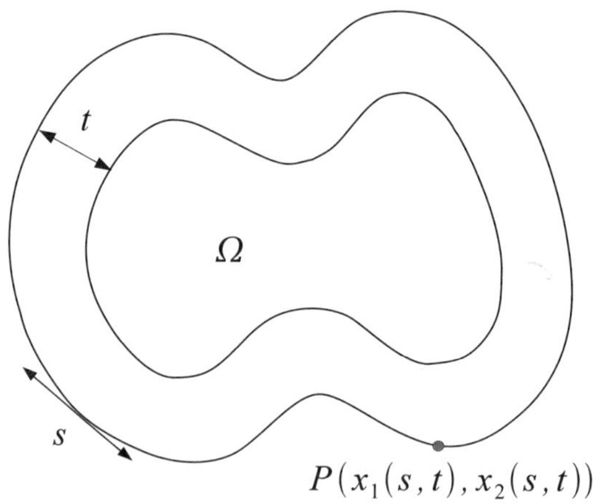

Figure 1.2. *The propagation of a front in the normal direction.*

where $u_0 : \mathbb{R}^d \to \mathbb{R}$ must be a *proper representation* of the front Γ_0, i.e., a continuous function u_0 which changes sign on $\Gamma_0 := \partial\Omega_0$. Considering for simplicity the case of two space dimensions, this would lead to the following definition.

Definition 1.3. *Assume that Γ_0 is a closed curve in $\mathbb{R}^2$, and denote by Ω_0 the region enclosed by Γ_0. A continuous function $u_0 : \mathbb{R}^2 \to \mathbb{R}$ is a* proper representation *of Γ_0 if and only if it satisfies the following conditions:*

$$\begin{cases} u_0(x) < 0, & x \in \Omega_0, \\ u_0(x) = 0, & x \in \Gamma_0, \\ u_0(x) > 0, & x \in \mathbb{R}^2 \setminus \overline{\Omega}_0. \end{cases} \tag{1.12}$$

In this way, the front Γ_0 at the initial time is identified with the zero-level set of u_0, as shown in Figure 1.1.

Assume now that we let Γ_0 evolve in the normal direction with velocity $c(x)$, as shown in Figure 1.2. To fix ideas, we can assume that the normal η is directed outwards with respect to Ω and that the velocity is positive (this situation corresponds to an expansion of the front, whereas an opposite sign would produce a shrinking).

The front at time t will be denoted by Γ_t and a generic point $P \in \Gamma_t$ will be represented as $P = (x_1(s,t), x_2(s,t))$, where s refers to the parametric representation of Γ_t (e.g., as an arc length parametrization). Since the velocity V at every point has magnitude $c(P)$ and is directed in the normal direction η, we can write

$$V(P) = (\dot{x}_1(s,t), \dot{x}_2(s,t)) = c(P)\eta(P). \tag{1.13}$$

After describing Γ_0 as the zero-level set of u_0, our plan is to describe the front Γ_t at a generic time t as the zero-level set of the solution of a Cauchy problem, having u_0 as its initial condition. In order to derive the equation characterizing Γ_t, we consider the total derivative of $u(x_1(s,t), x_2(s,t))$. Taking into account that this derivative vanishes on a level set of constant value, we have

$$\begin{aligned}
0 &= \frac{d}{dt} u(x_1(s,t), x_2(s,t), t) \\
&= u_t(x_1(s,t), x_2(s,t)) + Du(x_1(s,t), x_2(s,t)) \cdot (\dot{x}_1, \dot{x}_2) \\
&= u_t(P) + c(P) Du(P) \cdot \eta(P) \\
&= u_t(P) + c(P)|Du(P)|,
\end{aligned}$$

where the normal direction to a level set of u (computed at P) has been written in terms of u as $Du(P)/|Du(P)|$. In conclusion, solving (1.11) and looking at the zero-level set of the solution $u(x,t)$ we can recover the front Γ_t. Note that the choice of the zero level is arbitrary, and in fact all level sets of u are moving according to the same law.

Once we formulate the front evolution by means of (1.11), we can drop the smoothness assumption on Γ_t and apply the same method to more general initial configurations. Here, the possibility of defining weak (nondifferentiable) solutions for (1.11) plays a crucial role. For example, taking Γ_0 as the union of a number of separate curves in the plane, if the evolution is driven by a positive velocity, the various components of the front will collide in a finite time and we will have a *merging* of the fronts. It is clear that, at a point where two components of the front are tangent, we cannot expect to have a normal vector and a singularity appears in the evolution (and in the equation). However, as we will see later on, the level set method relies on a weak concept of solution which is robust enough to handle both the onset of a singularity and the change of topology.

This quick explanation of the level set method also shows its main drawback. In order to follow a $(d-1)$-dimensional front we are forced to embed the problem in $\mathbb{R}^d$, and this implies an extra computational cost. This remark has motivated the development of "local" versions of the level set method which reduce the computational cost by taking into account only the values in a small neighborhood of the front, the so-called *narrow band*. More details on such methods will be given in Chapter 9.

Optimal control and Dynamic Programming A second relevant application of the HJ equation is to characterize the value function of an optimal control problem. Let a controlled dynamical system be described by

$$\begin{cases} \dot{y}(t) = f(y(t), t, \alpha(t)), \\ y(t_0) = x_0, \end{cases} \tag{1.14}$$

where $y \in \mathbb{R}^d$ is the *state* of the system, $\alpha : (t_0, T) \to A \subseteq \mathbb{R}^m$ is the *control* function, and $f : \mathbb{R}^d \times (t_0, T) \times A \to \mathbb{R}^d$ is the controlled drift (or *dynamics*).

Consider the class of admissible controls

$$\mathcal{A} := \{\alpha : (t_0, T) \to A, \text{measurable}\},$$

where A is a compact subset of $\mathbb{R}^m$. It is known that for any given control $\alpha \in \mathscr{A}$ the solution of (1.14) is well defined (in a weak sense). Then, we can introduce the *cost functional* $J : \mathbb{R}^d \times \mathbb{R} \times \mathscr{A} \to \mathbb{R}$ which will be used to select the "optimal" (i.e., the minimizing) trajectory. For the *finite horizon problem*, the functional is usually defined as

$$J(x,t_0;\alpha) = \int_{t_0}^{T} g(y(s),s,\alpha(s))e^{-\lambda(s-t_0)}ds + e^{-\lambda(T-t_0)}\overline{g}(y(T)), \quad \lambda > 0, \tag{1.15}$$

where $g : \mathbb{R}^d \times (t_0, T) \times A \to \mathbb{R}$ is the *running cost* depending on the state of the system, on the time, and on the control, $\lambda \in \mathbb{R}_+$ is the *discount factor* which allows us to compare costs at different times by rescaling them at time t_0, and $\overline{g}$ is a terminal cost. The *value function* associated with this problem is defined as

$$u(x,t) = \inf_{\alpha} J(x,t;\alpha). \tag{1.16}$$

The value function represents the best value we can get from our cost functional and, as it will be shown in Chapter 8, enables us to construct the optimal solution. In all classical deterministic control problems, the *Dynamic Programming Principle* allows us to characterize the value function as the unique solution, once more in a suitably weak sense, of a PDE known as the Dynamic Programming (or Bellman) equation:

$$\begin{cases} u_t + \sup_{a \in A}\{-f(x,t,a)\cdot Du - g(x,t,a)\} = 0, & (x,t) \in \mathbb{R}^d \times (0,T), \\ u(x,T) = \overline{g}(x), & x \in \mathbb{R}^d, \end{cases} \tag{1.17}$$

where $a \in A$ is a parameter spanning the control set. Note that, although in backward form, (1.17) is of the HJ type (1.10). More details on this technique will be given in Chapter 8, whereas a more extensive presentation can be found in [BCD97].

1.3 ▪ Nonlinear stationary problems

We turn now to some examples of stationary HJ equations which will be analyzed in what follows. Problems of this kind may arise when dealing with applications related to control theory and image processing.

The eikonal equation The first classical example is given by the eikonal equation

$$\begin{cases} |Du| = f(x), & x \in \Omega, \\ u(x) = 0, & x \in \partial\Omega. \end{cases} \tag{1.18}$$

In a simple but relevant case, the right-hand side is constant (e.g., $f(x) \equiv 1$) on an external domain, so that the equation becomes

$$\begin{cases} |Du| = 1, & x \in \mathbb{R}^d \setminus \Omega, \\ u(x) = 0, & x \in \partial\Omega. \end{cases} \tag{1.19}$$

It is easy to see that, at least when Ω is a convex closed subset of $\mathbb{R}^d$ with a smooth boundary, the distance from Ω is a classical solution of (1.19). In fact, for every $x_0 \in \mathbb{R}^d \setminus \Omega$, we have

$$d(x_0, \Omega) = d(x_0, p_0), \tag{1.20}$$

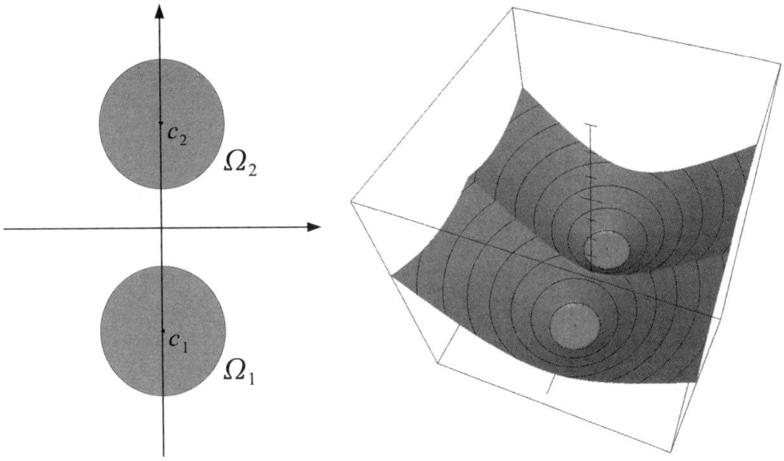

Figure 1.3. *A counterexample to the differentiability of the distance function $d(\cdot,\Omega)$.*

where p_0 is the unique projection of x_0 onto Ω. Now, the directional derivative along a generic direction $v \in \partial B(0,1)$ (where $B(0,1)$ denotes the unit ball in $\mathbb{R}^d$) reads as

$$Dd(x_0)\cdot v = \lim_{h\to 0} \frac{d(x_0+hv)-d(x_0)}{h}, \tag{1.21}$$

and we note that it attains its maximum value when v is aligned with $Dd(x_0)$ (this corresponds to a motion along the straight line connecting p_0 and x_0). Then, along this direction we obtain

$$Dd(x_0)\cdot \frac{x_0-p_0}{|x_0-p_0|} = \lim_{h\to 0} \frac{d\left(x_0+h\frac{x_0-p_0}{|x_0-p_0|}\right)-d(x_0)}{h} = \lim_{h\to 0}\frac{h}{h} = 1. \tag{1.22}$$

On the other hand, by the previous remark we must also have

$$Dd(x_0)\cdot \frac{x_0-p_0}{|x_0-p_0|} = Dd(x_0)\cdot \frac{Dd(x_0)}{|Dd(x_0)|} \tag{1.23}$$

and $d(x_0,\Omega)=0$ on $\partial\Omega$, and we conclude that the distance function is a smooth solution of (1.19).

As soon as we drop the convexity assumption on Ω, however, we no longer expect to have a differentiable distance function even if Ω has a smooth boundary. A simple counterexample is the following: take in $\mathbb{R}^2$ two disjoint circles Ω_1 and Ω_2, centered at c_1 and c_2, respectively, and with the same radius R, and set $\Omega := \Omega_1 \cup \Omega_2$. Consider the axis ρ of the segment c_1c_2, i.e., the locus of points having the same distance from both centers. Then, all points belonging to ρ are points at which the distance $d(\cdot,\Omega)$ is not differentiable, as shown in Figure 1.3. In this case, once more, we are naturally prompted to consider nonsmooth solutions of (1.19).

Control problems Another interesting example leading to an equation of eikonal type is the classical minimum time problem of control theory. In its simplest formulation, we consider a controlled system of the form

$$\begin{cases} \dot{y}(t) = \alpha(t), & t \in \mathbb{R}_+, \\ y(0) = x_0 \in \mathbb{R}^d \setminus \Omega, \end{cases} \tag{1.24}$$

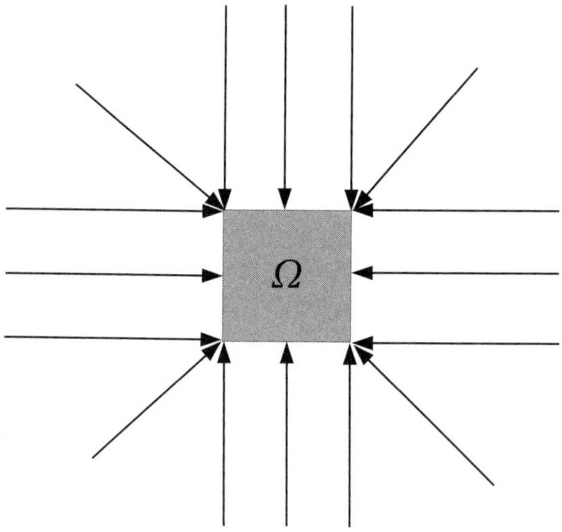

Figure 1.4. *Optimal trajectories of the minimum time problem for a square target Ω.*

where $\alpha : (0,+\infty) \to B(0,1)$ is a measurable control. Given a closed convex target set Ω, we want to find the minimal time necessary to drive the system from its initial position to the target (the minimal time represents the value function of this particular problem). Since the system can move in every direction, the optimal trajectory is a straight line connecting the initial position x_0 to its unique projection on Ω. Moreover, the velocity has at most unit norm, and therefore the minimum time of arrival from x_0 equals the distance $d(x_0,\Omega)$ and solves the eikonal equation (1.19). We show in Figure 1.4 an example with a square target Ω, whereas an approximation of the corresponding value function will be shown in Figure 1.8.

More generally (as we will see in Chapter 8, devoted to control problems and games), if the system is driven by a nonlinear controlled dynamical system of the form

$$\begin{cases} \dot{y}(t) = f(y(t),\alpha(t)), \\ y(0) = x_0 \in \mathbb{R}^d \setminus \Omega, \end{cases} \tag{1.25}$$

we end up with the following Bellman equation for the minimum time function:

$$\begin{cases} \max_{a\in A}\{-f(x,a)\cdot Du\} = 1, & x \in \mathbb{R}^d \setminus \Omega, \\ u(x) = 0, & x \in \partial\Omega. \end{cases} \tag{1.26}$$

The characterization of the value function in terms of a stationary Bellman equation also applies to other deterministic optimal control problems. For example, we mention that an equation of the form

$$\lambda u(x) + \max_{a\in A}\{-f(x,a)\cdot Du(x) - g(x,a)\} = 0 \tag{1.27}$$

characterizes the value function of an infinite horizon problem for the system (1.25), i.e., the function

$$u(x) := \inf_{\alpha\in\mathcal{A}} J(x,\alpha), \tag{1.28}$$

with a cost functional $J(x,\alpha)$ given by

$$J(x,\alpha) := \int_0^{+\infty} g(y(s),\alpha(s))e^{-\lambda s}\,ds. \tag{1.29}$$

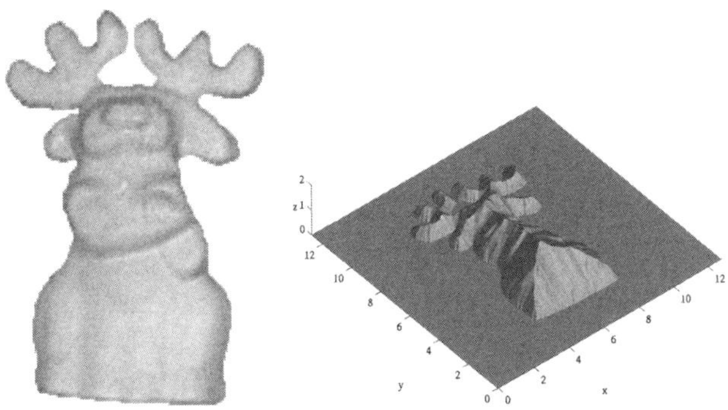

Figure 1.5. *A gray-level picture (left) and the corresponding surface (right).*

We refer the reader again to Chapter 8 for a more detailed presentation of Dynamic Programming techniques.

Shape-from-Shading An equation of eikonal type also arises in the Shape-from-Shading problem, a classical inverse problem in which one tries to reconstruct, using the shading information, the surface which has generated a certain monochromatic picture taken by a camera, as shown in Figure 1.5.

To be more precise, we introduce the basic assumptions and notation for this problem. We attach to the camera a three-dimensional coordinate system $(Oxyz)$, such that Oxy coincides with the image plane and Oz coincides with the optical axis. Under the assumption of orthographic projection, the visible part of the scene is, up to a scale factor, a graph $z = u(x)$, where $x = (x_1, x_2)$ is an image point. Horn has shown in [HB89] that the Shape-from-Shading problem can be modeled by the *image irradiance equation*

$$R(\nu(x)) = I(x), \tag{1.30}$$

where $I(x)$ is the grey level measured in the image at point x (in fact, $I(x)$ is the *irradiance* at point x, but the two quantities are proportional) and $R(\nu(x))$ is the *reflectance* function, giving the value of the light re-emitted by the surface as a function of its orientation, i.e., of the unit normal $\nu(x)$ to the surface at a point $(x, u(x))$. For the graph of a function u, this normal can be expressed as

$$\nu(x) = \frac{1}{\sqrt{1+p(x)^2+q(x)^2}}(-p(x), -q(x), 1), \tag{1.31}$$

where $p = \partial u/\partial x_1$ and $q = \partial u/\partial x_2$, so that $Du(x) = (p(x), q(x))$. The irradiance function (or image brightness) I is given as a measure of grey level (from 0 to 255 in digital images) at each pixel of the image, and in order to construct a continuous model, we will assume that it is renormalized in the interval $[0,1]$. We also assume that there is a unique light source at infinity, whose direction (supposed to be known) is given by the unit vector $\omega = (\omega_1, \omega_2, \omega_3) \in \mathbb{R}^3$.

The height function u, which is the unknown of the problem, has to be reconstructed on a compact domain $\Omega \subset \mathbb{R}^2$, called the *reconstruction domain*. Recalling that, for a Lambertian surface of uniform unit albedo, $R(\nu(x)) = \omega \cdot \nu(x)$, and using (1.31), then (1.30) can be rewritten for any $x \in \Omega$ as

$$I(x)\sqrt{1+|Du(x)|^2} + (\omega_1, \omega_2) \cdot Du(x) - \omega_3 = 0, \tag{1.32}$$

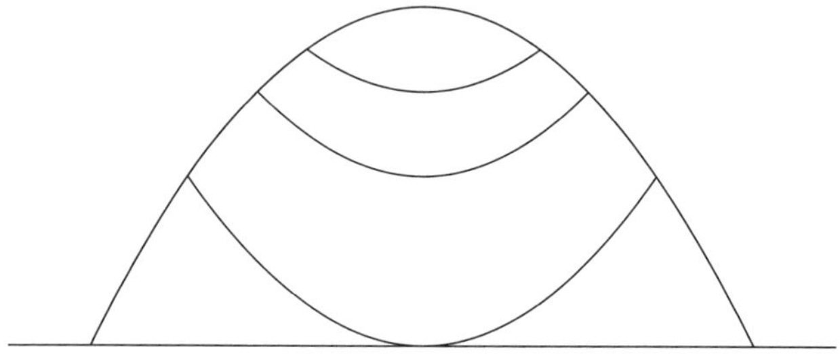

Figure 1.6. *One classical solution and some a.e. solutions of the Shape-from-Shading problem.*

which is a stationary first-order equation of HJ type. Points $x \in \Omega$ such that $I(x)$ is maximal correspond to the particular situation where ω and $\nu(x)$ point in the same direction: these points are usually called "singular points." The most widely studied case in the Shape-from-Shading literature corresponds to a frontal light source at infinity, i.e., $\omega = (0,0,1)$. Then, (1.32) becomes an eikonal equation of the form

$$|Du(x)| = \sqrt{\frac{1-I^2(x)}{I^2(x)}}. \tag{1.33}$$

It is known that the eikonal equation may not have a unique solution whenever the right-hand side vanishes at some point. So the existence of singular points corresponds to an intrinsic ill-posedness of the Shape-from-Shading problem. In fact, a simple example can show that uniqueness does not hold even for classical solutions. Consider a one-dimensional case in which the surface is given by $u_1(x) = 1 - x^2$ in the interval $[-1,1]$. This surface clearly satisfies the boundary conditions $u_1(1) = u_1(-1) = 0$, and, assuming a vertical light source with $\omega = (0,1)$, it also satisfies (1.33) for the corresponding irradiance, which is given by

$$I(x) = \frac{(0,1)\cdot(2x,1)}{\sqrt{1+4x^2}}. \tag{1.34}$$

At the point $x = 0$, I takes its maximum value $I(0) = 1$, and therefore the right-hand side of (1.33) vanishes. Note that the Lambertian assumption on the surface implies that I depends only on the angle between the normal $\nu(x)$ and ω, so that it is clear that the mirror image of the surface u_1 with respect to the x-axis, i.e., $u_2(x) = -1 + x^2$, is still a classical solution of (1.33). It is natural to observe (see Figure 1.6) that all mirror reflections of a part of both surfaces u_1 and u_2 with respect to a horizontal line satisfy (1.33) a.e. (although this does not imply that they could be acceptable weak solutions).

1.4 ▪ Simple examples of approximation schemes

In this section, we try to give some very basic ideas (in a single space dimension) about the approximation schemes for the model problems introduced so far. We start with the case of the *rigid transport problem*, i.e., the linear advection at constant speed c (to fix ideas, we assume that $c > 0$), which is described by the equation

$$\begin{cases} u_t + c u_x = 0, & (x,t) \in \mathbb{R} \times (0,T), \\ u(x,0) = u_0(x), & x \in \mathbb{R}. \end{cases} \tag{1.35}$$

The most classical method for constructing an approximation of (1.35) is to build a uniform grid in space and time (a lattice) with constant steps Δx and Δt, covering the domain of the solution:

$$\left\{(x_j,t_n)=(j\Delta x,n\Delta t),\quad j\in\mathbb{Z},n\in\mathbb{N},n\leq\frac{T}{\Delta t}\right\}. \tag{1.36}$$

The basic idea behind all finite difference approximation is to replace every derivative by an incremental ratio. Thus, one obtains a finite-dimensional problem whose unknowns are the values of the numerical solution at all the nodes of the lattice, so that the value u_j^n associated with the node (x_j,t_n) should be regarded as an approximation of $u(x_j,t_n)$. For the time derivative it is natural to choose the forward incremental ratio

$$u_t(x,t)\approx\frac{u(x,t+\Delta t)-u(x,t)}{\Delta t}, \tag{1.37}$$

which allows us, starting with the solution at the initial time t_0, to compute explicitly the approximations for increasing times $t_k > t_0$. Writing at (x_j,t_n) the forward incremental ratio in time, we get

$$u_t(x_j,t_n)\approx\frac{u_j^{n+1}-u_j^n}{\Delta t}. \tag{1.38}$$

For the approximation of u_x we have more choices, like

$$\begin{aligned} u_x(x_j,t_n)&\approx\frac{u_{j+1}^n-u_j^n}{\Delta x} &&\text{(right finite difference)},\\ u_x(x_j,t_n)&\approx\frac{u_j^n-u_{j-1}^n}{\Delta x} &&\text{(left finite difference)},\\ u_x(x_j,t_n)&\approx\frac{u_{j+1}^n-u_{j-1}^n}{2\Delta x} &&\text{(centered finite difference)}, \end{aligned} \tag{1.39}$$

which are all based on the values at the nodes (x_k,t_n).

In this case, the choice of the approximation for u_x crucially affects the convergence of the numerical solution to the exact solution. Although the centered finite difference would in principle provide a more accurate approximation of the space derivative, as we will see later on, given the time derivative approximation (1.38), the only choice in (1.39) which results in a convergent scheme is to use the left incremental ratio if $c > 0$ and the right incremental ratio if $c < 0$, which corresponds to the so-called *upwind scheme*. In conclusion, for $c > 0$, we obtain a scheme in the form

$$\frac{u_j^{n+1}-u_j^n}{\Delta t}+c\,\frac{u_j^n-u_{j-1}^n}{\Delta x}=0. \tag{1.40}$$

A different way to construct the approximation of (1.35) is to consider the advection term as a directional derivative and write

$$cu_x(x_j,t_n)\approx-\frac{u^n(x_j-c\delta)-u_j^n}{\delta}, \tag{1.41}$$

where δ is a "small" positive parameter, and u^n denotes an extension of the numerical solution (at time t_n) to be computed outside of the grid. Coupling the forward finite difference in time with this approximation we get

$$\frac{u_j^{n+1}-u_j^n}{\Delta t}-\frac{u^n(x_j-c\delta)-u_j^n}{\delta}=0,$$

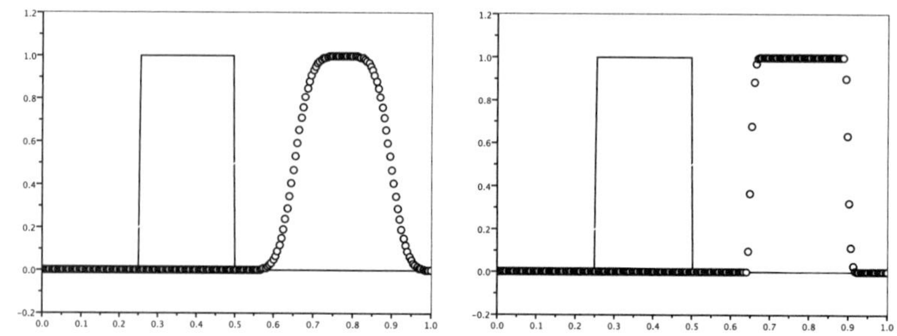

Figure 1.7. *Approximation of the rigid transport problem by the upwind scheme (left, $\Delta t = 0.025$) and by the SL scheme (right, $\Delta t = 0.05$).*

and finally, choosing $\delta = \Delta t$, we obtain the scheme

$$u_j^{n+1} = u^n(x_j - c\Delta t). \tag{1.42}$$

This is the Semi-Lagrangian (SL) scheme for (1.35), which could also be derived from a discretization of the representation formula (1.3). Opposite to finite difference schemes, SL schemes use points ($x_j - c\Delta t$ in our example) which may not be grid nodes, so the corresponding values of the numerical solution need to be reconstructed by some interpolation technique. Figure 1.7 shows two approximations of the rigid transport problem obtained via, respectively, the upwind and the SL schemes.

This technique can also be extended to the eikonal equation (1.18), via the identity

$$|Du(x)| = \max_{a\in B(0,1)} a \cdot Du(x). \tag{1.43}$$

In fact, the maximum in (1.43) is attained at $a^* = Du(x)/|Du(x)|$, and this shows that the right-hand side equals $|Du(x)|$. Using now (1.43) to rewrite the eikonal equation, we get

$$\max_{a\in B(0,1)} \{a \cdot Du(x)\} = f(x). \tag{1.44}$$

Since the term inside the square brackets is linear, we can apply the SL scheme, obtaining

$$u_j^{n+1} = \min_{a\in B(0,1)} \left\{u^n(x_j - a\Delta t)\right\} + \Delta t f(x). \tag{1.45}$$

Figure 1.8 shows the level set corresponding to the minimum time function for a square target and the simple dynamics $f(x,\alpha) = \alpha \in B(0,1)$.

1.5 ▪ Some difficulties arising in the analysis and in the approximation

After giving models, applications and motivations for the theory which will be the object of this book, it is worth pointing out a number of mathematical peculiarities of the problems under consideration.

1. The analysis of linear transport problems and the HJ equation leads us to drop classical solutions and introduce weak solutions in order to treat relevant applications. The analysis of weak solutions in this framework started in the 1960s with the pioneering works by S.N. Kružkov and has proceeded through the last decades of

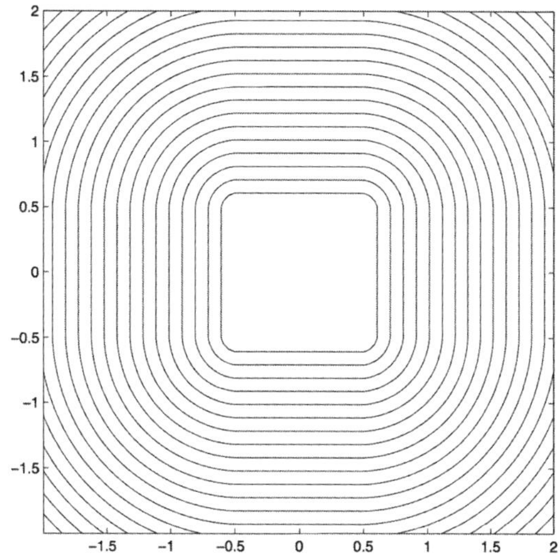

Figure 1.8. *Level sets of the solution of the eikonal equations approximated via the SL scheme* ($\Delta t = 0.05$).

the century when M.G. Crandall and P.L. Lions introduced the notion of viscosity solution. Typically, viscosity solutions are Lipschitz continuous, but, at least for convex Hamiltonians, the concept can be extended to discontinuous solutions. Although the theory is still undergoing technical developments, and new classes of problems are analyzed in this framework, this book will mainly refer to the basic theory for convex Hamiltonians, which is at the moment quite an established matter.

2. The lack of smoothness of viscosity solutions makes it difficult to develop efficient approximations. Starting from the 1980s, monotone finite difference methods have been proposed using the relationship between HJ equations and conservation laws (this relationship will be discussed in the next chapter). On this basis, monotone finite difference methods conceived for conservation laws have been adapted to the approximation of the HJ equation. However, monotone schemes are typically very diffusive, a serious drawback when treating nonsmooth problems. On the other hand, SL schemes may provide less diffusive, still monotone approximations through the use of large time steps. Construction and convergence theory of SL schemes, as well as their comparison with difference schemes, will be a major subject of this book.

3. In the last decades, the low efficiency of monotone schemes has motivated a large effort to develop less diffusive, high-order approximation schemes. This ongoing research, which can have a great impact on applications, has already produced a number of methods of great accuracy, although very few theoretical results are available. In this book, we will try to summarize the existing results for high-order SL schemes into a rigorous convergence analysis, at least on model cases.

4. The theory of viscosity solutions has been tested on a variety of applications, ranging from control and game theory to image processing, and from Fluid Dynamics to combustion and geophysics, and has proved to be particularly effective in characterizing the relevant solution. Some chapters of this book will be devoted to

these topics. However, these applications often require working in high dimensions. Along with their theoretical study, the development of fast and accurate numerical schemes for high-dimensional problems seems to be a critical, challenging task for future research.

1.6 ▪ How to use this book

We conclude this introductory chapter by giving, on the basis of our personal experience, some hints about the use of the material presented in the book. Such hints can also be useful for a reader approaching the subject for the first time.

- A basic introductory course on the analysis and approximation of HJ equations can use the first five chapters, which also provide an overview of the analytical aspects.
- An advanced numerical course on nonlinear hyperbolic differential equations will include, for example, Chapters 2, 3, 4, and 6.
- A course on applications to front propagation, level set methods, and Fluid Dynamics can use the material in Chapters 2, 3, 4, 5, and 9 or 7 depending on the emphasis to be given to different topics.
- A course on applications to control theory can use the material in Chapters 2, 3, 4, 5, and 8.

We will be glad to receive other suggestions for the use of this book for graduate and advanced undergraduate courses.

Chapter 2
Viscosity solutions of first-order PDEs

This chapter contains a general presentation of the theory of weak solutions in the viscosity sense for first-order PDEs, starting from the stationary case (which includes, e.g., the eikonal equation) and moving to evolutive models (which arise in control theory and front propagation).

The chapter reviews basic and essential results of existence and uniqueness for viscosity solutions. Moreover, it also presents other related topics, such as characteristics and Hopf–Lax formulae, regularity results (typically, based on the notion of *semiconcavity*), boundary conditions, and the relationship with entropic solutions of conservation laws. Most of these topics will prove useful in the construction and analysis of the approximation schemes presented in what follows. Finally, we briefly sketch the concept of discontinuous viscosity solutions.

2.1 ▪ The definition of viscosity solution

To introduce the theory of weak solutions in the viscosity sense, we start by considering a stationary problem of the general form

$$H(x,u,Du)=0, \qquad x\in\Omega. \tag{2.1}$$

Here, Ω is an open domain of $\mathbb{R}^d$ and the Hamiltonian $H:\mathbb{R}^d\times\mathbb{R}\times\mathbb{R}^d\to\mathbb{R}$ is a continuous real-valued function on $\Omega\times\mathbb{R}\times\mathbb{R}^d$. In this first discussion, we will avoid treating boundary conditions. In fact, boundary conditions need to be understood in a suitable, more delicate "weak sense" which will be examined later on in this chapter.

Typically, the main results in the theory of viscosity solutions assume that the Hamiltonian $H(x,u,p)$ satisfies the following basic set of assumptions:

(A1) $H(\cdot,\cdot,\cdot)$ is uniformly continuous on $\Omega\times\mathbb{R}\times\mathbb{R}^d$.

(A2) $H(x,u,\cdot)$ is convex on $\mathbb{R}^d$.

(A3) $H(x,\cdot,p)$ is monotone on $\mathbb{R}$.

The need for a notion of weak solution arises when observing that problem (2.1) is not expected to have a classical (i.e., $C^1(\Omega)$) solution, as the following counterexample shows.

A counterexample Consider the one-dimensional problem

$$\begin{cases} |u_x| = 1, & x \in \Omega = (-1,1), \\ u(-1) = u(1) = 0. \end{cases} \tag{2.2}$$

Obviously, $u_1(x) = x$ and $u_2(x) = -x$ are C^1 and satisfy the equation pointwise in $(-1,1)$, but they do not satisfy both boundary conditions.

On the other hand, if the assumption of differentiability is dropped, we might consider functions which satisfy the equation a.e. and satisfy the boundary conditions. Two such solutions would be given by the functions

$$u_3(x) = |x| - 1, \quad u_4(x) = 1 - |x|,$$

but in fact, by collecting piecewise affine functions with slopes ± 1, it is possible to construct infinitely many functions which solve the equation a.e. and satisfy the boundary conditions. Therefore, it is clear that the notion of "a.e. solution" gives too many solutions and is unsuitable for a uniqueness result. One possibility for recovering uniqueness is to perform an *elliptic regularization*, i.e., to regularize the problem by adding a second order term $-\varepsilon u_{xx}$, thus obtaining the elliptic problem

$$\begin{cases} -\varepsilon u_{xx} + |u_x| = 1, & x \in \Omega = (-1,1), \\ u(-1) = u(1) = 0. \end{cases} \tag{2.3}$$

By standard results on semilinear elliptic equations, for every positive ε this problem has a unique, regular solution u_ε, which can be explicitly written as

$$u_\varepsilon(x) = 1 - |x| + \varepsilon\left(e^{-\frac{1}{\varepsilon}} - e^{-\frac{|x|}{\varepsilon}}\right). \tag{2.4}$$

Since (2.2) is the limit problem of (2.3) as $\varepsilon \to 0$, it would be natural to define the limit of the family u_ε (if a limit exists) as a generalized solution for (2.2). In fact, on $(0,1)$, u_ε converges uniformly to $u(x) = 1 + |x|$, and we will see that this is precisely the viscosity solution of (2.2).

The following definition of viscosity solution improves this original idea by avoiding the limiting procedure and giving a local characterization of the weak solution.

The "classical" definition We first give the most popular definition of viscosity solution for $u \in BUC(\Omega)$ (the space of bounded and uniformly continuous functions over the open set Ω).

Definition 2.1. *Let $u \in BUC(\Omega)$. We say that u is a* viscosity solution *of* (2.1) *if and only if, for any $\varphi \in C^1(\Omega)$, the following conditions hold:*

(i) *at every point $x_0 \in \Omega$, local* maximum *for $u - \varphi$,*

$$H(x_0, u(x_0), D\varphi(x_0)) \le 0$$

(i.e., u is a viscosity subsolution*);*

(ii) *at every point $x_0 \in \Omega$, local* minimum *for $u - \varphi$,*

$$H(x_0, u(x_0), D\varphi(x_0)) \ge 0$$

(i.e., u is a viscosity supersolution*).*

Going back to the counterexample, we explain how such a definition allows to select a unique a.e. solution.

First, we show that any a.e. solution u which has a local minimum at x_0 cannot be a viscosity supersolution. In fact, choose $\varphi \equiv c$ (a constant). Clearly, x_0 is a local minimum point for $u-\varphi$, and therefore we should have

$$|\varphi_x(x_0)| \geq 1,$$

which is false since $\varphi_x \equiv 0$.

Note that the same argument works in a different way for subsolutions. Take an a.e. solution u which has a local maximum at x_0 and choose $\varphi \equiv c$. Clearly, x_0 is a local maximum point for $u-\varphi$ so that we should have

$$|\varphi_x(x_0)| \leq 1,$$

which is true, since $\varphi_x \equiv 0$. This implies that the viscosity solution can have maximum but not minimum points. The only viscosity solution of our problem is therefore

$$u(x) = 1 - |x|.$$

2.1.1 ▪ Some properties of viscosity solutions

We review here some properties of viscosity solutions which will be useful in what follows. In particular, we want to examine the link between classical and viscosity solutions, as well as some other properties which are relevant for the construction of approximation schemes.

Relationship with classical solutions The first result states the local character of viscosity solutions, along with their relationship with classical solutions.

Proposition 2.2. *The following statements hold true:*

(i) *If $u \in C(\Omega)$ is a viscosity solution of* (2.1) *in Ω, then u is a viscosity solution of the same equation in any open set $\Omega' \subset \Omega$.*
(ii) *If u is a classical $C^1(\Omega)$ solution, then it is also a viscosity solution.*
(iii) *Any viscosity solution that is also regular is a classical solution.*

Proof. (i) If x_0 is a local maximum point on Ω' for $u-\varphi$, $\varphi \in C^1(\Omega')$, then x_0 is a local maximum point on Ω for $u-\widetilde{\varphi}$, for any test function $\widetilde{\varphi} \in C^1(\Omega)$ such that $\widetilde{\varphi} \equiv \varphi$ on $\overline{B}(x_0, r)$ for some positive r. Then, by the definition of viscosity subsolutions,

$$H(x_0, u(x_0), D\varphi(x_0)) = H(x_0, u(x_0), D\widetilde{\varphi}(x_0)) \leq 0.$$

(ii) Assume $u \in C^1$ is a classical solution, so that $H(x, u(x), Du(x)) = 0$ for every $x \in \Omega$. Consider a test function φ and a local maximum point for $u-\varphi$. It is not restrictive to choose $\varphi(x_0) = u(x_0)$. Then, we have

$$D(u-\varphi)(x_0) = 0. \tag{2.5}$$

This implies that

$$D\varphi(x_0) = Du(x_0),$$

so u is viscosity subsolution. The proof that u is a supersolution is analogous.
(iii) We can always choose $\varphi = u$ so that every point is a local maximum and a local minimum for $u-\varphi$. Then, for any $x \in \Omega$ both inequalities will be satisfied:

$$H(x, u(x), Du(x)) \leq 0,$$
$$H(x, u(x), Du(x)) \geq 0.$$

This implies that the equation is satisfied pointwise. □

Min/Max of viscosity solutions Viscosity subsolutions (respectively, supersolutions) are stable with respect to the max (respectively, the min) operator. Introducing, for u, $v \in C(\Omega)$, the notations

$$(u \vee v)(x) = \max\{u(x), v(x)\},$$
$$(u \wedge v)(x) = \min\{u(x), v(x)\},$$

we have the following stability result.

Proposition 2.3. *The following statements hold true:*

(i) *Let u, $v \in C(\Omega)$ be viscosity subsolutions of the stationary equation in* (2.1). *Then, $u \vee v$ is a viscosity subsolution.*
(ii) *Let u, $v \in C(\Omega)$ be viscosity supersolutions of the stationary equation in* (2.1). *Then, $u \wedge v$ is a viscosity supersolution.*

Proof. Let x_0 be a local maximum point for $u \vee v - \varphi$ where $\varphi \in C^1(\Omega)$ is our test function. Without loss of generality, we can assume that $(u \vee v)(x_0) = u(x_0)$. Since x_0 is local maximum point for $u-\varphi$, we have

$$H(x_0, u(x_0), D\varphi(x_0)) \leq 0,$$

which proves (i). The reverse assertion (ii) can be proved is a similar way. □

An important property which follows from Proposition 2.3 is that the viscosity solution u can be characterized as the maximal subsolution of the equation, i.e.,

$$u \geq v \qquad \text{for any } v \in S, \tag{2.6}$$

where S is the space of subsolutions, i.e., $S = \{v \in C(\Omega):$ condition (i) is satisfied$\}$.

Proposition 2.4. *Let $u \in C(\Omega)$ be a viscosity subsolution of the stationary equation in* (2.1), *such that $u \geq v$ for any viscosity subsolution $v \in C(\Omega)$. Then, u is a viscosity supersolution and therefore a viscosity solution of* (2.1).

Proof. We will prove the result by contradiction. Assume that

$$d := H(x_0, u(x_0), D\varphi(x_0)) < 0$$

for some $\varphi \in C^1(\Omega)$ and $x_0 \in \Omega$ such that

$$u(x_0) - \varphi(x_0) \leq u(x) - \varphi(x) \qquad \forall x \in \overline{B}(x_0, \delta_0) \subset \Omega$$

for some $\delta_0 > 0$. Now, consider the function $w \in C^1(\Omega)$ defined as

$$w(x) := \varphi(x) - |x - x_0|^2 + u(x_0) - \varphi(x_0) + \frac{1}{2}\delta^2$$

for $0 < \delta < \delta_0$. It is easy to check that, by construction,

$$(u - w)(x_0) < (u - w)(x) \qquad \forall x \text{ such that } |x - x_0| = \delta. \tag{2.7}$$

We prove now that, for some sufficiently small δ,

$$H(x, w(x), Dw(x)) \leq 0 \qquad \forall x \in B(x_0, \delta). \tag{2.8}$$

For this purpose, a local uniform continuity argument shows that, for $0 < \delta < \delta_0$,

$$\begin{cases} |\varphi(x) - \varphi(x_0)| \leq \omega_1(\delta), \\ |D\varphi(x) - 2(x - x_0) - D\varphi(x_0)| \leq \omega_2(\delta) + 2\delta \end{cases} \tag{2.9}$$

for any $x \in \overline{B}(x_0, \delta)$, where the ω_i, $i = 1, 2$, are the moduli of continuity of, respectively, φ and $D\varphi$. Then, we have

$$|w(x) - u(x_0)| \leq \omega_1(\delta) + \delta^2, \qquad x \in \overline{B}(x_0, \delta).$$

Now,

$$\begin{aligned} H(x, w(x), Dw(x)) &= d + H(x, w(x), Dw(x)) \\ &= d + H(x, w(x), D\varphi(x) - 2(x - x_0)) - H(x_0, w(x_0), D\varphi(x_0)). \end{aligned} \tag{2.10}$$

Denoting by ω the modulus of continuity of H, we can write

$$H(x, w(x), Dw(x)) \leq d + \omega(\delta, \omega_1(\delta) + \delta^2, \omega_2(\delta) + 2\delta)$$

for all $x \in \overline{B}(x_0, \delta)$. Since d is negative, the above inequality proves (2.8) for $\delta > 0$ small enough. Let us fix such a δ and set

$$\widehat{v}(x) := \begin{cases} u \vee w, & x \in B(x_0, \delta), \\ u, & x \in \Omega \setminus B(x_0, \delta). \end{cases} \tag{2.11}$$

It is easy to check that, by (2.7), $\widehat{v} \in C(\Omega)$, so by Propositions 2.2 and 2.3(i) $\widehat{v}$ is a subsolution of (2.1). Since $\widehat{v}(x_0) > u(x_0)$, the statement is proved. □

Uniform convergence of viscosity solutions Viscosity solutions are also stable with respect to uniform convergence in $C(\Omega)$, as the following result shows. It is worth pointing out that this property does not hold for generalized a.e. solutions.

Proposition 2.5. *Let $u_n \in C(\Omega)$, $n \in \mathbb{N}$, be a viscosity solution of*

$$H_n(x, u_n(x), Du_n(x)) = 0, \qquad x \in \Omega. \tag{2.12}$$

Assume that

$$\begin{array}{ll} u_n \to u & \textit{locally uniformly in } \Omega, \\ H_n \to H & \textit{locally uniformly in } \Omega \times \mathbb{R} \times \mathbb{R}^d. \end{array} \tag{2.13}$$

Then, u is a viscosity solution of (2.1).

Proof. Let $\varphi \in C^1(\Omega)$, and let x_0 be a local maximum point for $u-\varphi$. It is not restrictive to assume that

$$u(x_0)-\varphi(x_0) > u(x)-\varphi(x)$$

for $x \neq x_0$, $x \in B(x_0, r)$ for some $r > 0$. By uniform convergence, and for n large enough, $u_n - \varphi$ attains a local maximum at a point x_n close to x_0. Then, we have

$$H_n(x_n, u_n(x_n), D\varphi_n(x_n)) \leq 0.$$

Since x_n tends to x_0 for $n \to +\infty$, passing to the limit we get

$$H(x_n, u(x_0), D\varphi(x_0)) \leq 0.$$

The proof that u is also a viscosity supersolution can be obtained by a similar argument. □

In addition, the viscosity solution u may be characterized as the uniform limit (for $\varepsilon \to 0$) of classical solutions u^ε of the regularized problem

$$-\varepsilon u^\varepsilon_{xx} + H(x, u^\varepsilon, Du^\varepsilon) = 0, \tag{2.14}$$

that is,

$$\lim_{\varepsilon \to 0+} u^\varepsilon = u.$$

This explains the name of *viscosity solutions*, since the term $-\varepsilon u_{xx}$ corresponds to the viscosity term in Fluid Dynamics. In principle, for a given ε, the regularization term $-\varepsilon u^\varepsilon_{xx}$ allows us to use the theory of elliptic equations to establish the existence of a regular solution $u^\varepsilon \in C^2(\Omega)$. Nevertheless, passing to the limit for $\varepsilon \to 0$ is quite a technical point, since the regularizing effect of the second-order term is cut down as ε vanishes. Moreover, for what we have seen in this chapter, we do not expect to have a regular solution for the limiting (first-order) equation. Without proving a precise result (which can be found in [Ba93]), we try give an idea of the relationship between the definition of viscosity solution and the limiting behavior of u^ε.

Assume that $u^\varepsilon \in C^2(\mathbb{R})$ converge uniformly as $\varepsilon \to 0$ to some $u \in C(\mathbb{R})$. Let us now take a regular function $\varphi \in C^2(\mathbb{R})$ and assume that x is a local maximum point for $u - \varphi$. By uniform convergence, $u^\varepsilon - \varphi$ attains a local maximum at some point x^ε and

$$\lim_{\varepsilon \to 0+} x^\varepsilon = x.$$

By the maximum properties, we get

$$D(u^\varepsilon - \varphi)(x^\varepsilon) = 0, \qquad -D^2(u^\varepsilon - \varphi)(x^\varepsilon) \geq 0,$$

where D^2 denotes the second derivative operator. Then, using the previous inequalities and the fact that u^ε is a classical solution of (2.14), we obtain

$$-\varepsilon\varphi_{xx}(x^\varepsilon) + H(x^\varepsilon, u(x^\varepsilon), D\varphi(x^\varepsilon)) \leq -\varepsilon u^\varepsilon_{xx}(x^\varepsilon) + H(x^\varepsilon, u(x^\varepsilon), Du(x^\varepsilon)) = 0.$$

Passing to the limit and using the continuity of φ_{xx}, $D\varphi$, and H we finally get

$$H(x, u(x), D\varphi(x)) \leq 0.$$

This computation explains why, from the theoretical point of view, it is reasonable to obtain viscosity solutions via an elliptic regularization of the first-order equation. Such a technique, however, is impractical for the numerical approach. In fact, the introduction of a second-order term has a smoothing effect on the singularities of solutions, which are definitely a point of interest in the approximation of HJ equations.

Before concluding the section, we give a result concerning changes of variables in (2.1). In what follows, this result will be useful for dealing with certain control problems.

Proposition 2.6. *Let $u \in C(\Omega)$ be a viscosity solution of (2.1), and let $\Phi : \mathbb{R} \to \mathbb{R}$ be a function in $C^1(\mathbb{R})$ such that $\Phi'(t) > 0$. Then $v := \Phi(u)$ is a viscosity solution of*

$$H\left(x, \Phi^{-1}(v(x)), \Phi^{-1}(v(x))Dv(x)\right) = 0, \qquad x \in \Omega. \tag{2.15}$$

We conclude this review with a warning about the fact that viscosity solutions are not preserved by a change of sign in the equation. This behavior, although unusual, is easily understood if we go back to the definition, which takes into account only the local properties at minimum and maximum points for $u - \varphi$. Indeed, the fact that any local maximum of $u-\varphi$ is a local minimum of $-(u-\varphi)$ implies that u is a viscosity subsolution of (2.1) if and only if $v = -u$ is a viscosity supersolution of $-H(x, -v(x), -Dv(x)) = 0$ in Ω. Similarly, u is a viscosity supersolution of (2.1) if and only if $v = -u$ is a viscosity subsolution of $-H(x, -v(x), -Dv(x)) = 0$ in Ω.

2.1.2 ▪ An alternative definition of viscosity solution

The definition of viscosity solutions based on test functions is probably the most popular and useful. However, the original definition was based on the notion of sub- and superdifferentials which are standard tools in convex analysis.

To introduce this notion, we first recall the elementary definition of a differentiable function.

Definition 2.7. *v is differentiable at x_0 if and only if there exists a linear operator A such that*

$$\lim_{y \to x_0} \frac{v(y) - v(x_0) - A(y - x_0)}{|y - x_0|} = 0.$$

If v is not differentiable at x_0 we can still have a generalized version of the differential, based on the following idea (sketched in one dimension for simplicity). In a univariate differentiable function the tangent at x_0 is uniquely defined by taking the limit of secants passing through points of the graph of v which converge to x_0, the limit slope defining the derivative v'. If v is not differentiable at x_0, this procedure does not select a unique tangent (and a unique derivative), but rather a set of "generalized tangents." For example, $v(x) = |x|$ has a unique tangent for every $x \neq 0$, whereas for $x = 0$ the "generalized tangents" are all the straight lines passing in $(0, 0)$ and not crossing the graph of v, i.e., the lines of the set

$$\{y = mx, m \in (-1, 1)\}.$$

The standard extension of derivatives used in nonsmooth analysis is based on this idea and corresponds to the following definition of *sub- and superdifferentials.*

Definition 2.8. *Let Ω be an open set of $\mathbb{R}^d$, and let $v : \Omega \to \mathbb{R}$.*

The superdifferential *$D^+v(x)$ of v at $x \in \Omega$ is defined as the set*

$$D^+v(x) := \left\{ p \in \mathbb{R}^d : \limsup_{y \to x\ y \in \Omega} \frac{v(y) - v(x) - p \cdot (y - x)}{|y - x|} \leq 0 \right\}.$$

The subdifferential $D^- v(x)$ *of* v *at* $x \in \Omega$ *is defined as the set*

$$D^- v(x) := \left\{ q \in \mathbb{R}^d : \liminf_{y \to x\ y \in \Omega} \frac{v(y) - v(x) - q \cdot (y - x)}{|y - x|} \geq 0 \right\}.$$

We remark that the sub- and superdifferentials are (possibly empty) convex and closed subsets of $\mathbb{R}^d$. Moreover, if at the point x we have

$$D^+ v(x) = D^- v(x),$$

then v is differentiable in the sense of Definition 2.7. Going back to the example of $v(x) = |x|$, we have, for $x \neq 0$,

$$D^+ v(x) = D^- v(x) = \begin{cases} -1 & \text{for } x < 0, \\ 1 & \text{for } x > 0, \end{cases}$$

so they coincide and give the standard derivative.

At $x = 0$, we have

$$\begin{aligned} D^- v(0) &= \left\{ q \in \mathbb{R}^d : \liminf_{y \to 0} \frac{|y| - q\, y}{|y|} \geq 0 \right\} \\ &= \left\{ q \in \mathbb{R}^d : \liminf_{y \to 0} (1 - q \operatorname{sgn}(y)) \geq 0 \right\}. \end{aligned}$$

Taking, for example, $q = 1$, we have

$$\liminf_{y \to 0} (1 - \operatorname{sgn}(y)) = \min\{2, 0\} = 0.$$

In the same way, we can show that every $q \in [-1, 1]$ belongs to the subdifferential. Then, $D^- v(0) = [-1, 1]$.

Considering now the superdifferential at $x = 0$, we have

$$\begin{aligned} D^+ v(0) &= \left\{ p \in \mathbb{R}^d : \limsup_{y \to 0} \frac{|y| - p\, y}{|y|} \leq 0 \right\} \\ &= \left\{ p \in \mathbb{R}^d : \limsup_{y \to 0} (1 - p \operatorname{sgn}(y)) \leq 0 \right\}. \end{aligned}$$

On the other hand, we have

$$\begin{aligned} \limsup_{y \to 0} (1 - p \operatorname{sgn}(y)) &= \lim_{y \to 0} (1 + \max(p, -p)) \\ &= \lim_{y \to 0} (1 + |p|) \geq 1. \end{aligned}$$

This shows that $D^+ v(0) = \emptyset$.

We can now give the definition of viscosity solution based on sub- and superdifferentials.

Definition 2.9. *Let* $u \in BUC(\Omega)$. *We say that* u *is a viscosity solution of* (2.1) *if and only if the following conditions hold:*

(i) $H(x, u(x), p) \leq 0$ *for all* $p \in D^+ u(x)$, $x \in \Omega$
(i.e., u is a viscosity subsolution*);*

(ii) $H(x,u(x),p) \geq 0$ *for all* $p \in D^{-}u(x)$, $x \in \Omega$
(i.e., u is a viscosity supersolution*).*

The following result explains the link between the two definitions.

Theorem 2.10. *Let Ω be an open subset of $\mathbb{R}^d$, and let $u \in C(\Omega)$; then the following hold:*

(i) $p \in D^{+}u(x)$, $x \in \Omega$, *if and only if there exists a function* $\varphi \in C^1(\Omega)$ *such that* $D\varphi(x) = p$ *and* $u - \varphi$ *has a local maximum point at* x.
(ii) $q \in D^{-}u(x)$, $x \in \Omega$, *if and only if there exists a function* $\varphi \in C^1(\Omega)$ *such that* $D\varphi(x) = q$ *and* $u - \varphi$ *has a local minimum point at* x.

Proof. A proof of this theorem can be found in [BCD97] (Lemma 1.7, p. 29). □

We conclude this section with a further result linking viscosity and a.e. solutions.

Theorem 2.11. *The following statements hold true:*

(i) *if* $u \in C(\Omega)$ *is a viscosity solution of* (2.1), *then*

$$H(x,u(x),Du(x)) = 0$$

at any point $x \in \Omega$ *where* u *is differentiable;*
(ii) *if* u *is locally Lipschitz continuous and is a viscosity solution of* (2.1), *then*

$$H(x,u(x),Du(x)) = 0 \qquad \text{a.e. in } \Omega.$$

Proof. Let u be a point of differentiability for u. Then, $D^{+}u(x) \cap D^{-}u(x) \neq \emptyset$ since this set contains $Du(x)$. Moreover, this intersection reduces to a singleton, so that

$$Du(x) = D^{+}u(x) = D^{-}u(x).$$

Hence, by Definition 2.9 we have

$$0 \geq H(x,u(x),Du(x)) \geq 0,$$

which proves (i). Statement (ii) follows immediately from (i) and Rademacher's theorem (which states that Lipschitz continuous functions are a.e. differentiable; see [E98, pp. 280–281] for the proof). □

2.1.3 ▪ Uniqueness of viscosity solutions

The crucial point in the theory of viscosity solution (and also the essential advantage over the concept of solutions a.e.) is to prove uniqueness. This is done via a *comparison principle*, also termed *maximum principle*.

Theorem 2.12. *Let* $u, v \in BUC(\Omega)$ *be, respectively, a sub- and a supersolution for* (2.1), *and let*

$$u(x) \leq v(x) \qquad \text{for any } x \in \partial\Omega.$$

Then,

$$u(x) \leq v(x) \qquad \text{for any } x \in \Omega.$$

This is enough to get uniqueness, since two viscosity solutions of the equation are (both) sub- and supersolutions. This implies that both $u(x) \leq v(x)$ and $u(x) \geq v(x)$ for any $x \in \Omega$, i.e., $u(x) = v(x)$ in Ω.

The following is a classical assumption for uniqueness.

(A4) Let $\omega(\cdot)$ be a modulus of continuity. We assume that

$$|H(x,u,p) - H(y,u,p)| \leq \omega(|x-y|(1+|p|))Q_R(x,y,u,p)$$

for any $x, y \in \Omega$, $u \in [-R, R]$, and $p \in \mathbb{R}^d$, where

$$Q_R(x,y,u,p) = \max(\gamma(H(x,u,p)), \gamma(H(y,u,p)))$$

and $\gamma : \mathbb{R} \to \mathbb{R}_+$ is continuous.

We can now give a sufficient condition for the comparison principle to hold.

Theorem 2.13. *Let assumptions* (A1)–(A4) *be satisfied. Then, the comparison principle (Theorem* 2.12*) holds for* (2.1)*; i.e., the viscosity solution is unique.*

Proof. We only sketch the proof, which is carried out via variable doubling.

The goal is to prove that $M = \max_{x \in \overline{\Omega}}(u - v)$ is negative. To this end, we introduce a test function depending on *two variables*:

$$\psi_\varepsilon(x,y) = u(x) - v(y) - \frac{|x-y|^2}{\varepsilon^2}.$$

Due to the penalization term, we can expect that the maximum points $(x_\varepsilon, y_\varepsilon)$ for ψ_ε should have x_ε and y_ε close enough for ε small. Moreover, for $\varepsilon \to 0^+$ we have

$$\begin{aligned} &M_\varepsilon \to M, \\ &\frac{|x-y|^2}{\varepsilon^2} \to 0, \\ &u(x_\varepsilon) - v(y_\varepsilon) \to M. \end{aligned}$$

This allows us to pass to the limit and get the comparison result. □

2.2 ▪ Viscosity solution for evolutive equations

Adapting the same basic ideas, the definition of viscosity solution can be easily extended to the evolutive case, by taking into account also the time derivative.

Definition 2.14. *$u \in BUC(\Omega \times (0,T))$ is a* viscosity solution *in $\Omega \times (0,T)$ of the equation*

$$u_t + H(x,t,u,Du) = 0 \tag{2.16}$$

if and only if, for any $\varphi \in C^1(\Omega \times (0,T))$, the following conditions hold:

(i) *at every point $(x_0,t_0) \in \Omega \times (0,T)$, local* maximum *for $u - \varphi$,*

$$\varphi_t(x_0,t_0) + (H(x_0,t_0,u(x_0,t_0),D\varphi(x_0,t_0)) \leq 0$$

(i.e., u is a viscosity subsolution*);*

(ii) *at every point* $(x_0,t_0)\in\Omega\times(0,T)$, *local* minimum *for* $u-\varphi$,

$$\varphi_t(x_0,t_0)+(H(x_0,t_0,u(x_0,t_0),D\varphi(x_0,t_0))\geq 0$$

(i.e., u is a viscosity supersolution*).*

This definition is very close to that used for stationary problems. In fact, it could also be derived from Definition 2.1 by setting the problem (2.1) in $d+1$ variables $(x,t)\in\mathbb{R}^d\times(0,T)$ and making the time derivative explicit.

The following comparison result refers to the evolutive problem and usually gives the uniqueness for the particular case $H(x,t,u,Du)=H(t,Du)$.

Theorem 2.15. *Assume* $H\in C((0,T)\times\mathbb{R}^d)$*. Let* $u_1,u_2\in C(\mathbb{R}^d\times(0,T))$ *be, respectively, the viscosity sub- and supersolutions in* $\mathbb{R}^d\times[0,T]$ *of the equation*

$$u_t(x,t)+H(t,Du(x,t))=0.$$

Then,

$$\sup_{\mathbb{R}^d\times(0,T)}(u_1-u_2)\leq\sup_{\mathbb{R}^d}(u_1(\cdot,0)-u_2(\cdot,0)).$$

Proof. The proof can be found in [BCD97, p. 56]. Note that comparison results for more general Hamiltonians can be found in [Ba98]. □

2.2.1 ▪ Representation formulae and Legendre transform

In some cases it is possible to derive representation formulae for viscosity solutions. These formulae have a great importance from both the analytical and the numerical points of view and will be derived here in two major cases: linear advection equations and convex HJ equations.

Linear advection equation

Let us start by the linear case in which the representation formula can be obtained via the *method of characteristics.*

Theorem 2.16. *Let* $u:\mathbb{R}^d\times(t_0,T)\to\mathbb{R}$ *be a viscosity solution of the initial value problem*

$$\begin{cases} u_t(x,t)+\lambda u(x,t)+f(x,t)\cdot Du(x,t)=g(x,t), & (x,t)\in\mathbb{R}^d\times(t_0,T),\\ u(x,t_0)=u_0(x), & x\in\mathbb{R}^d. \end{cases} \tag{2.17}$$

Assume that $f:\mathbb{R}^d\times(t_0,T)\to\mathbb{R}^d$ *and* $g:\mathbb{R}^d\times(t_0,T)\to\mathbb{R}$ *are continuous in* (x,t)*, and* f *is globally Lipschitz continuous with respect to* x*. Then,*

$$u(x,t)=e^{\lambda(t_0-t)}u_0(y(x,t;t_0))+\int_{t_0}^{t}e^{\lambda(s-t)}g(y(x,t;s),s)ds, \tag{2.18}$$

where $y(x,t;s)$ *is the position at time* s *of the solution trajectory passing through* x *at time* t*, i.e., solving the Cauchy problem*

$$\begin{cases} \dfrac{d}{ds}y(x,t;s)=f(y(x,t;s),s),\\ y(x,t;t)=x. \end{cases} \tag{2.19}$$

Proof. We give the proof under the additional assumption that $u \in C^1$.

Let (x,t) be fixed, and denote for shortness the solution of (2.19) as $y(s) = y(x,t;s)$. Writing the equation in (2.17) at a point $(y(s),s)$ and multiplying by $e^{\lambda s}$, we have

$$e^{\lambda s} u_s(y(s),s) + \lambda e^{\lambda s} u(y(s),s) + e^{\lambda s} f(y(s),s) \cdot Du(y(s),s) = e^{\lambda s} g(y(s),s).$$

Since u is differentiable, this may also be rewritten as

$$\frac{d}{ds}\left[e^{\lambda s} u(y(s),s)\right] = e^{\lambda s} g(y(s),s). \tag{2.20}$$

Integrating (2.20) over the interval $[t_0, t]$ we get

$$e^{\lambda t} u(y(t),t) = e^{\lambda t_0} u(y(t_0),t_0) + \int_{t_0}^{t} e^{\lambda s} g(y(s),s) ds. \tag{2.21}$$

Recalling that $y(t) = y(x,t;t) = x$ and $u(y(t_0),t_0) = u_0(y(t_0))$ and dividing by $e^{\lambda t}$, we get

$$u(x,t) = e^{\lambda(t_0 - t)} u_0(y(t_0)) + \int_{t_0}^{t} e^{\lambda(s-t)} g(y(s),s) ds,$$

which coincides with (2.18). □

We recall that a solution of (2.19) is called *characteristic curve*. Note that, in using (2.18), (2.19) is integrated backwards. Note also that if λ is strictly positive, f and g do not depend on t, and the source g is bounded, then $u(x,t)$ has a limit for $t - t_0 \to \infty$. Setting conventionally $t = 0$ and letting $t_0 \to -\infty$, we obtain in fact the limit

$$\begin{aligned} u(x) &= \int_{-\infty}^{0} e^{\lambda s} g(y(x,0;s)) ds \\ &= \int_{0}^{\infty} e^{-\lambda s} g(y(x,0;-s)) ds, \end{aligned} \tag{2.22}$$

which is a regime solution for problem (2.17) or, in other terms, solves the stationary equation

$$\lambda u(x) + f(x) \cdot Du(x) = g(x)$$

for $x \in \mathbb{R}^d$.

HJ equations

Concerning HJ equations, the representation formula is known as *Hopf–Lax formula* and is typically related to the problem

$$\begin{cases} u_t + H(Du) = 0, & (x,t) \in \mathbb{R}^d \times (0,T), \\ u(x,0) = u_0(x), & x \in \mathbb{R}^d, \end{cases} \tag{2.23}$$

where $H : \mathbb{R}^d \to \mathbb{R}$ is convex and satisfies the coercivity condition

$$\lim_{|p| \to +\infty} \frac{H(p)}{|p|} = +\infty. \tag{2.24}$$

Assumption (2.24) allows us to give the following definition.

Definition 2.17. *Let* (2.24) *be satisfied. We define the* Legendre–Fenchel conjugate *(or* Legendre–Fenchel transform*) of H for $q \in \mathbb{R}^d$ as*

$$H^*(q) = \sup_{p\in\mathbb{R}^d} \{p \cdot q - H(p)\}. \tag{2.25}$$

Note that the convexity assumption on H implies that H is continuous, and also being coercive in the sense of (2.24), the sup in (2.25) is in fact a maximum.

In general, the Legendre–Fenchel transform may not allow for an explicit computation—we will show in Chapter 3 a numerical procedure for approximating it. A few examples, anyway, can be computed analytically, among which is the quadratic Hamiltonian

$$H_2(p) = \frac{|p|^2}{2},$$

for which an easy computation gives

$$H_2^*(q) = \frac{|q|^2}{2}.$$

The definition of Legendre–Fenchel transform may also work for the noncoercive case, but in this case the conjugate function will not in general be defined everywhere and be bounded. For example, taking

$$H_1(p) = |p|$$

and using the definition (2.25), it is easy to check that

$$H_1^*(q) = \begin{cases} 0 & \text{for } |q| \le 1, \\ +\infty & \text{elsewhere.} \end{cases}$$

The main result of interest here concerns two important properties of the Legendre transform.

Theorem 2.18. *Let* (2.24) *be satisfied. Then, the function H^* has the following properties:*

(i) *$H^* : \mathbb{R}^d \to \mathbb{R}$ is convex and*

$$\lim_{|p|\to+\infty} \frac{H^*(p)}{|p|} = +\infty;$$

(ii) *$H(p) = H^{**}(p)$ for any $p \in \mathbb{R}^d$.*

Proof. The proof can be found in [E98, p. 122]. □

The above theorem says that by applying the Legendre transform to H twice we obtain back H itself. The definition of Legendre transform is very useful for characterizing the unique solution of (2.23) by means of the so-called *Hopf–Lax representation formula* (see [Ho65]). The following theorems provide the basic results concerning this characterization.

Theorem 2.19. *The function u defined by the following Hopf–Lax formula:*

$$u(x,t)=\inf_{y\in\mathbb{R}^d}\left\{u_0(y)+tH^*\left(\frac{x-y}{t}\right)\right\} \tag{2.26}$$

is Lipschitz continuous, is differentiable a.e. in $\mathbb{R}^d\times(0,+\infty)$, *and solves in the viscosity sense the initial value problem* (2.23).

Proof. The proof relies on the functional identity

$$u(x,t)=\min_{y\in\mathbb{R}^d}\left\{(t-s)H^*\left(\frac{x-y}{t-s}\right)+u(y,s)\right\} \tag{2.27}$$

(which is proved in [E98, p. 126]). This identity implies that u is Lipschitz continuous and, by Rademacher's theorem, differentiable a.e. Now, consider a point of differentiability (x,t).

Step 1. u is a subsolution at (x,t).
Fix $p\in\mathbb{R}^d$ and a positive parameter h. By (2.27), we have

$$u(x+hp,t+h)=\min_{y\in\mathbb{R}^d}\left\{hH^*\left(\frac{x+hp-y}{h}\right)+u(y,t)\right\}\le hH^*(p)+u(x,t),$$

which implies

$$\frac{u(x+hp,t+h)-u(x,t)}{h}\le H^*(p).$$

Now, for $h\to 0$, we get

$$p\cdot Du(x,t)+u_t(x,t)\le H^*(p).$$

This inequality holds for any $p\in\mathbb{R}^d$, so that we have

$$u_t(x,t)+H(Du(x,t))+u_t(x,t)+\max_{p\in\mathbb{R}^d}\{p\cdot Du(x,t)-H^*(p)\}\le 0,$$

where we have used the fact that $H=H^{**}$.

Step 2. u is a supersolution at (x,t).
Let us choose z such that

$$u(x,t)=t\,H^*\left(\frac{x-z}{t}\right)+u_0(z).$$

Fix a positive parameter h, and set $s=t-h$, $y=x\,s/t+(1-s/t)z$. This implies

$$\frac{x-s}{t}=\frac{y-z}{s},$$

and therefore

$$\begin{aligned}u(x,t)-u(y,s)&\ge tH^*\left(\frac{x-z}{t}\right)+u_0(z)-\left[sH^*\left(\frac{y-z}{s}\right)+u_0(z)\right]\\&=(t-s)H^*\left(\frac{x-z}{t}\right).\end{aligned}$$

As a consequence,

$$\frac{u(x,t)-u\left(\left(1-\frac{h}{t}\right)x+\frac{h}{t}z,t-h\right)}{h}\geq H^*\left(\frac{x-z}{t}\right),$$

and, for $h\to 0$, we obtain

$$\frac{x-z}{t}\cdot Du(x,t)+u_t(x,t)\geq H^*\left(\frac{x-z}{t}\right).$$

Then, we finally get

$$\begin{aligned}u_t(x,t)+H((Du(x,t))&=u_t(x,t)+\max_{p\in\mathbb{R}^d}\{p\cdot Du(x,t)-H^*(p)\}\\&\geq u_t(x,t)+\frac{x-z}{t}\cdot Du(x,t)-H^*\left(\frac{x-z}{t}\right)\geq 0,\end{aligned}$$

which concludes the proof. □

Theorem 2.20. *The unique viscosity solution of* (2.23) *is given by the Hopf–Lax representation formula.*

Proof. The proof of this result can be found in [E98, p. 561]. □

We finally note that an equivalent way of rewriting Hopf–Lax formula is to set

$$a:=\frac{x-y}{t},$$

so that $y=x+at$. Since the minimization with respect to y corresponds to a minimization with respect to a, we also obtain

$$u(x,t)=\inf_{a\in\mathbb{R}^d}\{u_0(x-at)+tH^*(a)\}. \tag{2.28}$$

This formula shows more clearly the link between the value of the solution at (x,t) and the values of the solution at points $(x-at,0)$, and it will be useful in the analysis of SL schemes.

2.2.2 ▪ Semiconcavity and regularity of viscosity solutions

As we have seen earlier in this chapter, we expect viscosity solutions to be in general only uniformly continuous. However, some further regularity results can be given. We will examine here the most typical of these results. A regularity property which plays an important role in HJ equations is *semiconcavity*, defined as follows.

Definition 2.21. *A function $u\in C(\Omega)$ is* semiconcave *in the open convex set Ω if there exists a constant $C>0$ such that, for any $x,z\in\Omega$ and $\mu\in[0,1]$,*

$$\mu u(x)+(1-\mu)u(y)\leq u(\mu x+(1-\mu)y)+\frac{1}{2}C\mu(1-\mu)|x-y|^2. \tag{2.29}$$

This definition corresponds to the request that $u(x)-\frac{1}{2}C|x|^2$ be concave, as it can be immediately verified. Note that an equivalent definition, when u is continuous, is to require that

$$u(x+h)-2u(x)+u(x-h)\leq C|h|^2 \tag{2.30}$$

for all $x\in\Omega$ and $h\in\mathbb{R}^d$.

Clearly, concave functions are also semiconcave. Moreover, C^1 functions having a Lipschitz continuous gradient are semiconcave. Another important example of nondifferentiable semiconcave functions is given by *marginal functions*, which are functions defined as

$$u(x) = \inf_{a \in A} F(x,a), \tag{2.31}$$

provided $F(\cdot,a)$ satisfies (2.29) uniformly with respect to a. Marginal functions are particularly important in control problems, since value functions belong to this class. We will see that semiconcavity is one of the properties that the value function inherits from the data in many control problems.

Example: Distance function from a set Given a domain $C \subset \mathbb{R}^d$, $C \neq \emptyset$, define as usual the distance $d(x,C)$ by

$$d(x,C) := \inf_{y \in C} |x-y|.$$

Then, d^2 is semiconcave in $\mathbb{R}^d$, since the application $x \to |x-y|^2$ is C^∞ and has constant second derivatives. Moreover, d itself is semiconcave on any compact set having strictly positive distance from C, because the mapping $x \to |x-y|$ also has bounded second derivatives in such a set, and that bound is uniform for y in a bounded set. Note that our counterexample (2.2) corresponds to the characterization of the distance function from the set $\mathbb{R} \setminus (-1,1)$ and that the viscosity solution $u(x) = 1 - |x|$ is semiconcave (and it is also the only semiconcave a.e. solution).

Properties of semiconcave functions We review here some properties of semiconcave functions which are of particular interest in relationship with viscosity solutions of HJ equations (see [CS04] for further details).

Theorem 2.22. *Let u be semiconcave in Ω. Then, u is locally Lipschitz continuous in Ω.*

Proof. For any fixed $x \in \Omega$ and any h such that $x+h \in \Omega$, we have

$$u(x+h) - u(x) = \psi(x+h) - \psi(x) + Cx \cdot h + \frac{C}{2}|h|^2,$$

where $\psi(x) = u(x) - \frac{1}{2}C|x|^2$ is concave and, therefore, locally Lipschitz continuous in Ω. Since the sum of the last two terms is quadratic in h (and, hence, locally Lipschitz too), this proves the assertion. □

Theorem 2.23. *Assume that the Hamiltonian H is convex and that the initial condition u_0 is semiconcave. Then, the viscosity solution of* (2.23) *is semiconcave.*

Proof. Assume that u_0 satisfies (2.30), and let x and h be fixed. By the Hopf–Lax formula, the solution $u(x,t)$ of (2.23) may be written at x and $x \pm h$ as

$$\begin{aligned}
u(x,t) &= \inf_{a \in \mathbb{R}^d} \{u_0(x-at) + tH^*(a)\} = u_0(x - \bar{a}t) + tH^*(\bar{a}), \\
u(x+h,t) &= \inf_{a \in \mathbb{R}^d} \{u_0(x+h-at) + tH^*(a)\} = u_0(x+h-\bar{a}_+ t) + tH^*(\bar{a}_+), \\
u(x-h,t) &= \inf_{a \in \mathbb{R}^d} \{u_0(x-h-at) + tH^*(a)\} = u_0(x-h-\bar{a}_- t) + tH^*(\bar{a}_-),
\end{aligned}$$

where we have denoted by $\overline{a}$, $\overline{a}_+$, and $\overline{a}_-$ the minimizers for, respectively, x, $x-h$, and $x+h$ in the Hopf–Lax formula. On the other hand, replacing $\overline{a}_\pm$ by $\overline{a}$, we obtain the inequalities

$$u(x+h,t) \le u_0(x+h-\overline{a}t) + tH^*(\overline{a}),$$
$$u(x-h,t) \le u_0(x-h-\overline{a}t) + tH^*(\overline{a}),$$

so that at last we get

$$\begin{aligned} u(x+h)-2u(x)+u(x-h) &\le u_0(x+h-\overline{a}t)-2u_0(x-\overline{a}t)+u_0(x-h-\overline{a}t) \\ &\le C|h|^2 \end{aligned}$$

and, by semiconcavity of u_0, we obtain semiconcavity of u. □

Let now $u \in W^{1,\infty}_{loc}(\Omega)$, and define the set

$$D^*u(x) := \left\{ p \in \mathbb{R}^d : p = \lim_{n\to\infty} Du(x_n), x_n \to x \right\}.$$

Then, D^* is nonempty and closed for any $x \in \Omega$. Denote by $\operatorname{co} D^*u(x)$ its convex hull. A classical result in nonsmooth analysis (see [Cl83] for the proof) states that

$$\operatorname{co} D^*u(x) = \partial u(x) \qquad \forall x \in \Omega,$$

where $\partial u(x)$ is the *generalized gradient* of u at x, defined by

$$\begin{aligned} \partial u(x) &:= \left\{ p \in \mathbb{R}^d : \overline{u}(x,q) \ge p\cdot q \ \forall q \in \mathbb{R}^d \right\} \\ &= \left\{ p \in \mathbb{R}^d : \underline{u}(x,q) \le p\cdot q \ \forall q \in \mathbb{R}^d \right\}, \end{aligned}$$

and $\overline{u}$, $\underline{u}$ are, respectively, the generalized directional derivatives defined by

$$\overline{u}(x;q) := \limsup_{y\to x, t\to 0+} \frac{u(y+tq)-u(y)}{t},$$
$$\underline{u}(x;q) := \liminf_{y\to x, t\to 0+} \frac{u(y+tq)-u(y)}{t}.$$

An important property is that, under the assumption of semiconcavity, some generalized and classical derivatives coincide.

Proposition 2.24. *Let u be semiconcave in Ω. Then, for all $x \in \Omega$, the following hold:*

(i) $D^+u(x) = \partial u(x) = \operatorname{co} D^*u(x)$.
(ii) *Either $D^-u(x) = \emptyset$ or u is differentiable at x.*
(iii) *If $D^+u(x)$ is a singleton, then u is differentiable at x.*
(iv) $\frac{\partial u}{\partial q}(x) = \min_{p\in D^+u(x)} p\cdot q$ *for all unit vectors q.*

Proof. The proof can be found in [BCD97, p. 66]. □

Remark 2.25. *Claim* (ii) *has an important consequence. Assume that the Legendre transform H^* is coercive and has bounded second derivatives. Then, writing the Hopf–Lax representation formula for u,*

$$u(x,t) = \inf_{a\in\mathbb{R}^d} \{u_0(x-at)+tH^*(a)\},$$

we note that the infimum is actually a minimum and that the function to be minimized is semiconcave. At a minimum point $\overline{a}$, its subdifferential cannot be empty (in fact, it must contain at least the origin), and by the regularity of H^*, claim (ii) *implies that u must be differentiable at the point $x-\overline{a}t$.*

Moreover, semiconcavity allows us to clarify the link between a.e. solutions and viscosity supersolutions.

Proposition 2.26. *Let u be semiconcave and such that*

$$H(x,u(x),Du(x))\geq 0 \qquad \text{a.e. in } \Omega,$$

where H is continuous. Then, u is a viscosity supersolution of

$$H(x,u(x),Du(x))=0 \qquad \text{in } \Omega.$$

Proof. The proof can be found in [BCD97]. □

Note that this proposition gives a sharper interpretation of counterexample (2.2). In fact, all a.e. solutions are subsolutions in the viscosity sense, but only the semiconcave one $u(x)=1-|x|$ is also a supersolution.

Under suitable assumptions, it is possible to prove the semiconcavity of viscosity solutions for HJ equations of more general structure, as the following theorem shows.

Theorem 2.27. *Let $\lambda>0$, and let $u\in W^{1,\infty}(\mathbb{R}^d)$ be a viscosity solution of*

$$\lambda u(x)+H(x,Du(x))=0, \qquad x\in\mathbb{R}^d,$$

with Lipschitz constant L_u. Assume that H satisfies

$$|H(x,p)-H(x,q)|\leq \omega(|p-q|) \qquad \forall x,p,q\in\mathbb{R}^d \tag{2.32}$$

and that, for some $C>0$ and $\widehat{L}>2L_u$,

$$H(x+h,p+Ch)-2H(x,p)+H(x-h,p-Ch)\geq -C|h|^2 \tag{2.33}$$

holds for all $x,h\in\mathbb{R}^d$, $p\in B(0,\widehat{L})$. Then, u is semiconcave on $\mathbb{R}^d$.

Proof. The proof can be found in [BCD97, p. 69]. □

A final remark is that combining convexity of the Hamiltonian with semiconcave regularity one can obtain an unexpected differentiability result for the viscosity solution.

Proposition 2.28. *Assume that $u\in C(\Omega)$ is a viscosity solution in Ω of*

$$\lambda u(x)+H(x,Du(x))=0$$

with $\lambda\geq 0$. Assume also that $H(x,\cdot)$ is strictly convex for any fixed $x\in\Omega$ and that $-u$ is semiconcave. Then $u\in C^1(\Omega)$.

Proof. The proof can be found in [BCD97]. □

2.3 ▪ Problems in bounded domains

Another peculiar point of the theory of viscosity solutions is the way boundary conditions are satisfied. It is not surprising that boundary conditions cannot be arbitrarily assigned in first-order equations—even for advection equations we know that Dirichlet data can be imposed only at points of the boundary where the drift is directed inwards. The technical point for nonlinear equations is that the equation plays a role up to the boundary.

2.3.1 ▪ Boundary conditions in a weak sense

In the theory of viscosity solutions, the typical compact form for boundary conditions is

$$\min(H(x,u(x),Du(x)),B(x,u(x),Du(x))) \le 0,$$
$$\max(H(x,u(x),Du(x)),B(x,u(x),Du(x))) \ge 0,$$

where $x \in \partial\Omega$, and B represents a suitably defined *boundary operator*. In order to explain this setting, we will consider in detail the main situations of interest. For example, the Dirichlet condition would be formulated in a classical form as

$$u(x) = b(x), \qquad x \in \partial\Omega.$$

In its weak formulation, this corresponds to a boundary operator defined by

$$B(x,u(x),Du(x)) = u(x) - b(x).$$

We still remark that not all the boundary conditions are compatible with the equation. In this section, we briefly analyze the effect of Dirichlet, Neumann, and "state constraint" boundary conditions, posed on parts of the boundary. First, note that boundary conditions should be imposed in a weak sense. The condition which defines u as a viscosity subsolution for (2.1) requires that for any test function $\varphi \in C^1(\overline{\Omega})$ and $x \in \partial\Omega$ local maximum point for $u - \varphi$,

$$\min\{H(x,u(x),D\varphi(x)),B(x,u,D\varphi(x))\} \le 0 \tag{2.34}$$

for a given boundary operator B. Similarly, the condition for supersolutions requires that for any test function $\varphi \in C^1(\overline{\Omega})$ and $x \in \partial\Omega$ local minimum point for $u - \varphi$,

$$\max\{H(x,u(x),D\varphi(x)),B(x,u,D\varphi(x))\} \ge 0. \tag{2.35}$$

The effect of the Dirichlet condition is to impose a value on u according to the above conditions; in particular, the value $u(x) = b(x)$ is set at every point where $H(x,u(x),D\varphi(x)) \ge 0$ (for subsolutions) and $H(x,u(x),D\varphi(x)) \le 0$ (for supersolutions).

The Neumann condition, which is classically stated as

$$\frac{\partial u}{\partial \nu}(x) = m(x), \qquad x \in \partial\Omega,$$

uses in its weak formulation the boundary operator

$$B(x,u(x),Du(x)) = \frac{\partial u}{\partial \nu} - m,$$

where $\nu(\cdot)$ represents the outward normal to the domain Ω. A typical use of this condition occurs when we know (or presume) that the level curves of the solution are orthogonal to the boundary $\partial\Omega$ or to a part of it, in which case we simply choose $m(x)=0$.

On the other hand, in the "state constraints" boundary condition neither a value for u nor a value for its normal derivative $\partial u/\partial\nu(x)$ is imposed. In this respect, it has been sometimes interpreted as a "no boundary condition" choice, although this interpretation is quite sloppy. In fact, a bounded and uniformly continuous function u is said to be a *state constrained viscosity solution* if and only if it is a subsolution in Ω and a supersolution in $\overline{\Omega}$ (i.e., up to the boundary). Whenever a bound on the L^∞ norm of the solution is known, this condition can be also stated as a Dirichlet boundary condition by simply setting $b(x)\equiv\overline{C}$, provided the constant $\overline{C}$ satisfies

$$\overline{C} > \max_{x\in\Omega} u(x).$$

By this choice (2.34) is trivially satisfied, whereas (2.35) requires (strictly)

$$H(x,u(x),D\varphi(x))\geq 0. \tag{2.36}$$

Representation formulae in bounded domains We treat separately the linear case in a bounded domain Ω. Define by

$$\Gamma_{in} := \{x\in\partial\Omega : f(x,t)\cdot\nu(x)<0\} \tag{2.37}$$

the subset of boundary points where the vector field f is pointing inwards to Ω. We consider again the linear evolutive problem (1.1), which is rewritten here as

$$\begin{cases} u_t+f(x,t)\cdot Du=g(x,t), & (x,t)\in\Omega\times(t_0,T),\\ u(x,0)=u_0(x), & x\in\overline{\Omega}, \end{cases}$$

and complemented with the Dirichlet boundary condition

$$u(x,t)=b(x,t), \qquad x\in\Gamma_{in}. \tag{2.38}$$

Note that, in terms of characteristics, the existence of an inflow boundary Γ_{in} means that the characteristic $y(x,t;s)$ may have a finite exit time from Ω for s going backwards. More precisely, we will set

$$\theta(x,t) := \sup\left\{s\leq t : y(x,t;s)\in\mathbb{R}^N\setminus\Omega\right\}. \tag{2.39}$$

By its definition, $\theta(x,t)$ represents the time at which the characteristic passing at (x,t) starts from the boundary. The representation formula (written at (x,t)) may be changed accordingly, obtaining

$$u(x,t)=\int_{t_0\vee\theta(x,t)}^{t} g(y(x,t;s),s)ds+u(y(x,t;t_0\vee\theta(x,t)),t_0\vee\theta(x,t)), \tag{2.40}$$

where $a\vee b=\max(a,b)$, finite or $-\infty$. Despite its formal complexity, (2.40) has a natural interpretation. If $\theta(x,t)<t_0$, that is, if the characteristic has run in the interior of Ω during the time interval (t_0,t), the solution $u(x,t)$ has the usual representation, i.e.,

$$u(x,t)=\int_{t_0}^{t} g(y(x,t;s),s)ds+u(y(x,t;t_0),t_0).$$

On the other hand, if $\theta(x,t) \geq t_0$, it happens that the characteristic passing at (x,t) has started from the boundary, so the value which is propagated is the boundary condition, computed at the intersection between the characteristic and the boundary itself:

$$u(x,t) = \int_{\theta(x,t)}^{t} g(y(x,t;s),s)ds + b(y(x,t;\theta(x,t)),\theta(x,t)).$$

This generalized version of the representation formula can be useful in assigning a Dirichlet boundary condition and in fact will be used in the numerical implementation of SL schemes.

2.4 ▪ Viscosity solutions and entropy solutions

A strong link exists between viscosity solutions and entropy solutions, which are the usual analytical tools in the framework of conservation laws. Although this link is valid only in one space dimension and for a particular class of Hamiltonians, it gives some interesting insight and also has an important role in the construction of numerical schemes for HJ equations.

Consider the two problems, an evolutive HJ equation

$$\begin{cases} v_t + H(v_x) = 0, & (x,t) \in \mathbb{R} \times (0,T), \\ v(x,0) = v_0(x), & x \in \mathbb{R} \end{cases} \tag{2.41}$$

(where the Hamiltonian H is assumed to be convex), and the associated conservation law

$$\begin{cases} u_t + H(u)_x = 0, & (x,t) \in \mathbb{R} \times (0,T), \\ u(x,0) = u_0(x), & x \in \mathbb{R}. \end{cases} \tag{2.42}$$

Assume now that

$$v_0(x) = \int_{-\infty}^{x} u_0(\xi)d\xi.$$

Then, it turns out that this relationship between u and v is preserved also for $t > 0$; that is, if u is the entropy solution of (2.42), then

$$v(x,t) = \int_{-\infty}^{x} u(\xi,t)d\xi$$

is the unique viscosity solution of (2.41). Vice versa, if v is the viscosity solution of (2.41), then $u = v_x$ is the unique entropy solution for (2.42).

Note that v is a.e. differentiable, and the singular points for its derivative v_x correspond to shocks for u. Note also that semiconcavity of v correspond to a one-sided (positive) bound on the derivative of u. This link will be also useful for numerical purposes.

We briefly review the main results.

Proposition 2.29. *Let $H \in C(\mathbb{R})$ be convex, and assume that $v \in W^{1,\infty}(\mathbb{R} \times (0,T))$ is a solution a.e. of* (2.41). *Then, $u := v_x$ is a weak solution of* (2.42).

Proof. The proof can be found in [Li82, p. 268]. □

Theorem 2.30. *Let $H \in C^1(\mathbb{R})$ be convex, and let $v_0 \in W^{1,\infty}(\mathbb{R})$. If $v \in W^{1,\infty}(\mathbb{R} \times (0,T))$ is the unique viscosity solution of* (2.41), *then $u := v_x$ is the unique entropy solution of* (2.42).

Proof. As we know, the viscosity solution v is the limit in $L^\infty(\mathbb{R}\times(0,T))$, as $\varepsilon\to 0^+$, of regular solutions v^ε of the following problem:

$$\begin{cases} v_t^\varepsilon(x,t)+H(v_x^\varepsilon)=\varepsilon v_{xx}^\varepsilon, & (x,t)\in\mathbb{R}\times(0,T),\\ v^\varepsilon(x,0)=v_0(x), & x\in\mathbb{R}. \end{cases}$$

Hence we have, for any $\varphi\in C^\infty(\mathbb{R}\times(0,T))$,

$$\begin{aligned} \lim_{\varepsilon\to 0}\int_0^T\int_{\mathbb{R}} v_x^\varepsilon(x,t)\varphi(x,t)\,dx\,dt &= -\lim_{\varepsilon\to 0}\int_0^T\int_{\mathbb{R}} v^\varepsilon(x,t)\varphi_x(x,t)\,dx\,dt\\ = -\lim_{\varepsilon\to 0}\int_0^T\int_{\mathbb{R}} v(x,t)\varphi(x,t)\,dx\,dt &= \int_0^T\int_{\mathbb{R}} v_x(x,t)\varphi(x,t)\,dx\,dt. \end{aligned}$$

Obviously, the function $u^\varepsilon := v_x^\varepsilon$ solves the problem

$$\begin{cases} u_t^\varepsilon(x,t)+H(u^\varepsilon)_x=\varepsilon u_{xx}^\varepsilon, & (x,t)\in\mathbb{R}\times(0,T),\\ u^\varepsilon(x,0)=v_{0x}(x), & x\in\mathbb{R}. \end{cases}$$

Since the sequence u^ε converges in $L^1_{loc}(\mathbb{R}\times(0,T))$ for $\varepsilon\to 0^+$, it must converge to the entropy solution u of (2.42). This implies that, for any $\varphi\in C^\infty(\mathbb{R}\times(0,T))$,

$$\lim_{\varepsilon\to 0}\int_0^T\int_{\mathbb{R}} u^\varepsilon(x,t)\varphi(x,t)\,dx\,dt = \int_0^T\int_{\mathbb{R}} u(x,t)\varphi(x,t)\,dx\,dt.$$

As a consequence,

$$\int_0^T\int_{\mathbb{R}} v_x(x,t)\varphi(x,t)\,dx\,dt = \int_0^T\int_{\mathbb{R}} u(x,t)\varphi(x,t)\,dx\,dt,$$

and $v_x = u$ a.e. in $\mathbb{R}\times(0,T)$. □

A converse of this result is also true. The first step is an intermediate result.

Proposition 2.31. *Let $H\in C(\mathbb{R})$ be convex, and assume that $u\in L^\infty_{loc}(\mathbb{R}\times(0,T))$ is a weak solution of* (2.42). *Define*

$$v(x,t) := \int_\alpha^x u(\xi,t)\,d\xi \tag{2.43}$$

for a fixed $\alpha\in\mathbb{R}$. Then, $v\in W^{1,\infty}_{loc}(\mathbb{R}\times(0,T))$ and v is a solution of (2.41) *a.e.*

Proof. Since $u\in L^\infty_{loc}(\mathbb{R}\times(0,T))$, there exists a set $A\subseteq(0,T)$ of zero Lebesgue measure, such that for any $t\in(0,T)\setminus A$, u is defined a.e. on $\mathbb{R}$ and $u(\cdot,t)\in L^\infty_{loc}(\mathbb{R})$. Then, for such values of t, $v(\cdot,t)\in L^\infty_{loc}(\mathbb{R})$. Moreover, for any $t\in(0,T)\setminus A$ and any $\varphi\in C_0^\infty(\mathbb{R}\times(0,T))$,

$$\begin{aligned} \int_{\mathbb{R}} v(x,t)\varphi_x(x,t)\,dx &= \int_{\mathbb{R}}\left[\int_\alpha^x u(\xi,t)\,d\xi\right]\varphi_x(x,t)\,d\xi\\ &= -\int_{\mathbb{R}} u(x,t)\varphi(x,t)\,dt. \end{aligned}$$

Thus, integrating on $(0,T)$, one has $u = v_x$ in the sense of distributions and a.e. Since u is a weak solution of (2.42), we have, for any $\varphi \in C_0^\infty(\mathbb{R}\times(0,T))$,

$$\int_0^T \int_{\mathbb{R}} H(u)\, \varphi_x(x,t)\, dx\, dt = -\int_0^T \int_{\mathbb{R}} u(x,t)\varphi_t(x,t)\, dx\, dt$$
$$= -\int_0^T \int_{\mathbb{R}} v_x(x,t)\varphi_t(x,t)\, dx\, dt = \int_0^T \int_{\mathbb{R}} v(x,t)\varphi_{tx}(x,t)\, dx\, dt.$$

So there exists v_t in the sense of distributions, and $v_t = -H(u) = -H(v_x)$. Therefore, $v \in W^{1,\infty}_{loc}(\mathbb{R}\times(0,T))$, and v is a solution a.e. of (2.41). □

Now, for any $u \in C([0,T];L^1(\mathbb{R}))$, we define

$$v(x,t) := \int_{-\infty}^x u(\xi,t)\, d\xi\,. \tag{2.44}$$

Clearly, for any $t \in (0,T)$, the function $v(\cdot,t)$ is absolutely continuous and $v_x = u$ a.e. We can now prove the main result.

Theorem 2.32. *Let $H \in C^1(\mathbb{R})$ and $u_0 \in L^\infty(\mathbb{R})\cap L^1(\mathbb{R})$. Assume that $u \in L^\infty(\mathbb{R}\times[0,T])\cap C([0,T];L^1(\mathbb{R}))$ is the unique entropy solution of* (2.42)*. Then, the function v given by* (2.44) *is the unique viscosity solution of* (2.41) *for the initial condition*

$$v_0(x) := \int_{-\infty}^x u_0(\xi)\, d\xi\,.$$

Proof. Since $u \in C([0,T];L^1(\mathbb{R}))$, we have that $v \in L^\infty(\mathbb{R}\times[0,T])$. As in the previous proposition, it is easy to show that $v \in W^{1,\infty}(\mathbb{R}\times[0,T])$ and that v is a solution a.e. of (2.41). Moreover,

$$\lim_{t\to 0}|v(x,t)-v_0(x)| \le \lim_{t\to 0}\int_{-\infty}^x |u(\xi,t)-u_0(\xi)|\, d\xi = 0.$$

Now, suppose that v is not the viscosity solution of (2.41), and denote by $\overline{v}$ the unique viscosity solution. Then, by Theorem 2.30, $\overline{v}_x$ is the unique entropy solution of (2.42), and therefore, for any $\varphi \in C_0^\infty(\mathbb{R}\times(0,T))$,

$$\int\int (v-\overline{v})\varphi_x\, dx\, dt = 0.$$

Since φ can be arbitrarily chosen, the conclusion follows. □

2.5 ▪ Discontinuous viscosity solutions

We conclude this chapter giving some ideas about the extension of the "classical" theory of viscosity solutions to the discontinuous case. This extension has been strongly motivated by a number of relevant applications, e.g., to image processing and to games, which may lead to discontinuous solutions.

Let $w : \mathbb{R}^d \to \mathbb{R}$ be a bounded function. We recall two definitions which will play a key role in defining a bounded, but possibly discontinuous, viscosity solution. The *lower semicontinuous envelope* of w is defined as

$$\underline{w}(x) := \liminf_{y \to x} w(y),$$

whereas the *upper semicontinuous envelope* is defined as

$$\overline{w}(x) := \limsup_{y \to x} w(y).$$

With this additional tool, we can extend the notion of viscosity solutions as follows.

Definition 2.33. *A function $u \in L^\infty(\Omega)$ is a viscosity solution in Ω of*

$$H(x, u, Du) = 0$$

if and only if, for any $\varphi \in C^1(\Omega)$, the following conditions hold:

(i) *at every point $x_0 \in \Omega$, local* maximum *for $\overline{u} - \varphi$,*

$$H(x_0, \overline{u}(x_0), D\varphi(x_0)) \leq 0$$

(i.e., $\overline{u}$ is a viscosity subsolution*);*

(ii) *at every point $x_0 \in \Omega$, local* minimum *for $\underline{u} - \varphi$,*

$$H(x_0, \underline{u}(x_0), D\varphi(x_0)) \geq 0;$$

i.e., $\underline{u}$ is a viscosity supersolution.

It is important to note that the extension of the comparison principle to this new setting is not trivial. However, in many relevant problems one can get a uniqueness result for a lower semicontinuous viscosity solution. A comprehensive introduction to these results goes beyond the scope of this book, but it can be found in [Ba93].

2.6 ▪ Commented references

The theory of viscosity solutions has started with the papers by Crandall and Lions [CL84] on the solution of first-order HJ equations. However, a very important source of inspiration for this theory had come from some earlier papers by Kružkov [Kr60, Kr64], in which the solution for first-order problems was obtained via the vanishing viscosity methods (i.e., by elliptic/parabolic regularization). These papers were strongly related to the analysis of conservation laws, where Kružkov gave important contributions (see the monographs by Leveque [L92] and Serre [Se99a, Se99b] for more information on this topic). Kružkov was also interested in the numerical approximation via finite difference schemes which he studied in [Kr66]. The main contribution by Crandall and Lions was finding an intrinsic definition of the weak solution which would not require passing to the limit in the regularized problem and could allow obtaining uniqueness results for a large class of stationary and evolutive nonlinear PDEs.

At the very beginning the definition was based on sub- and superdifferentials, and only at a later time was the definition based on test functions introduced ([CEL84]; see also [Is89]). Starting from these seminal papers the theory had increasing success in the following years, as witnessed by the books by Lions [Li82] and Barles [Ba98]. In particular, the latter contains a chapter on discontinuous viscosity solutions. More details on discontinuous viscosity solutions can be found in the original papers by Barron and Jensen [BJ90, BJ91], Frankowska [Fr93], Barles [Ba93], and Soravia [Sor93b]. In the framework of discontinuous Hamiltonians one should also look at the papers by Camilli and Siconolfi [CS03] and Soravia [Sor02].

The theory has naturally been extended to second-order fully nonlinear problems. For these extensions of the theory, we refer the reader to the survey paper by Crandall, Ishii, and Lions [CIL92] and to the book by Fleming and Soner [FS93].

Finally, it is worth saying that the development of this theoretical framework has been widely driven by the applications to control problems, image processing, and Fluid Dynamics. For the applications to control problems we refer the reader to the monographs by Bardi and Capuzzo Dolcetta [BCD97] for deterministic problems and Fleming and Soner [FS93] for stochastic problems. Moreover, in the books by Sethian [Se99] and Osher and Fedkiw [OF03] it is possible to find a number of applications of HJ equations, in particular in the framework of level set methods.

Chapter 3

Elementary building blocks

SL schemes require an assembly of different (and partly independent) conceptual blocks, in particular a strategy for moving along characteristics and a technique for reconstructing the numerical solution at the foot of a characteristic. In the case of HJ equations, a numerical minimization technique is also needed. Here, different choices for all such blocks are considered and explained, and their basic theoretical results are reviewed, with a greater emphasis on less established topics.

3.1 ▪ A review of ODE approximation schemes

The first typical operation performed by SL schemes is to move along characteristics to locate the value to be propagated. This operation amounts to approximating over a single time step a system of ODEs like (1.7), which will be rewritten as

$$\begin{cases} \dfrac{d}{dt} y(x_0, t_0; t) = f(y(x_0, t_0; t), t), & t \in \mathbb{R}, \\ y(x_0, t_0; t_0) = x_0, \end{cases} \tag{3.1}$$

with an initial condition $x_0 \in \mathbb{R}^d$ given at the time t_0, and with a vector field $f : \mathbb{R}^d \times \mathbb{R} \to \mathbb{R}^d$ globally Lipschitz continuous with respect to its first argument. Such a task can be accomplished by different techniques, usually borrowed from the ODE literature. This section will review the basic ideas, which will later be applied to the specific situations of interest, whereas for a more extensive treatment of the related techniques and theoretical results, we refer the reader to specialized textbooks. Throughout the section, we will assume that a set of nodes $t_k = t_0 + k\Delta t$ (with $k \in \mathbb{Z}$) is given on the time axis, with y_k denoting the corresponding approximations of $y(x_0, t_0; t_k)$.

The usual concepts of consistency and stability are involved in the convergence theory of methods for ODEs, as the basic convergence theorem states. Conceptually, consistency means that the scheme is a "small perturbation" of the exact equation, whereas stability means that the scheme satisfies (uniformly in Δt) a principle of continuous dependence upon the initial data. Note that, among the various concepts of stability which have been developed for this problem, we will refer here to *zero-stability*, which is that required for convergence.

Rather than outlining a general theory, we will sketch how this theory applies to the two main classes of schemes to approximate (3.1), which are referred to as one-step and multistep methods. We will briefly review their general philosophy, as well as the

Table 3.1. *First- and second-order one-step schemes for the system* (3.1).

Scheme	Form	Order
FE	$\Phi(\Delta t; y_k, t_k, y_{k+1}) = f(y_k, t_k)$	$p=1$
BE	$\Phi(\Delta t; y_k, t_k, y_{k+1}) = f(y_{k+1}, t_k + \Delta t)$	$p=1$
H	$\Phi(\Delta t; y_k, t_k, y_{k+1}) = \frac{1}{2}[f(y_k, t_k) + f(y_k + \Delta t f(y_k, t_k), t_k + \Delta t)]$	$p=2$
CN	$\Phi(\Delta t; y_k, t_k, y_{k+1}) = \frac{1}{2}[f(y_k, t_k) + f(y_{k+1}, t_k + \Delta t)]$	$p=2$

main results of consistency, stability, and convergence, and give some practical examples, including cases of use in SL schemes.

Remark 3.1. *While practical schemes for* (3.1) *are necessarily stable, we will see that, in the convergence theory of SL schemes, stability is not technically required. At the level of convergence, the choice of a particular scheme rather results in a different consistency error, although it may also affect the qualitative behavior of the solution.*

3.1.1 ▪ One-step schemes

In one-step schemes the system (3.1) is discretized in the form

$$\begin{cases} y_{k+1} = y_k + \Delta t \Phi(\Delta t; y_k, t_k, y_{k+1}), \\ y_0(x_0, t_0) = x_0. \end{cases} \tag{3.2}$$

In (3.2), the approximation y_{k+1} is constructed using only the informations available at the time step t_k. If the function Φ has a genuine dependence on y_{k+1}, then the scheme is termed as *implicit* and the computation of y_{k+1} itself requires solving (3.2) as a system of nonlinear equations (or a scalar equation if the solution is scalar).

Examples

We show in Table 3.1 some of the simplest examples of explicit and implicit one-step schemes of first and second order, namely Forward Euler (FE), Backward Euler (BE), Heun (H), and Crank–Nicolson (CN). More generally, this class of schemes includes all Runge–Kutta-type methods. Note that, although their use is clear when the dynamics $f(\cdot,\cdot)$ has an explicit expression, care should be taken when applying them to stepping back along characteristics, especially in nonlinear cases, since the information might not be available at any time step or at any point. This issue will be discussed in Chapter 5.

Theoretical results

Consistency The scheme (3.2) is said to be *consistent* with order $p \geq 1$ if, once we replace y_k, y_{k+1} with the exact values $y(t_k) = y(x_0, t_0; t_k)$, $y(t_{k+1}) = y(x_0, t_0; t_{k+1})$ of a smooth solution, the scheme satisfies

$$y(t_{k+1}) = y(t_k) + \Delta t \Phi(\Delta t; y(t_k), t_k, y(t_{k+1})) + O(\Delta t^{p+1}). \tag{3.3}$$

In general, we expect that (3.3) would result in obtaining an approximation error of order $O(\Delta t^p)$, and this is also true when using (3.2) for approximating characteristics in SL schemes (in this case, this consistency error affects the term related to time discretization alone).

Stability The definition of zero-stability in the case of one-step schemes requires that if (3.2) is used to approximate (3.1) starting from two different initial conditions x_0 and x_0' (with the corresponding numerical solutions denoted by y_k and y_k'), then

$$\|y_k - y_k'\| \le C \|x_0 - x_0'\|$$

for any k such that $t_k \in [t_0, t_0 + T]$, and with a constant C independent of Δt.

A general result states that if the function Φ is Lipschitz continuous with respect to its second and fourth arguments, the one-step scheme is zero-stable. Note that all the examples of Table 3.1 satisfy this condition.

Convergence Finally, we give the convergence theorem for one-step schemes.

Theorem 3.2. *If the scheme* (3.2) *is consistent and zero-stable, then, for any* k *such that* $t_k \in [t_0, t_0 + T]$,

$$\|y_k - y(x_0, t_0; t_k)\| \to 0$$

as $\Delta t \to 0$. *Moreover, if* (3.2) *is consistent with order* p *and the solution* y *is smooth enough, then*

$$\|y_k - y(x_0, t_0; t_k)\| \le C \Delta t^p.$$

3.1.2 ▪ Multistep schemes

Linear multistep schemes discretize the system (3.1) in the form

$$\begin{cases} y_{k+1} = \sum_{j=0}^{n_s-1} \alpha_j y_{k-j} + \Delta t \sum_{j=-1}^{n_s-1} \beta_j f(y_{k-j}, t_{k-j}), \\ y_0 = x_0, \\ y_1, \ldots, y_{n_s-1} \quad \text{given.} \end{cases} \tag{3.4}$$

In (3.4), the approximation y_{k+1} is constructed using the information available at the n_s time steps t_{k-n_s+1} to t_{k+1}. The startup values $y_1, \ldots, y_{n_s-1}$ are chosen so as to be an approximation (with suitable accuracy) of the corresponding values of the exact solution. As for the case of one-step schemes, if $\beta_{-1} \neq 0$ so that the right-hand side of (3.4) depends on y_{k+1}, then the computation of y_{k+1} requires solving (3.4) as a system of nonlinear equations.

Examples

In Table 3.2 we report two specific choices of the coefficients, both resulting in a method of order $p = 2$ and correspond to the midpoint scheme (MP) and to the second-order Adams–Bashforth scheme (AB2). Such cases turn out to be of common use in SL schemes for environmental Fluid Dynamics.

Table 3.2. *Examples of multistep schemes for the system* (3.1).

Scheme	α_0	α_1	β_0	β_1	Order
MP	0	1	2	0	$p=2$
AB2	1	0	$\frac{3}{2}$	$-\frac{1}{2}$	$p=2$

Theoretical results

Consistency Consistency is checked in multistep schemes according to the same principle used in one-step schemes, namely to replace the values y_k with the exact values $y(t_k) = y(x_0, t_0; t_k)$ of a smooth solution. Therefore, the multistep scheme (3.4) is said to be *consistent* with order $p \geq 1$ if

$$y(t_{k+1}) = \sum_{j=0}^{n_s-1} \alpha_j y(t_{k-j}) + \Delta t \sum_{j=-1}^{n_s-1} \beta_j f(y(t_{k-j}), t_{k-j}) + O(\Delta t^{p+1}). \tag{3.5}$$

It is possible to derive algebraic conditions on the coefficients α_j, β_j which ensure that (3.5) is satisfied. However, it is also possible to avoid these technical aspects and rather work directly on the formulation (3.5).

Stability Although similar to the definition adopted for one-step schemes, zero-stability of multistep schemes is defined in a way which takes into account not only the two initial points $y_0 = x_0$ and $y'_0 = x'_0$ but all the startup values $y_1, \ldots, y_{n_s-1}$ and $y'_1, \ldots, y'_{n_s-1}$. We require therefore that, for any k such that $t_k \in [t_0, t_0 + T]$,

$$\|y_k - y'_k\| \leq C \max_{0 \leq i \leq n_s-1} (\|y_i - y'_i\|)$$

with a constant C independent of Δt.

The endpoint of zero-stability analysis for linear multistep schemes is the so-called *root condition*, which gives a necessary and sufficient condition in algebraic form.

Theorem 3.3. *A multistep method in the form* (3.4) *is zero-stable if and only if, once we denote by ζ_i ($i = 1, \ldots, n_s$) the roots of the polynomial*

$$r(\zeta) = \zeta^{n_s} - \sum_{j=0}^{n_s-1} \alpha_j \zeta^{n_s-j-1},$$

one has $|\zeta_i| \leq 1$ for any i, and in addition, all roots such that $|\zeta_i| = 1$ are simple.

Convergence The convergence theorem for multistep schemes, besides the standard assumptions of consistency and stability, requires convergence of the startup values to the corresponding values for the solution.

Theorem 3.4. *If the scheme* (3.4) *is consistent and zero-stable, and if, for $1 \leq k \leq n_s - 1$,*

$$\|y_k - y(x_0, t_0; t_k)\| \to 0,$$

then, for any k such that $t_k \in [t_0, t_0 + T]$,

$$\|y_k - y(x_0, t_0; t_k)\| \to 0$$

as $\Delta t \to 0$. *Moreover, if* (3.4) *is consistent with order* p, $\|y_k - y(x_0, t_0; t_k)\| = O(\Delta t^p)$ *for* $1 \le k \le n_s - 1$, *and the solution* y *is smooth enough, then*

$$\|y_k - y(x_0, t_0; t_k)\| \le C \Delta t^p.$$

3.2 ▪ Reconstruction techniques in one and multiple space dimensions

The role of the reconstruction step in SL schemes is to recover the value of the numerical solution at the feet of characteristics, which are not in general grid points themselves. In their basic version, SL schemes do not use cell averages, and the reconstruction is rather based on the interpolation of pointwise values of the solution.

We assume therefore that a grid (not necessarily structured) with nodes x_j is set and that a function $v(x)$ is represented by the vector V of its corresponding nodal values v_j. The space discretization parameter Δx, in the basic case of a single space dimension with sequentially numbered nodes, is defined as

$$\Delta x = \sup_k (x_{k+1} - x_k), \tag{3.6}$$

whereas in multiple dimensions and unstructured cases, it requires more complex definitions which will be recalled when necessary.

In the simplest cases (symmetric Lagrange interpolation, Lagrange finite elements), the reconstruction is set in the form of a linear combination of suitable basis functions:

$$I[V](x) = \sum_k v_k \psi_k(x). \tag{3.7}$$

We will mostly consider functions ψ_k in piecewise polynomial form of degree r, and if necessary, the notation $I_r[V]$ will be used to state this degree more explicitly. Also, the ψ_k are usually chosen to be *cardinal functions*, that is, to satisfy $\psi_k(x_i) = \delta_{ik}$. When nonoscillatory reconstructions are considered, the linearity with respect to the values v_k is lost and the reconstruction takes a more complex form.

Whenever useful, we will apply in what follows the idea of *reference basis functions*, meaning a basis of cardinal functions which is defined on a reference grid (e.g., a grid with unity distance between nodes) and generates the actual basis of the reconstruction by an affine transformation depending on the geometry of the specific grid used.

In the remainder of the section, we will review three reconstruction strategies (symmetric Lagrange interpolation, ENO, WENO) which are usually implemented with constant step and extended to multiple dimensions by separation of variables. Moreover, a genuinely multidimensional and unstructured strategy (finite elements) will also be considered.

3.2.1 ▪ Symmetric Lagrange interpolation in $\mathbb{R}^1$

This kind of interpolation is usually applied with evenly spaced nodes (by (3.6), Δx will denote the constant step), and the samples of V used for the reconstruction are taken

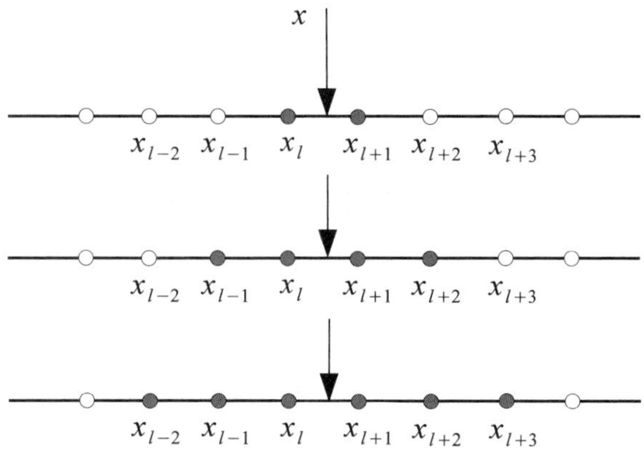

Figure 3.1. *Reconstruction stencil for linear, cubic, and quintic Lagrange interpolation.*

from a stencil surrounding the interval which contains the point x, resulting in a piecewise polynomial interpolation. Note that, in principle, Lagrange interpolation can be implemented on nonuniform grids as well as in globally polynomial form. In the SL framework, however, the former choice would lead in general to an unstable scheme and the latter to an undue numerical dispersion in the form of Gibbs oscillations. Lagrange interpolation is typically implemented with an odd degree, thus using an equal number of nodes on both sides of x. In fact, with such a choice, the interpolation error is related to an even derivative of the function, this resulting in a symmetric behavior of the SL scheme around singularities, as it will turn out from the numerical dispersion analysis.

Construction

Taking into account that the number of nodes must be one unity larger than the degree r of the interpolation, we obtain, for $x \in [x_l, x_{l+1}]$, that the value $I[V](x)$ is computed by a Lagrange polynomial constructed on a set of nodes which includes $\frac{r+1}{2}$ nodes on each side of the interval. This stencil of nodes will be denoted by

$$\mathscr{S} = \left\{ x_k : l - \frac{r-1}{2} \le k \le l + \frac{r+1}{2} \right\}$$

and is depicted in Figure 3.1 for the lowest odd orders of interpolation. As a consequence, the Lagrange polynomial reconstruction takes the form

$$I_r[V](x) = \sum_{x_k \in \mathscr{S}} v_k \prod_{x_i \in \mathscr{S} \setminus \{x_k\}} \frac{x - x_i}{x_k - x_i}. \tag{3.8}$$

Since the Lagrange basis used to interpolate changes as x moves from one interval to the other, at first glance it could seem that (3.7) would not be satisfied with a unique basis of cardinal functions. Nevertheless, the reconstruction being linear with respect to the values to be interpolated, such a unique basis can still be suitably defined. In fact, the interpolation operator $I[\cdot]$ is a linear map from the space l^∞ of bounded sequences into the space of continuous functions on $\mathbb{R}$:

$$I[\cdot] : l^\infty \to C^0(\mathbb{R}), \tag{3.9}$$

so that, by elementary linear algebra arguments, any basis function ψ_k can be defined to be nothing but the image of the base element e_k (that is, the interpolation of a sequence such that $v_k = 1$, $v_i = 0$ for $i \neq k$).

On the other hand, since the grid is uniform and the reconstruction is self-similar and invariant by translation, any such basis function may be written as

$$\psi_k(x) = \psi\left(\frac{x - x_k}{\Delta x}\right). \tag{3.10}$$

Here, the function ψ plays the role of *reference basis function* and may be obtained by applying the interpolation operator to the sequence e_0, with $\Delta x = 1$. In what follows, y will represent the variable in the reference space, and we will possibly use the notation $\psi^{[r]}$ to denote the reference function for a Lagrange interpolation of order r. Although the interpolation procedure, as performed by (3.8), does not require knowing the explicit expression of ψ, this expression will be useful later for the theoretical analysis of the schemes.

Examples

We sketch below the construction of the Lagrange reconstruction, as well as of the corresponding reference basis function corresponding to the cases ($r = 1, 3, 5$) shown above.

Linear ($\mathbb{P}_1$) interpolation We first show the simple situation of linear interpolation, which also corresponds to the $\mathbb{P}_1$ case in the finite element setting. The expression of the interpolating polynomial for the first order ($r = 1$) is

$$\begin{aligned} I_1[V](x) &= \frac{x - x_{l+1}}{x_l - x_{l+1}} v_l + \frac{x - x_l}{x_{l+1} - x_l} v_{l+1} \\ &= \frac{x_{l+1} - x}{\Delta x} v_l + \frac{x - x_l}{\Delta x} v_{l+1}, \end{aligned} \tag{3.11}$$

whereas the interpolation of the sequence e_0 in the reference space results in a reference function of the well-known structure

$$\psi^{[1]}(y) = \begin{cases} 1 + y & \text{if } \; -1 \leq y \leq 0, \\ 1 - y & \text{if } \; 0 \leq y \leq 1, \\ 0 & \text{elsewhere.} \end{cases} \tag{3.12}$$

Note that the value at a certain node affects the reconstruction only in the adjacent intervals.

Lagrange interpolation of higher order Take now as a second example cubic reconstruction, which performs an interpolation using the four nearest nodes (two on the left and two on the right of the interval containing x). The interpolating polynomial is computed as

$$\begin{aligned}
I_3[V](x) &= \sum_{k=l-1}^{l+2} v_k \prod_{i\neq k} \frac{x-x_i}{x_k-x_i} \\
&= -\frac{(x-x_l)(x-x_{l+1})(x-x_{l+2})}{6\Delta x^3} v_{l-1} \\
&\quad + \frac{(x-x_{l-1})(x-x_{l+1})(x-x_{l+2})}{2\Delta x^3} v_l \\
&\quad - \frac{(x-x_{l-1})(x-x_l)(x-x_{l+2})}{2\Delta x^3} v_{l+1} \\
&\quad + \frac{(x-x_{l-1})(x-x_l)(x-x_{l+1})}{6\Delta x^3} v_{l+2}
\end{aligned}$$

(note that the value at a node x_j affects the reconstruction in the interval (x_{j-2}, x_{j+2})). Then, referring again to the case in which e_0 is interpolated and $\Delta x = 1$, the explicit form of the reference base function $\psi^{[3]}$ is

$$\psi^{[3]}(y) = \begin{cases} \frac{1}{2}(y+1)(y-1)(y-2) & \text{if } 0 \le y \le 1, \\ -\frac{1}{6}(y-1)(y-2)(y-3) & \text{if } 1 \le y \le 2, \\ 0 & \text{if } y > 2, \\ \psi^{[3]}(-y) & \text{if } y < 0. \end{cases} \tag{3.13}$$

In turn, a similar computation for the quintic interpolation would yield for the interpolant the expression

$$\begin{aligned}
I_5[V](x) &= \sum_{k=l-2}^{l+3} v_k \prod_{i\neq k} \frac{x-x_i}{x_k-x_i} \\
&= -\frac{(x-x_{l-1})(x-x_l)(x-x_{l+1})(x-x_{l+2})(x-x_{l+3})}{120\Delta x^5} v_{l-2} \\
&\quad + \frac{(x-x_{l-2})(x-x_l)(x-x_{l+1})(x-x_{l+2})(x-x_{l+3})}{24\Delta x^5} v_{l-1} \\
&\quad - \frac{(x-x_{l-2})(x-x_{l-1})(x-x_{l+1})(x-x_{l+2})(x-x_{l+3})}{12\Delta x^5} v_l + \cdots \\
&\quad + \frac{(x-x_{l-2})(x-x_{l-1})(x-x_l)(x-x_{l+1})(x-x_{l+2})}{120\Delta x^5} v_{l+3}.
\end{aligned}$$

Accordingly, the reference basis function would be in the form

$$\psi^{[5]}(y) = \begin{cases} -\frac{1}{12}(y+2)(y+1)(y-1)(y-2)(y-3) & \text{if } 0 \le y \le 1, \\ \frac{1}{24}(y+1)(y-1)(y-2)(y-3)(y-4) & \text{if } 1 \le y \le 2, \\ -\frac{1}{120}(y-1)(y-2)(y-3)(y-4)(y-5) & \text{if } 2 \le y \le 3, \\ 0 & \text{if } y > 3, \\ \psi^{[5]}(-y) & \text{if } y < 0. \end{cases} \tag{3.14}$$

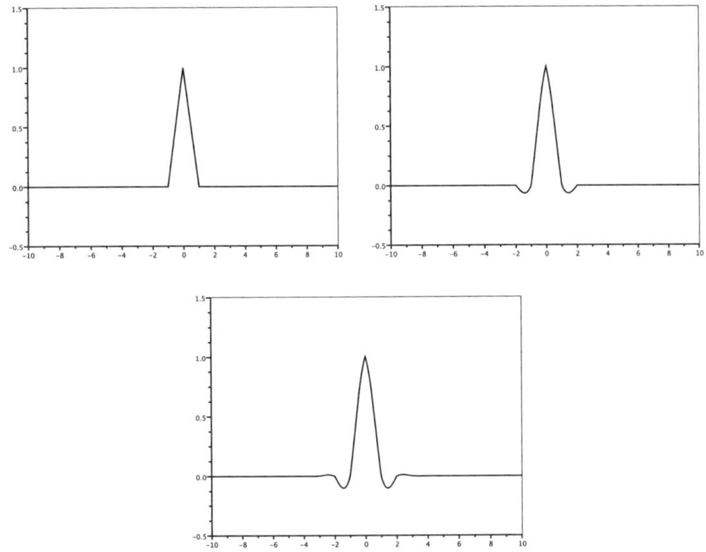

Figure 3.2. *The basis functions for linear, cubic, and quintic interpolation* [Fe13a].

More generally for an arbitrary odd interpolation degree r written in the form (3.8), the reference basis function $\psi^{[r]}$ is

$$\psi^{[r]}(y) = \begin{cases} \displaystyle\prod_{k\neq 0, k=-[r/2]}^{[r/2]+1} \frac{y-k}{-k} & \text{if } 0 \leq y \leq 1, \\ \vdots & \\ \displaystyle\prod_{k=1}^{r} \frac{y-k}{-k} & \text{if } [r/2] \leq y \leq [r/2]+1, \\ 0 & \text{if } y > [r/2]+1, \\ \psi^{[r]}(-y) & \text{if } y < 0. \end{cases} \tag{3.15}$$

Note that this construction can also be performed for an even degree of interpolation, but the result would not be a symmetric function.

We show in Figure 3.2 the reference basis functions for interpolation, $\psi^{[r]}(y)$, for $r = 1, 3, 5$.

Theoretical results

We recall here the approximation result concerning the interpolation error for Lagrange polynomial approximations. The proof is usually contained in any textbook in basic numerical analysis.

Theorem 3.5. *Let the interpolation $I_r[V]$ be defined by* (3.7), *with ψ_j defined by* (3.10) *and ψ defined by* (3.15). *Let $v(x)$ be a uniformly continuous function on $\mathbb{R}$, and let V be the vector of its nodal values. Then, for $\Delta x \to 0$,*

$$\|v - I_r[V]\|_\infty \to 0. \tag{3.16}$$

If, moreover, $v \in W^{s,\infty}(\mathbb{R})$, then

$$\|v - I_r[V]\|_\infty \leq C\Delta x^{\min(s,r+1)} \tag{3.17}$$

for some positive constant C depending on the degree of regularity s.

Remark 3.6. *Estimate* (3.17) *can be improved by taking into account the dependence on x. For our purposes, it suffices to point out that the interpolation error decreases linearly when approaching a node. Using the more general form of the interpolation error estimate, it is easy to prove that*

$$|v(x) - I_r[V](x)| \leq C\Delta x^{\min(s,r+1)} \min_{m\in\mathbb{Z}} \frac{x - x_m}{\Delta x}, \tag{3.18}$$

which clearly gives (3.17) *when passing to the ∞-norm but also accounts for a reduction of the error in the neighborhood of a node.*

Remark 3.7. *As we mentioned, globally polynomial interpolations show Gibbs oscillations when treating solutions with singularities, and therefore they are usually avoided in non-smooth contexts (like viscosity or entropy solutions), whereas piecewise polynomial approximations (e.g., Lagrange and finite element) have a better performance. However, although confined to a neighborhood of the singularity, high-order Lagrange reconstructions can still exhibit an oscillatory behavior, due to the changes of sign in the basis functions (see Figure* 3.2*). This motivates the search for* nonoscillatory *(ENO/WENO) reconstruction strategies which could reduce this drawback.*

3.2.2 ▪ Essentially Nonoscillatory interpolation in $\mathbb{R}^1$

Essentially Nonoscillatory (ENO) interpolation is a first strategy of reconstruction which has been developed in order to reduce Gibbs oscillations. ENO interpolation *essentially* cuts such oscillations (that is, reduces over- or undershoots of the reconstruction to the order of some power of Δx), still retaining a high order of accuracy in smooth regions of the domain. The basic tool for this operation is a strategy which selects, among all candidate stencils, the stencil where the function is smoother. In this procedure, different functions may require different stencils, so that linearity of the reconstruction (expressed by (3.7)) is lost.

Construction

We define the procedure for constructing a reconstruction of ENO type based on the one-dimensional mesh $\{x_j\}$, in order to approximate the value $v(x)$ for $x \in [x_l, x_{l+1}]$. Most of what follows does not need a uniform mesh, although this is the typical situation. In any case, Δx will be defined by (3.6).

We have to construct an interpolating polynomial of degree r, with a stencil of $r+1$ points, which must include x_l and x_{l+1} and be chosen in order to avoid singularities of v. This will be performed using Newton's form of the interpolating polynomial,

$$I_r[V](x) = V[x_{l_0}] + \sum_{k=1}^{r} V[x_{l_0}, \ldots, x_{l_k}] \prod_{m=0}^{k-1} (x - x_{l_m}) \tag{3.19}$$

with the divided differences $V[\cdot]$ defined as

$$V[x_{l_0}] = v(x_{l_0}),$$
$$V[x_{l_0},\dots,x_{l_k}] = \frac{V[x_{l_1},\dots,x_{l_k}] - V[x_{l_0},\dots,x_{l_{k-1}}]}{x_{l_k} - x_{l_0}}.$$

Note that, in order to reconstruct v in the interval $[x_l, x_{l+1}]$, we set $x_{l_0} = x_l$, $x_{l_1} = x_{l+1}$ and proceed to extend the stencil by adding further adjacent points.

Now, from the basic results concerning divided differences, it is known that if $v \in C^s(I)$ (I being the smallest interval containing $x_{l_0},\dots,x_{l_s}$), then there exists a point $\xi \in I$ such that

$$V[x_{l_0},\dots,x_{l_s}] = \frac{v^{(s)}(\xi)}{s!}. \tag{3.20}$$

On the other hand, if v has a discontinuity in I, then

$$V[x_{l_0},\dots,x_{l_s}] = O\left(\frac{1}{\Delta x^s}\right),$$

and more in general, if the discontinuity occurs at the level of the kth derivative $v^{(k)}$, with $k < s$, then

$$V[x_{l_0},\dots,x_{l_s}] = O\left(\frac{1}{\Delta x^{s-k}}\right). \tag{3.21}$$

Hence, as long as it affects the construction of the interpolating polynomial, the magnitude of a divided difference is a measure of the regularity of the function on the related stencil.

Let therefore the fixed initial stencil be denoted by

$$\mathcal{S}_1 = \{x_l, x_{l+1}\} = \{x_{l_0}, x_{l_1}\}. \tag{3.22}$$

At the kth step of the procedure, we pass from a stencil $\mathcal{S}_{k-1}$ to the stencil

$$\mathcal{S}_k = \{x_{l_0},\dots,x_{l_k}\} \tag{3.23}$$

by adding the point x_{l_k}. Then, if we plan to extend the stencil towards points where $v(x)$ is smoother, x_{l_k} will be chosen (between the two possible choices) as the one which minimizes the magnitude of the new divided difference. More formally, setting

$$l_k^- = \min(l_0,\dots,l_{k-1}) - 1, \quad l_k^+ = \max(l_0,\dots,l_{k-1}) + 1,$$

then the following hold:

- If $|V[x_{l_0},\dots,x_{l_{k-1}},x_{l_k^-}]| < |V[x_{l_0},\dots,x_{l_{k-1}},x_{l_k^+}]|$, then $l_k = l_k^-$ (the stencil is extended one point to the left).
- If $|V[x_{l_0},\dots,x_{l_{k-1}},x_{l_k^-}]| \geq |V[x_{l_0},\dots,x_{l_{k-1}},x_{l_k^+}]|$, then $l_k = l_k^+$ (the stencil is extended one point to the right).

After the selection, the stencil is updated and the algorithm proceeds to the next step, until $k = r$. The interpolation may be computed by adding each term of the Newton polynomial during the process or by performing a Lagrange interpolation on the final stencil $\mathcal{S} = \mathcal{S}_r$.

Theoretical results

We review some basic results about ENO interpolation.

Accuracy in smooth regions First, it's obvious that if the function to be interpolated is smooth, then the accuracy coincides with the accuracy of Lagrange interpolation.

Theorem 3.8. *Let the interpolation $I_r[V]$ be defined by* (3.19), *with the ENO procedure of selection for the $V[x_{l_0},\ldots,x_{l_k}]$. Let $v \in W^{s,\infty}(\mathbb{R})$. Then,*

$$\|v - I_r[V]\|_\infty \le C\Delta x^{\min(s,r+1)} \tag{3.24}$$

for some positive constant C.

Note that ENO interpolation is constructed on a polynomial basis, and therefore the sharper estimate (3.6) also applies to this case.

Bounds on the total variation The second property is related to the total variation of the reconstruction. We start with a preliminary result, which does not require the ENO procedure for computing I_r but is suitable for any piecewise polynomial interpolation.

Proposition 3.9. *Let v be discontinuous in the interval $[x_l, x_{l+1}]$. Then, $I_r[V]$ has no extrema in the same interval.*

Proof. We give only a sketch of the proof, based on the case of $v(x)$ defined as a step function:

$$v(x) = \begin{cases} 0, & x \le 0, \\ 1, & x > 0. \end{cases}$$

Let $p \in \mathbb{P}_r$ be the polynomial interpolating v on the stencil $\mathcal{S} = \mathcal{S}_r$, and assume the discontinuity point $x = 0$ satisfies

$$x_l < 0 < x_{l+1}$$

with $x_l, x_{l+1} \in \mathcal{S}_r$. On any interval $[x_j, x_{j+1}]$ not containing the discontinuity point we have

$$p(x_j) = u(x_j) = u(x_{j+1}) = p(x_{j+1}),$$

and therefore there exists a point $\xi_j \in (x_j, x_{j+1})$ such that

$$p'(\xi_j) = 0.$$

Then, we can find $r-1$ different roots for $p'(x)$, that is, one root for any interval (x_j, x_{j+1}) not containing the point $x = 0$. Since $p' \in \mathbb{P}_{r-1}$, it follows that $p'(x)$ cannot have further roots in the shock interval $[x_l, x_{l+1}]$; that is, $p(x)$ is monotone in this interval. □

Last, we give the result on the total variation of ENO interpolation. Basically, this results states that ENO interpolation is TVB; that is, it has bounded total variation. In addition, the increase in variation introduced is of the order of the interpolation error in smooth regions.

Theorem 3.10. *Let $v(x)$ be a piecewise continuous function with derivative $v^{(s)}$ bounded in the continuity set. Then, there exists a function $z(x)$ such that*

$$z(x) = I_r[V](x) + O(\Delta x^{\min(s,r+1)})$$

and satisfies

$$TV(z) \leq TV(v).$$

Proof. The statement directly follows from the two last results, taking $z(x) = v(x)$ in intervals where v is smooth and $z(x) = p(x)$ in intervals containing discontinuities. □

3.2.3 ▪ Weighted Essentially Nonoscillatory interpolation in $\mathbb{R}^1$

Weighted Essentially Nonoscillatory (WENO) interpolation is another strategy for nonoscillatory reconstruction. It is derived from the ENO strategy, but unlike ENO it uses all the computed information as much as possible.

In practice, while the ENO strategy computes twice the required number of divided differences to build a Newton polynomial, but at each step discards one divided difference among two, WENO strategy computes all the candidate interpolating polynomials. Then, if the function is smooth enough on the given stencil, candidate polynomials are combined in order to have a further increase of the accuracy. If the function is nonsmooth, then the reconstruction selects only the smoothest candidate approximation.

Although the construction of a WENO interpolation is a well-established matter, we will review here the details of the procedure, along with the main theoretical results, since it is usually treated and used in a different framework. In this part, we do not necessarily assume a uniform mesh spacing.

Construction

General structure To construct a WENO interpolation of degree $r = 2n-1$ on the interval $[x_l, x_{l+1}]$, we start from the Lagrange polynomial built on the stencil $\mathscr{S} = \{x_{l-n+1}, \ldots, x_{l+n}\}$ and written in the form

$$Q(x) = \sum_{k=1}^{n} C_k(x) P_k(x), \tag{3.25}$$

where the "linear weights" C_k are polynomials of degree $n-1$ and the P_k are polynomials of degree n interpolating V on the stencil $\mathscr{S}_k = \{x_{l-n+k}, \ldots, x_{l+k}\}$, $k = 1, \ldots, n$ (note that all the stencils $\mathscr{S}_k$ overlap on the interval $[x_l, x_{l+1}]$ and that the dependence on l has been dropped in this general expression). The nonlinear weights are then constructed so as to obtain an approximation of the highest degree if suitable smoothness indicators β_k give the same result on all stencils. This leads to defining

$$\alpha_k(x) = \frac{C_k(x)}{(\beta_k + \varepsilon)^2} \tag{3.26}$$

(with ε a properly small parameter, usually of the order of 10^{-6}), and then the nonlinear weights are defined as

$$w_k(x) = \frac{\alpha_k(x)}{\sum_h \alpha_h(x)}. \tag{3.27}$$

The final form of WENO interpolation is then

$$I_r[V](x) = \sum_{k=1}^{n} w_k(x)P_k(x), \tag{3.28}$$

where w_k is defined by (3.26)–(3.27).

The general concept of (3.28) is the following. If all the smoothness indicators β_k would have the same value, then $w_k(x) = C_k(x)$ and the reconstruction is given by the Lagrange polynomial (3.25), which has the highest possible degree. If some of the smoothness indicators have a "large" value with respect to the others (this meaning that the partial stencil contains a singularity), then $w_k(x) \ll C_k(x)$, and therefore the corresponding term is discarded via the weighting process.

Smoothness indicators Some degrees of freedom are left in the choice of the smoothness indicators. The general idea is that in smooth regions all indicators β_k should basically estimate the same quantity, and vanish at a certain rate when $\Delta x \to 0$, that is,

$$\beta_k = D_j \Delta x^{2p}(1 + O(\Delta x^q)), \tag{3.29}$$

where D_j is some constant depending on the function and on the interval index j. If condition (3.29) is satisfied, then in smooth conditions we would have $w_k(x) \approx C_k(x)$ (in a form depending on the exponent q, to be made more precise later), and this enables that scheme to increase the accuracy by merging the information from different stencils. On the other hand, if β_k is associated with a stencil containing a discontinuity of the function, then we require that

$$\beta_k = O(1), \tag{3.30}$$

so that asymptotically such a stencil would be discarded by the weighting procedure.

Following [Sh98], a typical way of defining the smoothness indicators is based on successive derivatives $P_k^{(h)}$ of the polynomials P_k, and more precisely

$$\beta_k = \sum_{h=1}^{n} \int_{x_l}^{x_{l+1}} \Delta x^{2h-1} P_k^{(h)}(x)^2 dx, \tag{3.31}$$

for which (3.29), (3.30) hold with $p = 1$, $q = 2$.

Remark 3.11. *We have shown here the most classical definition for the smoothness indicators, which enforces condition* (3.30) *in intervals containing discontinuities. However, if the solution is supposed to be smoother, different forms of the smoothness indicators could be chosen. For example, in HJ equations, where discontinuities may occur for the gradients (but not for the solution), the term related to the first derivative could be removed from* (3.31). *Accordingly, the scaling with respect to Δx should be changed to obtain* (3.30) *at singular points.*

Linear weights Let us consider now the linear weights C_k, which have not yet been given a precise expression. Due to (3.25) and to the definition of the polynomials P_k, the C_k are characterized by the fact that $Q(x)$ should be the interpolating polynomial of $V(x)$ on the stencil S. Since P_k interpolates V on the stencil $\mathcal{S}_k$, it would be natural to require that C_k should vanish at the nodes outside $\mathcal{S}_k$ and that in the nodes of $\mathcal{S}$ the nonzero weights should have unit sum. In this way, if P_k is the polynomial that interpolates the function on $\mathcal{S}_k$, then (3.25) is the polynomial that interpolates V on $\mathcal{S}$.

Taking into account that the number of nodes belonging to $\mathcal{S}$ but not to $\mathcal{S}_k$ is precisely n, we thus infer that the linear weights must necessarily have the form

$$C_k(x) = \gamma_k \prod_{x_h \in \mathcal{S} \setminus \mathcal{S}_k} (x - x_h) = \gamma_k \widetilde{C}_k(x) \tag{3.32}$$

with γ_k to be determined in order to have unit sum. Note that the conditions

$$\sum_{k=1}^{n} C_k(x_i) = 1 \tag{3.33}$$

for each $x_i \in \mathcal{S}$ apparently constitute a set of $2n$ equations in n unknowns. Actually, the correct perspective to look at these conditions is the following. The left-hand side of (3.33) is a polynomial of degree $n-1$ (computed at x_i), and on the right-hand side we have the polynomial $p(x) \equiv 1$. Therefore, imposing that the two polynomials coincide on more than $n-1$ points is equivalent to imposing that they be identical. Now, it is easy to show that the monic polynomials $\widetilde{C}_k$, $k = 1, \ldots, n$, are independent and form a basis of the space $\mathbb{P}_{n-1}$ of polynomials of degree $n-1$. Condition

$$\sum_{k=1}^{n} C_k(x) = 1 \quad \forall x \in \mathbb{R} \tag{3.34}$$

uniquely determines the constants γ_k as a polynomial identity.

Condition (3.34) can be conveniently satisfied by imposing it on a suitable set of points. Consider first the node x_{l-n+1}. Then, plugging (3.32) into (3.25), we note that, among all the linear weights computed at x_{l-n+1}, the only nonzero weight is C_1, so that

$$Q(x_{l-n+1}) = C_1(x_{l-n+1})P_1(x_{l-n+1}) = C_1(x_{l-n+1})v_{l-n+1}, \tag{3.35}$$

and this implies that $C_1(x_{l-n+1}) = 1$ and allows us to compute γ_1. On the other hand, in the following node x_{l-n+2}, the nonzero weights are C_1 and C_2, and with a similar argument we obtain the condition

$$\begin{aligned} Q(x_{l-n+2}) &= C_1(x_{l-n+2})P_1(x_{l-n+2}) + C_2(x_{l-n+2})P_2(x_{l-n+2}) \\ &= [C_1(x_{l-n+2}) + C_2(x_{l-n+2})]v_{l-n+2}, \end{aligned} \tag{3.36}$$

which implies that $C_1(x_{l-n+2}) + C_2(x_{l-n+2}) = 1$, whence γ_2 can be computed. Following this guideline, we finally obtain the set of conditions

$$\sum_{i=1}^{k} C_i(x_{l-n+k}) = 1 \qquad (k = 1, \ldots, n), \tag{3.37}$$

that is, more explicitly,

$$\sum_{i=1}^{k} \gamma_i \prod_{x_h \in \mathcal{S} \setminus \mathcal{S}_i} (x_{l-n+k} - x_h) = 1 \qquad (k = 1, \ldots, n), \tag{3.38}$$

which is a linear triangular system in the unknowns γ_i.

Examples

We give two examples of this construction (on a uniform mesh), including the expressions for the smoothness indicators (3.31). A general procedure for computing linear weights will be given in the next subsection.

Second/third-order WENO interpolation To construct a third-order interpolation we start from two polynomials of second degree, so that

$$I_3[V](x) = w_L P_L(x) + w_R P_R(x), \tag{3.39}$$

where $P_L(x)$ and $P_R(x)$ are second-order polynomials constructed, respectively, on the nodes x_{l-1}, x_l, x_{l+1} and on the nodes x_l, x_{l+1}, x_{l+2}. The two linear weights C_L and C_R are first degree polynomials in x, and according to the general theory outlined so far, they read as

$$C_L = \frac{x_{l+2} - x}{3\Delta x}, \quad C_R = \frac{x - x_{l-1}}{3\Delta x}, \tag{3.40}$$

and the expressions of α_L, α_R, w_L, and w_R may be easily recovered from the general form.

According to (3.31), the smoothness indicators have the explicit expressions

$$\beta_L = \frac{13}{12} v_{l-1}^2 + \frac{16}{3} v_l^2 + \frac{25}{12} v_{l+1}^2 - \frac{13}{3} v_{l-1} v_l + \frac{13}{6} v_{l-1} v_{l+1} - \frac{19}{3} v_l v_{l+1}, \tag{3.41}$$

$$\beta_R = \frac{13}{12} v_{l+2}^2 + \frac{16}{3} v_{l+1}^2 + \frac{25}{12} v_l^2 - \frac{13}{3} v_{l+2} v_{l+1} + \frac{13}{6} v_{l+2} v_l - \frac{19}{3} v_l v_{l+1}. \tag{3.42}$$

Third/fifth-order WENO interpolation To construct a fifth-order interpolation we start from three polynomials of third degree:

$$I_5[V](x) = w_L P_L(x) + w_C P_C(x) + w_R P_R(x), \tag{3.43}$$

where the third-order polynomials $P_L(x)$, $P_C(x)$, and $P_R(x)$ are constructed, respectively, on $x_{l-2}, x_{l-1}, x_l, x_{l+1}$, on $x_{l-1}, x_l, x_{l+1}, x_{l+2}$, and on $x_l, x_{l+1}, x_{l+2}, x_{l+3}$. The weights C_L, C_C, and C_R are second degree polynomials in x, and have the forms

$$C_L = \frac{(x - x_{l+2})(x - x_{l+3})}{20\Delta x^2},$$
$$C_C = \frac{(x - x_{l-2})(x - x_{l+3})}{10\Delta x^2},$$
$$C_R = \frac{(x - x_{l-2})(x - x_{l-1})}{20\Delta x^2},$$

while the smoothness indicators β_C and β_R have the expressions

$$\begin{aligned}\beta_C = {} & \frac{61}{45} v_{l-1}^2 + \frac{331}{30} v_l^2 + \frac{331}{30} v_{l+1}^2 + \frac{61}{45} v_{l+2}^2 - \frac{141}{20} v_{l-1} v_l + \frac{179}{30} v_{l-1} v_{l+1} \\ & + \frac{293}{180} v_{l-1} v_{l+2} - \frac{1259}{60} v_l v_{l+1} + \frac{179}{30} v_l v_{l+2} - \frac{141}{20} v_{l+1} v_{l+2},\end{aligned} \tag{3.44}$$

$$\begin{aligned}\beta_R = {} & \frac{407}{90} v_l^2 + \frac{721}{30} v_{l+1}^2 + \frac{248}{15} v_{l+2}^2 + \frac{61}{45} v_{l+3}^2 - \frac{1193}{60} v_l v_{l+1} + \frac{439}{30} v_l v_{l+2} \\ & - \frac{683}{180} v_l v_{l+3} - \frac{2309}{60} v_{l+1} v_{l+2} + \frac{309}{30} v_{l+1} v_{l+3} - \frac{553}{60} v_{l+2} v_{l+3},\end{aligned} \tag{3.45}$$

and β_L can be obtained using the same set of coefficients of β_R in a symmetric way (that is, replacing the indices $l-2, \ldots, l+3$ with $l+3, \ldots, l-2$).

Theoretical results

It is easy to check that, due to their structure, the linear weights for the two examples above are always positive in the interval $[x_l, x_{l+1}]$. We prove now that this result holds in general. In addition, we give an explicit expression for the linear weights in the case of evenly spaced nodes and prove an interpolation error estimate.

Positivity of the weights We first give a general proof of positivity for the linear weights $C_k(x)$. We point out that, due to the general structure of nonlinear weights, positivity of the former set of weights implies positivity of the latter.

Theorem 3.12. *Let $\{x_i\}$ be a family of consecutively numbered points of $\mathbb{R}$. Then, once the Lagrange polynomial $Q(x)$ of a given function $v(x)$ built on the stencil $\mathcal{S} = \{x_{l-n+1}, \dots, x_{l+n}\}$ is written in the form (3.25), the linear weights $C_k(x)$ $(k = 1, \dots, n)$ are nonnegative for any $x \in (x_l, x_{l+1})$. Moreover, $\sum_k C_k(x) \equiv 1$.*

Proof. Let us denote by $Q^{(h,m)}(x)$ (with $j - n + 1 \le h \le j < j + 1 \le m \le j + n$ and $m - h \ge n$) the interpolating polynomial constructed on the stencil $\{x_h, \dots, x_m\}$, so that

$$Q(x) = Q^{(j-n+1,j+n)}(x), \tag{3.46}$$

and, moreover,

$$P_k(x) = Q^{(j-n+k,j+k)}(x). \tag{3.47}$$

We will also write a generic $Q^{(h,m)}(x)$ as

$$Q^{(h,m)}(x) = \sum_k C_k^{(h,m)}(x) P_k(x), \tag{3.48}$$

where the summation may be extended to all $k = 1, \dots, n$ by setting $C_k^{(h,m)} = 0$ whenever the stencil of P_k is not included in $\{x_h, \dots, x_m\}$. Therefore, the final linear weights will be

$$C_k(x) = C_k^{(j-n+1,j+n)}(x). \tag{3.49}$$

We proceed by induction on the degree of $Q^{(h,m)}$, given by $i = \deg Q^{(h,m)} = m - h$, starting with $i = n$ up to $i = 2n - 1$. First we prove that the claim of the theorem extends to all polynomials $Q^{(h,m)}(x)$ defined in (3.48). The claim is obviously true for $Q^{(j-n+k,j+k)}(x)$ by (3.47). In this case the coefficients read as $C_k^{(h,m)} = 1$, $C_s^{(h,m)} \equiv 0$, $s \ne k$. Then, we assume the claim is true for the set of linear weights $C_k^{(h,m)}(x)$ of a generic $Q^{(h,m)}(x)$ such that $m - h = i$ and add a node to the right (adding a node to the left leads to an analogous computation). By inductive assumption we have, for any h and m such that $h \le j < j + 1 \le m$ and $m - h = i$, that (3.48) holds with $C_k^{(h,m)}(x) \ge 0$. On the other hand, by elementary interpolation theory arguments (Neville's recursive form of the interpolating polynomial),

$$Q^{(h,m+1)}(x) = \frac{x_{m+1} - x}{x_{m+1} - x_h} Q^{(h,m)}(x) + \frac{x - x_h}{x_{m+1} - x_h} Q^{(h+1,m+1)}(x). \tag{3.50}$$

Note that $\deg Q^{(h,m)} = \deg Q^{(h+1,m+1)} = i$, and that both fractions which multiply $Q^{(h,m)}(x)$ and $Q^{(h+1,m+1)}(x)$ are positive as long as $x \in (x_l, x_{l+1})$. Therefore, plugging (3.48) into

(3.50), we get

$$Q^{(h,m+1)}(x) = \sum_k \left[\frac{x_{m+1}-x}{x_{m+1}-x_h} C_k^{(h,m)}(x) + \frac{x-x_h}{x_{m+1}-x_h} C_k^{(h+1,m+1)}(x) \right] P_k(x). \qquad (3.51)$$

The term in square brackets is $C_k^{(h,m+1)}(x)$, and it is immediate to check that it is nonnegative. Iterating this argument up to $Q^{(j-n+1,j+n)}(x)$ completes the proof of nonnegativity for all the linear weights $C_k(x)$.

Last, setting $v(x) \equiv 1$, we also have $Q(x) \equiv 1$, and, for any k, $P_k(x) \equiv 1$. Plugging these identities into (3.25) we obtain $\sum_k C_k(x) \equiv 1$. □

Explicit form of the linear weights We turn now to the problem of giving an explicit expression to the linear weights in the situation of evenly spaced nodes. The expression of the weights for the linear scheme is

$$C_k = \gamma_k \prod_{x_h \in \mathscr{S} \setminus \mathscr{S}_k} (x - x_h),$$

so that the problem is providing the expression of the constants γ_k by applying to the case of a uniform grid the general procedure already outlined. The endpoint of this analysis is given by the following theorem.

Theorem 3.13. *Assume the space grid is evenly spaced with step Δx. Then, the linear weights (3.32) have the explicit form*

$$C_k(x) = \frac{(-1)^{n+k}}{\Delta x^{n-1}} \frac{n!(n-1)!}{(n-k)!(k-1)!(2n-1)!} \prod_{x_h \in S \setminus S_k} (x - x_h) \qquad (3.52)$$

and satisfy the positivity condition $C_k(x) > 0$ for any $k \in \{1, \dots, n\}$, $x \in [x_l, x_{l+1}]$.

Proof. First observe that the polynomial $C_k(x)$ does not change sign in the interval $[x_l, x_{l+1}]$. More precisely, the sign of it, in this interval, is given by $\mathrm{sgn}(\gamma_k)(-1)^{n+k}$. Therefore

$$C_k(x) > 0 \quad \forall x \in [x_l, x_{l+1}], \forall k, n \quad \Leftrightarrow \quad \mathrm{sgn}(\gamma_k) = (-1)^{n+k}. \qquad (3.53)$$

For simplicity, set $l = n$ in the definition of the stencil $\mathscr{S}$, so that $\mathscr{S} = \{x_1, \dots, x_{2n}\}$. Let us introduce the reference stencils

$$\widetilde{\mathscr{S}} = \mathscr{S}/\Delta x, \; \widetilde{\mathscr{S}_k} = \mathscr{S}_k/\Delta x.$$

Then the polynomials C_k can be written as

$$C_k(\Delta x \eta) = \widetilde{\gamma}_k \prod_{h \in \widetilde{\mathscr{S}} \setminus \widetilde{\mathscr{S}_k}} (\eta - h),$$

where η is the variable in the reference space, $\widetilde{\gamma}_k = \gamma_k \Delta x^{n-1}$. The dimensionless constants $\widetilde{\gamma}_k$ satisfy the condition

$$\sum_{i=1}^{k} \widetilde{\gamma}_i \prod_{\substack{h=1 \\ h \notin \{i, \dots, i+n\}}}^{2n} (i - h) = 1.$$

Introducing now the matrix

$$a_{ki} = \frac{\prod_{h=1}^{2n}(k-h)}{\prod_{h=i}^{i+n}(k-h)} = \frac{(k-1)!(2n-k)!}{(n+i-k)!(k-i)!}(-1)^{n+i}, \tag{3.54}$$

the system can be written as

$$\sum_{i=1}^{k} a_{ki}\widetilde{\gamma}_i = 1.$$

Let $\mu_i \equiv (-1)^{n+i}\widetilde{\gamma}_i$. Then, such constants satisfy the triangular system

$$\sum_{i=1}^{k} |a_{ki}|\mu_i = 1 \tag{3.55}$$

and, according to (3.53), $C_i(x) > 0$, $x \in [x_n, x_{n+1}]$, if and only if $\mu_i > 0$.

The solution to system (3.55) can be given explicitly. In fact, we will prove now that

$$\mu_i = \frac{n!(n-1)!}{(n-i)!(i-1)!(2n-1)!} \qquad (i = 1, \dots, n). \tag{3.56}$$

Actually, plugging (3.56) and (3.54) into (3.55) we obtain the set of conditions that we want to check:

$$\sum_{i=1}^{k} \frac{(k-1)!(2n-k)!n!(n-1)!}{(n+i-k)!(k-i)!(n-i)!(i-1)!(2n-1)!} = 1.$$

These relations can be rearranged in the form

$$\frac{(k-1)!(2n-k)!}{(2n-1)!}\sum_{i=1}^{k}\binom{n}{k-i}\binom{n-1}{i-1} = 1,$$

and hence, rewriting the first term and shifting the summation index as $j = i-1$,

$$\sum_{j=0}^{k-1}\binom{n}{k-1-j}\binom{n-1}{j} = \binom{2n-1}{k-1},$$

which is in turn a special case of the identity (see [AS64])

$$\sum_{j=0}^{N}\binom{n}{N-j}\binom{m}{j} = \binom{n+m}{N},$$

with $N = k-1$, and $m = n-1$. □

Remark 3.14. *We can further prove here that the μ_i are inverses of integers. In fact, note first that by* (3.56), $\mu_i = \mu_{n-i+1}$, *so that it suffices to prove the claim for $i \le (n+1)/2$. Then, we rewrite the μ_i as*

$$\mu_i = \begin{cases} \dfrac{n!}{(2n-1)!} & \text{if} \quad i = 1, \\[2ex] \dfrac{(n-i+1)\cdots(n-1)n(n-1)!}{(i-1)!(2n-1)!} & \text{if} \quad i > 1. \end{cases} \tag{3.57}$$

Table 3.3. *Values of $1/\mu_i$ for the first 8 values of n.*

n	$1/\mu_i$
1	1
2	3, 3
3	20, 10, 20
4	210, 70, 70, 210
5	3024, 756, 504, 756, 3024
6	55440, 11088, 5544, 5544, 11088, 55440
7	1235520, 205920, 82368, 61776, 82368, 205920, 1235520
8	32432400, 4633200, 1544400, 926640, 926640, 1544400, 4633200, 32432400

The claim is obvious if $i = 1$, whereas for $i > 1$ we derive from (3.57)

$$\mu_i = \frac{(2n-2i+2)\cdots(2n-2)}{2^{i-1}(i-1)!(n+1)\cdots(2n-1)}.$$

Now, if $i \le (n+1)/2$, then $2n-2i+2 \ge n+1$, and therefore any term of the product $(2n-2i+2)\cdots(2n-2)$ can be simplified with a term of the product $(n+1)\cdots(2n-1)$. This completes the proof of the claim.

We report in Table 3.3 the value of $1/\mu_i$ for values of n up to $n = 8$.

Interpolation error estimates We finally prove here the convergence result for WENO interpolations, restricting ourselves to the smooth case.

Theorem 3.15. *Let $v(x)$ be a C^∞ function, and let the interpolation $I_r[V]$ be defined by the WENO form* (3.28) *with the smoothness indicators β_k satisfying* (3.29). *Then,*

$$\|v - I_r[V]\|_\infty \le C\Delta x^{\min(2n,n+q+1)}. \tag{3.58}$$

Proof. First, note that if (3.29) is satisfied, then

$$w_k(x) = C_k(x) + \omega_k(x, \Delta x^q), \tag{3.59}$$

where for any k, $\omega_k(x, \Delta x^q) = O(\Delta x^q)$, and, since both the w_k and the C_k have unity sum, we also have $\sum_k \omega_k(x, \Delta x^q) = 0$. Then,

$$|v(x) - I_r[V](x)| \le \left| v(x) - \sum_k C_k(x) P_k(x) \right| + \left| \sum_k C_k(x) P_k(x) - \sum_k w_k(x) P_k(x) \right|.$$

For the first term it is clear that

$$\left| v(x) - \sum_k C_k(x) P_k(x) \right| \le C\Delta x^{2n}, \tag{3.60}$$

whereas for the second, using (3.59) and the fact that the P_k are themselves interpolating polynomials of degree n,

$$\begin{aligned}\left|\sum_k C_k(x)P_k(x) - \sum_k w_k(x)P_k(x)\right| &= \left|\sum_k [C_k(x) - w_k(x)]P_k(x)\right| \\ &= \left|\sum_k \omega_k(x,\Delta x^q)\left[v(x) + O(\Delta x^{n+1})\right]\right| \\ &= \left|v(x)\sum_k \omega_k(x,\Delta x^q) + O(\Delta x^{n+q+1})\right| \\ &= O(\Delta x^{n+q+1}), \end{aligned} \tag{3.61}$$

where in the last display we have used the fact that the $\omega_k(x,\Delta x^q)$ have zero sum. Finally, (3.58) follows from (3.60), (3.61). □

3.2.4 ▪ Handling multiple dimensions by separation of variables

Lagrange interpolation When passing to multiple space dimensions, the extension of Lagrange interpolation (that is, in practice, the definition of suitable cardinal basis functions) can be performed in different forms. In the simplest case, cardinal basis functions are defined by the product of one-dimensional functions in the form that follows. Let the space grid of points x_j be uniform and orthogonal, and let $j = (j_1,\dots,j_d)$ be the multi-index associated with a given node. Now, once we ensure that the ψ_j are cardinal functions, the form (3.7) still defines an interpolant of v. On the other hand, on a uniform orthogonal grid, a basis of Lagrange cardinal functions can be defined by

$$\psi_j(\xi) = \prod_{k=1}^{d} \psi_{j_k}(\xi_k), \tag{3.62}$$

where, in order to avoid ambiguities, we have denoted the space variable by $\xi = (\xi_1,\dots,\xi_d)$. In (3.62), it should be understood that ψ_j (which is indexed by a multi-index and has an argument in $\mathbb{R}^d$) refers to the d-dimensional grid, whereas ψ_{j_k} (which has a simple index and a scalar argument) refers to the grid on the kth variable. It is clear that (3.62) defines a cardinal function, and in fact

$$\psi_j(x_i) = \begin{cases} 1 & \text{if} \quad i_1 = j_1,\dots,i_d = j_d, \\ 0 & \text{else.} \end{cases}$$

Figure 3.3 shows the bidimensional version of the cubic interpolation reference basis function of Figure 3.2.

A different interpretation of this procedure can be given by splitting the operation implied by (3.7). Working for simplicity in two dimensions and plugging (3.62) into (3.7), we have

$$\begin{aligned} I[V](\xi) &= \sum_{j_1}\sum_{j_2} v(x_{j_1,j_2})\psi_{j_1}(\xi_1)\psi_{j_2}(\xi_2) \\ &= \sum_{j_2}\left[\sum_{j_1} v(x_{j_1,j_2})\psi_{j_1}(\xi_1)\right]\psi_{j_2}(\xi_2). \end{aligned} \tag{3.63}$$

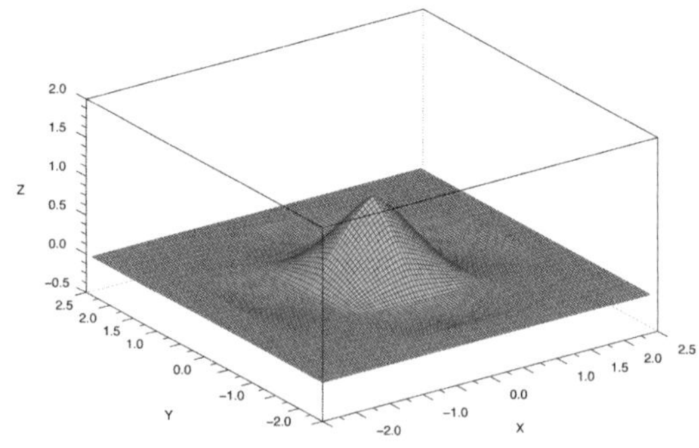

Figure 3.3. *Reference basis function for cubic interpolation in* $\mathbb{R}^2$.

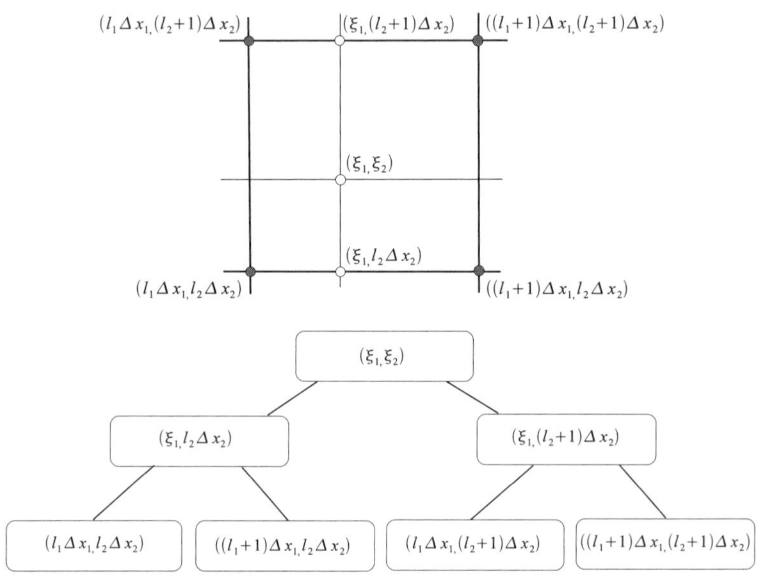

Figure 3.4. *Dimensional splitting for linear interpolation in* $\mathbb{R}^2$.

In practice, the process appears to be the result of two (in general, d) nested one-dimensional interpolations. The upper diagram of Figure 3.4 shows the points involved in this process for a linear Lagrange interpolation. First, the point ξ is located in a rectangle of the grid, say $[l_1\Delta x_1, (l_1+1)\Delta x_1] \times [l_2\Delta x_2, (l_2+1)\Delta x_2]$. Then, in the first phase of the interpolation, two one-dimensional linear reconstructions are performed to recover the interpolated values at $(\xi_1, l_2\Delta x_2)$ and $(\xi_1, (l_2+1)\Delta x_2)$. In the second phase, a linear interpolation is performed between these two values to recover the interpolated value at (ξ_1, ξ_2).

Once the same points in a tree are arranged (as shown in the lower diagram), the interpolation is computed by advancing from the leaves towards the root (in terms of data structures, the interpolated value is computed as the tree is visited in *postorder*).

In general, this procedure can be applied to any dimension d, as well as any order of interpolation. The dimension appears as the depth of the tree, whereas the number of nodes required for one-dimensional reconstruction coincides with the number of sons of a given node of the tree.

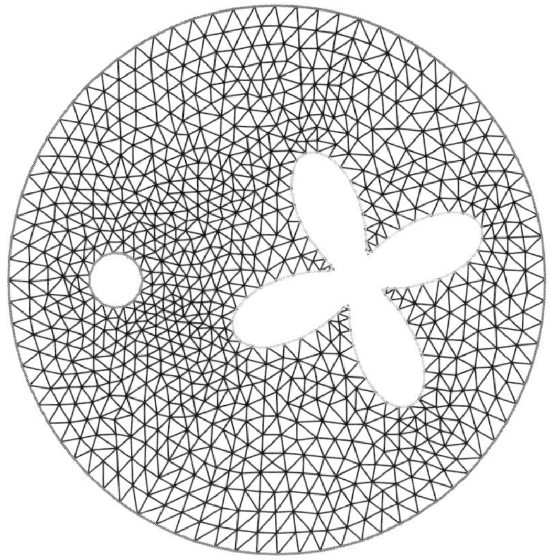

Figure 3.5. *Tessellation of a complex geometry into triangles.*

ENO/WENO interpolation The nonlinear nature of nonoscillatory reconstructions prevents them from being extended to $\mathbb{R}^d$ in the forms (3.7), (3.62). On the other hand, the mechanism of successive one-dimensional interpolations may also be applied to ENO or WENO (or, in general, to nonlinear) reconstructions, although in this case, due to the existence of various candidate stencils, the total number of points involved is considerably higher. It is also possible, however, to define nonoscillatory interpolants based on genuinely multidimensional stencils, possibly on unstructured grids.

3.2.5 ▪ Finite element interpolation

Finite element interpolation is a reconstruction strategy which, still interpolating a given function in a piecewise polynomial form, does not need any particular structure of the space grid and is intrinsically multidimensional, although it also includes the one-dimensional case.

A definite advantage of finite element interpolation is the ability to treat complex geometries, which in general do not allow for a structured grid. We show in Figure 3.5 an example, which also illustrates the possibility of performing local refinements for accuracy purposes.

Construction

By definition, a finite element is a triple $(K, \Sigma, \mathbb{V}_r)$, where the following hold:

- K is the reference domain of the element and also in practice the kind of elementary figure into which the computational domain is split. In one space dimension K is necessarily an interval; in $\mathbb{R}^2$ typical choices are triangles, and less frequently rectangles, and so on.

- $\mathbb{V}_r$ is the space of polynomials in which the interpolation is constructed. Two typical choices we will consider here are $\mathbb{P}_r$ (the space of polynomials in d variables of degree no larger than r) and $\mathbb{Q}_r$ (the space of polynomials of degree no larger than r with respect to each variable).

- Σ is a set of functionals $\gamma_i : \mathbb{P}_r \to \mathbb{R}$ (usually referred to as *degrees of freedom*) which completely determine a polynomial of $\mathbb{P}_r$. In the case of Lagrange finite elements, which is the one of interest here, the degrees of freedom are the values of the polynomial on a unisolvent set of nodes in K (with some abuse of notation, they are identified with the nodes), and the polynomial is expressed on the basis of Lagrange functions associated with such a set of nodes.

We will not explain in detail the construction of a Lagrange finite element interpolation in its greatest generality (this topic is treated in the specialized literature), but rather sketch the basic ideas by working on some one- and two-dimensional examples. As a starting point, the computational domain should be (approximately) tessellated with nonoverlapping elements obtained by affine transformations of the reference element. Interpolation nodes are placed on each element by mapping the nodes of the reference element. As outlined for symmetric Lagrange interpolation, the basis function ψ_k used in (3.7) can be conceptually obtained by interpolating the sequence e_k.

Examples

To make the main ideas clear, we present here the construction of the $\mathbb{P}_1$, $\mathbb{P}_2$, $\mathbb{Q}_1$, and $\mathbb{Q}_2$ Lagrange finite element spaces in one and two space dimensions. For a general treatment of the topic, we refer the reader to classical monographs on finite element schemes.

$\mathbb{P}_1$ and $\mathbb{P}_2$ finite elements in $\mathbb{R}$ In one dimension, the only possibility is to define the reference element as an interval. In order to interpolate with a $\mathbb{P}_1$ or $\mathbb{P}_2$ basis, respectively, two and three nodes must be placed in the reference interval. To obtain a continuous interpolant, two of them must be placed at the interface between elements, that is, in the extreme points of the reference element. Therefore, if the reference interval keeps unity distance between nodes, in the $\mathbb{P}_1$ case it could be defined as the interval $[0,1]$ with the nodes $\eta_0 = 0$, $\eta_1 = 1$ (here and in what follows, we use η as a variable in the reference space). As a result, the Lagrange reference basis functions (shown in the left plot of Figure 3.6) will be given by

$$\ell_0(\eta) = 1 - \eta, \qquad \ell_1(\eta) = \eta,$$

while a generic point $x \in [x_l, x_{l+1}]$ will be carried in the reference interval $[0,1]$ by the transformation

$$\eta = \frac{x - x_l}{x_{l+1} - x_l}.$$

The cardinal basis function ψ_k associated with a node x_k of a one-dimensional grid reads then as

$$\psi_k(x) = \begin{cases} \dfrac{x - x_{k-1}}{x_k - x_{k-1}} & \text{if } x \in (x_{k-1}, x_k], \\[2ex] 1 - \dfrac{x - x_k}{x_{k+1} - x_k} & \text{if } x \in (x_k, x_{k+1}), \\[2ex] 0 & \text{elsewhere.} \end{cases}$$

Note that, except for allowing a nonuniform grid, $\mathbb{P}_1$ finite element interpolation coincides with first-order Lagrange interpolation. This correspondence, however, is lost for higher interpolation orders.

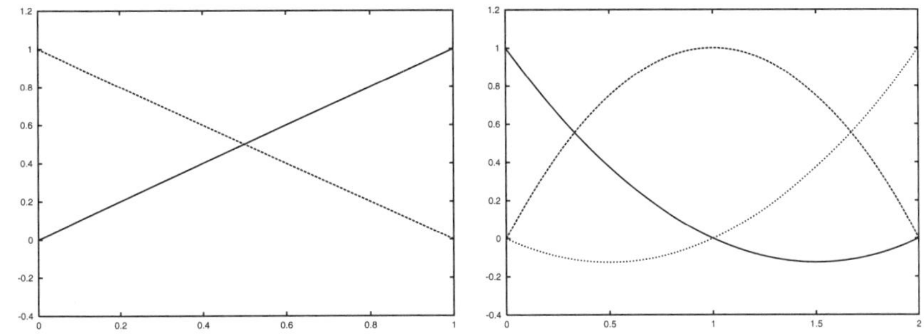

Figure 3.6. $\mathbb{P}_1$ *and* $\mathbb{P}_2$ *reference elements and basis functions in* $\mathbb{R}$.

In the case of $\mathbb{P}_2$ interpolation, we can choose the reference interval as $[0,2]$ and place the three nodes at $\eta = 0, 1, 2$. The three Lagrange reference basis functions (shown in the right plot of Figure 3.6) are now given by

$$\ell_0(\eta) = \frac{1}{2}(\eta-1)(\eta-2), \qquad \ell_1(\eta) = 1-(\eta-1)^2, \qquad \ell_2(\eta) = \frac{1}{2}\eta(\eta-1),$$

and, if $x \in [x_l, x_{l+2}]$, the transformation from the variable x to the reference variable η is

$$\eta = 2\frac{x-x_l}{x_{l+2}-x_l}.$$

The cardinal basis function ψ_k is constructed in a way which parallels the $\mathbb{P}_1$ case; in particular,

$$\psi_k(x) = \begin{cases} 1-\left(\dfrac{x-x_{k-1}}{x_k-x_{k-1}}-1\right)^2 & \text{if } x \in (x_{k-1}, x_{k+1}), \\ 0 & \text{elsewhere} \end{cases}$$

if x_k is internal to an element, and

$$\psi_k(x) = \begin{cases} \dfrac{1}{2}\left(\dfrac{x-x_{k-2}}{x_k-x_{k-1}}\right)\left(\dfrac{x-x_{k-2}}{x_k-x_{k-1}}-1\right) & \text{if } x \in (x_{k-2}, x_k], \\ \dfrac{1}{2}\left(\dfrac{x-x_k}{x_{k+1}-x_k}-1\right)\left(\dfrac{x-x_k}{x_{k+1}-x_k}-2\right) & \text{if } x \in (x_k, x_{k+2}), \\ 0 & \text{elsewhere} \end{cases}$$

if x_k is at the interface between two elements.

Note that in the $\mathbb{P}_2$ case the situations in which x_k is internal to or an extreme point of an element leads to different forms, even on an uniformly spaced mesh. More generally, in the $\mathbb{P}_r$ case (with $r > 1$), the form of ψ_k depends on the position of x_k within the element. Thus, in any of these cases the interpolation fails in general to be translation invariant.

We show in Figure 3.6 two examples of reference intervals and basis functions for $\mathbb{P}_1$ and $\mathbb{P}_2$ finite element interpolation in one space dimension.

$\mathbb{P}_1$, $\mathbb{P}_2$, $\mathbb{Q}_1$, and $\mathbb{Q}_2$ finite elements in $\mathbb{R}^2$ At the increase of dimensions, the geometry of the reference element allows for an increasing number of options. In $\mathbb{R}^2$, two main situations of interest can be recognized: the situations in which the reference element is,

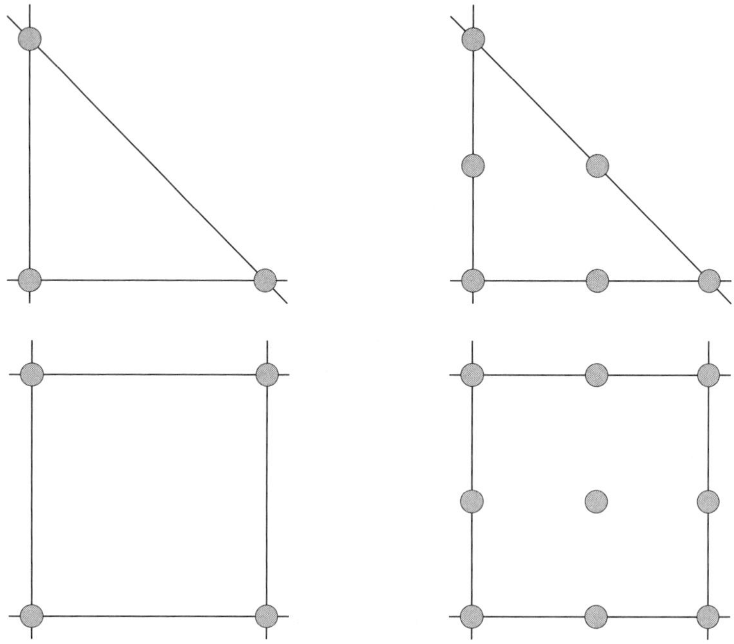

Figure 3.7. $\mathbb{P}_1$-$\mathbb{P}_2$ *(upper) and* $\mathbb{Q}_1$-$\mathbb{Q}_2$ *(lower) reference elements in* $\mathbb{R}^2$.

respectively, a triangle (leading to $\mathbb{P}_r$ finite element spaces) and a rectangle (leading to $\mathbb{Q}_r$ finite element spaces).

On triangular elements, it is necessary to construct the space of polynomials of degree r, with $r = 1,2$ in our example. We will not give the explicit expression of the reference basis functions, but rather show the choice of the nodes. In the $\mathbb{P}_1$ case, the polynomial space has dimension three, and the nodes are placed in the vertices of the element. In the $\mathbb{P}_2$ case, the space has dimension six, and the nodes are typically placed at vertices and midpoints. The upper line of Figure 3.7 shows the position of nodes in the two situations. Clearly, the kth Lagrange reference basis function is a polynomial of degree r taking the value $\ell_k(\eta_m) = \delta_{km}$ at a generic node η_m. Note that the restriction of a Lagrange basis function to one side of the triangle gives a polynomial of degrees $r = 1$ and $r = 2$, respectively, and the number of nodes on the side is enough to determine univocally this polynomial. Thus, the interpolation remains continuous passing from one element to the other.

On quadrilateral elements, the Lagrange polynomials are constructed as tensor products of one-dimensional Lagrange bases, as it has been discussed concerning multidimensional Lagrange reconstruction. Therefore, the $\mathbb{Q}_1$ finite element has four nodes at the vertices of the quadrilateral, whereas the $\mathbb{Q}_2$ finite element has nine nodes (vertices, midpoints, and center of the quadrilateral element). The lower line of Figure 3.7 reports the position of nodes in this case. Again, the number of nodes on each side of the quadrilateral implies that the interpolation is continuous at the interface of two elements.

Theoretical results

The approximation result concerning finite element interpolation follows the general result for polynomial approximations. However, when dealing with nonuniform space grids, care should be taken to avoid that triangular elements could degenerate when refining the grid, typically by flattening towards segments. While we leave a formal theory to specialized literature, we simply remark here that grid refinement in a finite element setting requires that the largest size Δx of the elements vanish and that the grid be *nondegenerate*. This means that, once we denote by R_K the radius of the element K and by r_K the radius of the inscribed circle, their ratio remains bounded for all K:

$$\frac{R_K}{r_K} \leq C. \tag{3.64}$$

Last, we give the approximation result for finite element interpolations.

Theorem 3.16. *Let the interpolation $I_r[V]$ be defined by a $\mathbb{P}_r$ finite element space on a space grid satisfying* (3.64). *Let $v(x)$ be a uniformly continuous function on $\mathbb{R}$, and let V be the vector of its nodal values. Then, for $\Delta x \to 0$,*

$$\|v - I_r[V]\|_\infty \to 0. \tag{3.65}$$

If, moreover, $v \in W^{s,\infty}(\mathbb{R})$, then

$$\|v - I_r[V]\|_\infty \leq C\Delta x^{\min(s,r+1)} \tag{3.66}$$

for some positive constant C depending on the degree of regularity s.

3.3 ▪ Function minimization

As a building block of SL schemes for nonlinear problems, we will be concerned with the problem of minimization of a function of N variables,

$$f(x^*) = \min_{x \in \mathbb{R}^N} f(x),$$

and with the related numerical techniques. Since in our case the function to be minimized has no explicit form, our interest here is mainly focused on *derivative-free methods*, i.e., methods which minimize a function without making use of the analytic expressions of the gradient or the hessian. Without aiming at completeness, a list of the main approaches used to perform this operation includes

- direct search methods,
- descent methods (adapted to work without derivatives),
- Powell's method and its modifications, and
- trust-region methods based on quadratic models.

We will briefly review the general philosophy of the various classes of methods. The huge amount of related studies cannot be condensed in such a brief survey; rather than giving implementation details, we will sketch the general ideas and refer the reader to suitable, specialized literature and software packages.

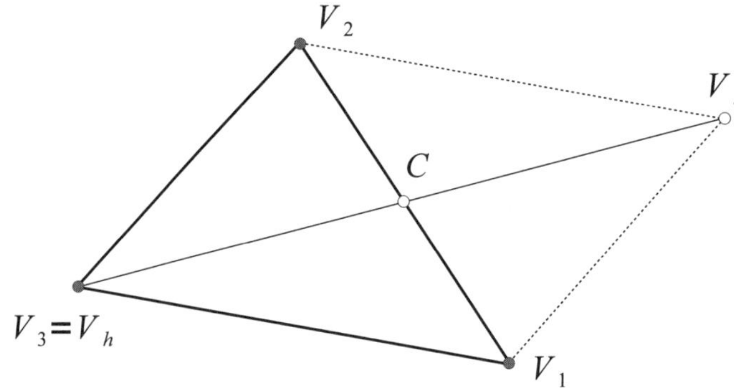

Figure 3.8. *Updating of a simplex in the basic version of simplex method.*

3.3.1 ▪ Direct search methods

The name of *direct search methods* is generally used to collect methods of heterogeneous nature, which share the feature of minimizing a function without using, either in exact or approximate form, its gradient. Direct search methods rather proceed by computing the function at a discrete, countable set of points.

The most common schemes of this class are based on the *simplex* technique (not to be confused with the algorithm used in linear programming), whose basic idea dates back to the 1960s and is shown in Figure 3.8.

Let a simplex of vertices $V_1, \ldots, V_m$ be constructed in $\mathbb{R}^N$. Once we compute the corresponding values $y_1, \ldots, y_m$ of f at the vertices, the highest value y_h is selected, and the corresponding vertex V_h is reflected with respect to the centroid C of the simplex made by the remaining vertices. Then, the new simplex obtained by replacing the node V_h with V_r is accepted if the value of the function at V_r is not the highest value in the new simplex. Otherwise, the second highest value of the original simplex is selected, and the procedure is repeated. If all possible updates of the simplex lead to finding the highest value on the reflected node, the algorithm stops. The size of the simplex should then be considered as a tolerance in locating the minimum, and to refine the search the algorithm should be restarted with a simplex of smaller size.

In a later (and more widely used) version, the algorithm has been restated replacing the reflection of V_h with a discrete line search. In this version the geometry of simplices may change, and while this allows the scheme to adapt to the geometry of the function, it can also cause the simplices to degenerate, thus stalling the algorithm. Specific tools have been devised to handle this situation, the most obvious one being the restart.

3.3.2 ▪ Descent methods

In descent methods, an initial estimate of the minimum point x_0 is given, and the updating of the approximation x_k is performed via the formula

$$x_{k+1} = x_k + \beta_k d_k; \tag{3.67}$$

that is, the scheme moves from x_k to x_{k+1} along the direction d_k with step β_k. As a rule (which also gives the name to this class of schemes), the direction d_k should be a descent direction, i.e., $(d_k, \nabla f(x_k)) < 0$, β_k should be positive, and $f(x_{k+1}) < f(x_k)$. Neglecting for the moment the problem of avoiding the computation of derivatives, the first and most typical scheme of this class is the *method of steepest descent* (or *gradient method*), in which $d_k = -\nabla f(x_k)$, so that

$$x_{k+1} = x_k - \beta_k \nabla f(x_k), \tag{3.68}$$

and β_k is chosen on the basis of an *exact line search*, that is,

$$f(x_{k+1}) = \min_{\beta}[x_k - \beta \nabla f(x_k)] \tag{3.69}$$

(we will not discuss here the algorithms used for this one-dimensional minimization). Despite the fact that such a choice of the direction d_k gives locally the largest negative directional derivative, the method of steepest descent in its original formulation requires a high complexity to determine β_k by the one-dimensional minimum problem (3.69), but has nevertheless a slow convergence to the minimum. Since the precision in solving the exact line search has little influence on the convergence speed, a first improvement consists in relaxing the definition (3.69), so that a certain β becomes acceptable if it ensures that the function "decreases enough" passing from x_k to x_{k+1} (once more, we give up describing such strategies in detail). This generates the so-called *inexact line search*. A further improvement is to choose more efficient search directions d_k. The most common choices are the following:

- Newton's method: it consists in setting $d_k = -H_f(x_k)^{-1}\nabla f(x_k)$. This choice ensures quadratic convergence for smooth functions but has $O(N^3)$ complexity for each iteration.

- Quasi-Newton methods: here, d_k is defined as $d_k = -H_k \nabla f(x_k)$, where H_k is a matrix constructed at each iteration so as to mimic (in a suitable sense) the behavior of $H_f(x_k)^{-1}$ but whose construction requires $O(N^2)$ operations. On the other hand, the speed of convergence is superlinear.

- Conjugate direction methods: initially conceived for positive definite quadratic functions,

  $$f(x) = \frac{1}{2}(Ax, x) - (b, x),$$

  they are based on search directions d_k which are *A-conjugate*, that is,

  $$(Ad_i, d_i) > 0, \qquad (Ad_i, d_j) = 0 \quad (i \neq j).$$

  In the case of quadratic functions, it can be proved that such a scheme converges at most in n steps. Its adaptation to nonquadratic function (the so-called *conjugate gradient*, which requires one to compute ∇f at each iteration) has superlinear convergence, while the cost per iteration is $O(N^2)$.

To turn back to the problem of implementing schemes of this class without computing derivatives, we remark that in principle it is possible to replace the gradient and, possibly, the hessian of f by their finite difference approximations, although the complexity of this operation is $N+1$ computations of f for the gradient and even higher for the hessian. Whenever the computation of f has a critical complexity (and this is definitely the case when using minimization algorithms within an SL scheme), the minimization is better accomplished by cheaper techniques.

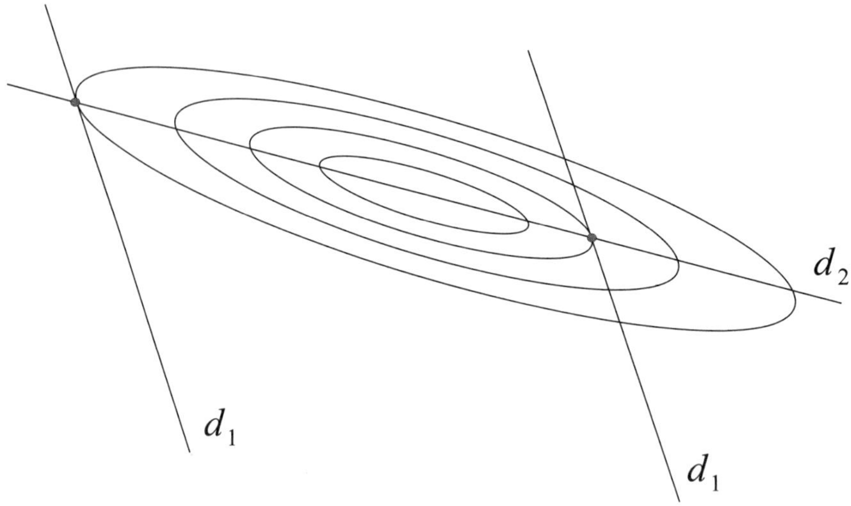

Figure 3.9. *Construction of conjugate directions in Powell's method.*

3.3.3 ▪ Powell's method and its modifications

This method is genuinely derivative-free, but search directions are rather computed exploiting a geometric property of quadratic functions, which allows one to construct conjugate directions without computing the gradient or the hessian. The basic idea is sketched in Figure 3.9 in the case of two space dimensions.

Let a quadratic function f (shown through its elliptical level curves) and a search direction d_1 be given. Consider the minima of f along two parallel lines with direction d_1 (we recall that at a line-constrained minimum the search direction is tangent to the level curve). Then, the direction d_2 through the two minima is conjugate with d_1. In Figure 3.9, this is recognized by the fact that d_2 passes through the center of the ellipses—this agrees with the fact that the second conjugate direction brings the global minimum, as it happens for a quadratic function in two dimensions.

The technique can be extended to generate conjugate directions for a generic dimension N and produces a scheme in the form (3.67) of a descent method. Being a conjugate direction method, Powell's method converges in N iterations for quadratic functions of N variables and is expected to be superlinear for smooth nonquadratic functions.

Clearly, a certain number of technical details (which will not be reviewed here) should be set up to make the idea work in practice, but this method has inspired a large number of efficient algorithms. Among the most successful related software, we mention the routine PRAXIS, which is freely available as an open-source Fortran code.

3.3.4 ▪ Trust-region methods based on quadratic models

In this class of schemes, the iterative minimization of the function f is carried out by working at each step on an approximate (typically, quadratic) model of the function, which is associated with a so-called *trust region*, a set in which the model is considered to be a good approximation of the function.

Let x_k denote the current iterate. We write the quadratic model for the function f at x_k as

$$m_k(x_k+s)=f(x_k)+(g_k,s)+\frac{1}{2}(H_k s,s) \tag{3.70}$$

and the trust region as the ball

$$\mathscr{B}_k = \{x \in \mathbb{R}^N : ||x - x_k|| \le \Delta_k\}.$$

Although the vector g_k and the (symmetric) matrix H_k play the roles of, respectively, the gradient and the hessian of f, their construction should not use any information on derivatives. In practice, they are typically constructed by interpolating a certain set of previous iterates where the function f has already been computed.

Once the model (3.70) has been constructed at the iterate x_k, a new candidate point x_k^+ is computed as the minimum point for m_k in the trust region. If the value $f(x_k^+)$ represents a sufficient improvement of $f(x_k)$, then the iterate is updated as $x_{k+1} = x_k^+$, and typically a former point is removed from the set of points used in the interpolation. If not, the trust region is reduced by decreasing Δ_k, and/or new points are added to improve the quality of the approximation.

The robustness of the algorithm is crucially related to the geometric properties of the set of points used to construct the model (3.70), so various recipes have been proposed to generate this set, to iteratively update it, and to improve it whenever the algorithm is unable to proceed. Once more, we will not present the technical details, but rather point out that one such algorithm, NEWUOA, has available open-source implementations.

3.4 ▪ Numerical computation of the Legendre transform

In the construction of SL schemes for HJ equations, a central role is played by the Lax–Hopf formula and therefore by the Legendre transform H^* of the Hamiltonian function H. The possibility of an explicit computation of H^* holds only for a small class of functions, so we will address here the problem of its numerical approximation, focusing in particular on "fast" algorithms.

We recall that the Legendre transform of the function $H(p)$ is defined by

$$H^*(q) = \sup_{p \in \mathbb{R}^d} \{p \cdot q - H(p)\}, \tag{3.71}$$

it takes finite values if the coercivity condition (2.24) is satisfied, and it is a strictly convex function of q whenever H is a strictly convex function of p. Moreover, in the specific framework of interest here, the sup is in fact a maximum and can be searched for on a bounded set. In multiple dimensions, a "factorization formula" holds, so that

$$H^*(q) = \sup_{p_1 \in \mathbb{R}} \left\{ p_1 q_1 + \sup_{p_2 \in \mathbb{R}} \left\{ p_2 q_2 + \cdots + \sup_{p_d \in \mathbb{R}} \{p_d q_d - H(p)\} \cdots \right\}\right\}.$$

On the basis of this decomposition, it is possible to construct multidimensional algorithms for the Legendre transform by successive increases in the dimension. For this reason, we will sketch the basic algorithm in a single space dimension, that is, using (3.71) with $d = 1$.

Since the set in which we look for the max is bounded, it can be discretized by setting up a grid of points $p_1, \dots, p_N$, with corresponding values $H(p_1), \dots, H(p_N)$ (note that in this case, the subscript refers to the index of a point, the problem being one-dimensional). A parallel grid $q_1, \dots, q_M$ (typically, with $M \sim N$) is set up in the variable q. The Discrete Legendre Transform algorithm consists in computing (3.71) at the discrete points q_k, with the maximum obtained over the discrete set $\{p_1, \dots, p_N\}$, that is,

$$H^*(q_k) = \max_{i \in \{1,\dots,N\}} \{p_i q_k - H(p_i)\} \quad (k = 1, \dots, M). \tag{3.72}$$

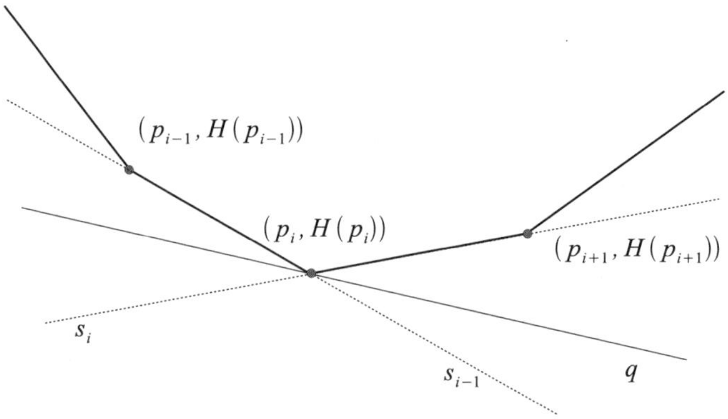

Figure 3.10. *Computation of the Discrete Legendre Transform with piecewise linear approximation.*

In this form, the algorithm is convergent under natural assumptions but has the drawback of a quadratic complexity. This has prompted the search for faster implementations, and in particular with complexity $O(N \log N)$ or even $O(N)$ as in the version outlined here.

The ordered set of vertices $(p_i, H(p_i))$ defines a piecewise linear approximation of $H(p)$ as shown in Figure 3.10, where the approximation between the nodes p_i and p_{i+1} has the slope

$$s_i = \frac{H(p_{i+1}) - H(p_i)}{p_{i+1} - p_i}. \tag{3.73}$$

Note that, H being convex, this sequence is nondecreasing.

Given q, the problem of computing $H^*(q)$ is solved once one knows the argmax of $pq - H(p)$. On the other hand, a certain value $\overline{p}$ is in this argmax if and only if q is included in the subdifferential of H at $\overline{p}$. In the discretized version, depending on the fact that $\overline{p}$ is a node or is internal to an interval of the grid, this subdifferential may be written as

$$D^- H(\overline{p}) = \begin{cases} [s_{i-1}, s_i] & \text{if } \overline{p} = p_i, \\ s_i & \text{if } \overline{p} \in (p_i, p_{i+1}) \end{cases}$$

(this is shown in Figure 3.10), and consequently the argmax (as a function of q) has the form

$$\operatorname{argmax}\{pq - H(p)\} = \begin{cases} p_i & \text{if } q \in [s_{i-1}, s_i], \\ (p_i, p_{i+1}) & \text{if } q = s_i. \end{cases} \tag{3.74}$$

It suffices therefore to create the table of slopes (3.73) and to compare a given value of q with the values of s_i to find the argmax and hence $H^*(q)$. Since a grid is also set up in the q-domain, the comparison of the q_k with the s_i in (3.74) simply requires a merging of the two vectors. The final algorithm performs only operations of linear complexity.

3.5 ▪ Commented references

Most of the numerical techniques reviewed in this chapter are analyzed in detail in dedicated monographs. Classical textbooks on the approximation of ODEs are [HNW93] and [But03], but many others are available. The basic theory of polynomial interpolation is contained in any textbook on basic numerical analysis, and any specific reference is useless. Finite element approximations are a somewhat more specialized (although well-established) issue, and among the many monographs dedicated to this topic, we can quote [Cia02] and [BS08].

On the contrary, to our knowledge no monograph has been devoted yet to nonoscillatory schemes. A classical review on the basic ideas of ENO and WENO interpolation is [Sh98] (from which we have reported most results concerning ENO interpolation). WENO interpolation was first studied in detail in [CFR05], proving in particular the results concerning positivity of the weights. Moreover, in addition to what has been presented in this chapter, other reconstruction techniques have been applied to SL schemes. In particular, we mention here the so-called *shape preserving interpolations* (see [RW90] and the references therein) and, more recently, RBF interpolation (see [Buh03] for a general review on RBF interpolation and, for example, [Is03] for its application to SL schemes). The algorithm for tensorized multidimensional interpolations based on a tree structure has been proposed in [CFF04], whereas a different approach to polynomial interpolation in high dimensions is provided by the so-called *sparse grids*, for which an extensive review is given in [BG04].

A general, up-to-date reference on optimization is [NW06]. Concerning direct search methods, a more detailed review can be found in [KLT03], while a parallel (although less recent) review devoted to trust-region methods is presented in [CST97]. Documentation and theory about two public domain optimization codes, PRAXIS and NEWUOA, can be found in, respectively, [Bre73] and [Pow08].

Finally, the algorithm of Fast Legendre Transform (of complexity $O(N \log N)$) was first proposed in [Br89] and further developed in [Co96]. More recently, the possibility of a linear-time (i.e., $O(N)$) algorithm has been shown in [Lu97]. This algorithm corresponds to what has been presented in this chapter.

Chapter 4

Convergence theory

This chapter presents the main results of convergence for numerical schemes. In the case of the Lax–Richtmeyer equivalence theorem for linear problems, the notions of consistency and stability are reviewed, with a special emphasis on monotone and L^2 stability, which play a central role for, respectively, the low-order and the high-order schemes. In the case of convex HJ equations, we present the convergence results of Crandall and Lions, Barles and Souganidis, and Lin and Tadmor, which require ad hoc stability concepts (monotonicity in the first two cases, uniform semiconcavity for the third).

4.1 ▪ The general setting

We will carry out the discretization in the usual framework of difference schemes. Time is discretized with a (fixed) time step Δt, so that $t_k = k\Delta t$, whereas discretization with respect to space variables will require us to set up a space grid in the computational domain (we make the standing assumption that this grid is *nondegenerate*, in a sense to be made explicit). We write a generic node as x_j, $j \in \mathcal{I}$, for a given set $\mathcal{I}$ of indices. Possible choices include

- unbounded structured uniform meshes in one space dimension, for which $j \in \mathcal{I} = \mathbb{Z}$ and $x_j = j\Delta x$;
- unbounded structured uniform meshes in multiple space dimensions, for which the node index $j = (j_1, \ldots, j_d)$ is a multi-index, $j \in \mathcal{I} = \mathbb{Z}^d$, and $x_j = (j_1\Delta x_1, \ldots, j_d\Delta x_d)$ for a set of space discretization parameters $\Delta x_1, \ldots, \Delta x_d \le \Delta x$ (these parameters are usually required to be linearly related to one another to avoid degeneracy of the grid);
- bounded structured uniform meshes in one or multiple space dimensions, for which $\mathcal{I}$ is the product of finite sets of indices, but nodes remain evenly spaced;
- bounded unstructured meshes in one or more space dimensions (typically based on triangulations), for which $\mathcal{I} = \{1, \ldots, n_n\}$, and each node is located with its own coordinates. Suitable geometric conditions on triangles ensure in this case that the grid does not degenerate (see Chapter 3).

Note that, in the unbounded case, it is usual to assume that the analytical solution is compactly supported. This makes it easier to develop a convergence theory in Hölder norms different from $\|\cdot\|_\infty$, and this will be implicitly done in what follows.

We denote by v_j^n the desired approximation of $u(x_j, t_n)$ and by V^n and U (respectively, $U(t)$) the sets of nodal values for the numerical solution at time t_n and for the exact solution $u(x)$ (respectively, $u(x,t)$). W also denote by W and Φ (respectively, $W(t)$ and $\Phi(t)$) the sets of nodal values of generic functions $w(x)$ and $\phi(x)$ (respectively, $w(x,t)$ and $\phi(x,t)$). In general, we will refer to the set of nodal values as a (possibly infinite) vector.

Convergence of numerical schemes will be analyzed in a specific norm, and throughout the book we will always use normalized Hölder norms, defined in the structured uniform case by

$$||W||_\alpha := \begin{cases} \left(\Delta x_1 \cdots \Delta x_d \sum_j |w_j|^\alpha\right)^{1/\alpha} & \text{if } \alpha < \infty, \\ \max_j |w_j| & \text{if } \alpha = \infty. \end{cases} \tag{4.1}$$

The natural matrix norm associated with the vector norm $||\cdot||_\alpha$ is defined as usual by

$$||B||_\alpha := \sup_{W \neq 0} \frac{||BW||_\alpha}{||W||_\alpha}, \tag{4.2}$$

which leads, in particular, for $\alpha = 1, 2, \infty$, to the forms

$$\begin{aligned} ||B||_1 &= \sup_{j \in \mathcal{I}} \sum_{i \in \mathcal{I}} |b_{ij}|, \\ ||B||_2 &= \left(\sup_{m \in \mathcal{I}} \rho_m(B^t B)\right)^{1/2}, \\ ||B||_\infty &= \sup_{i \in \mathcal{I}} \sum_{j \in \mathcal{I}} |b_{ij}| \end{aligned}$$

(where $\rho_m(\cdot)$ is a generic eigenvalue of the matrix inside the brackets). Note that natural matrix norms are insensitive to the scaling factor $(\Delta x_1 \cdots \Delta x_d)^{1/\alpha}$ appearing in (4.1).

In the unstructured case, once we write the numerical solution as a linear combination of basis function in the form

$$\sum_j w_j \psi_j(x), \tag{4.3}$$

a proper way of defining a Hölder norm is

$$||W||_\alpha := \begin{cases} \left(\int_{\mathbb{R}^d} \left|\sum_j w_j \psi_j(x)\right|^\alpha dx\right)^{1/\alpha} & \text{if } \alpha < \infty, \\ \max_j |w_j| & \text{if } \alpha = \infty. \end{cases} \tag{4.4}$$

Once we fix the initial condition V^0 by setting $v_j^0 = u_0(x_j)$, the numerical scheme will be defined for $n \geq 0$ by the iteration

$$\begin{cases} V^{n+1} = S(\Delta; t_n, V^n), \\ V^0 = U_0, \end{cases} \tag{4.5}$$

where $\Delta = (\Delta x, \Delta t)$ denotes the discretization parameters, with possible constraints posed by stability and/or consistency requirements. If the dependence on the discretization parameters need not be explicitly taken into account, we will simply write $S(t_n, V^n)$ instead of $S(\Delta; t_n, V^n)$; furthermore, if the evolution operator does not depend on t, we will write the scheme as $S(\Delta; V^n)$ or in the simplified form $S(V^n)$. Whenever useful, we will adopt the notation $S^*(\cdot)$ in which the asterisk, if specified, stands for a particular choice of the scheme. We will also denote by $S_j(\cdot)$ the jth component of S, that is, the scheme computed at x_j.

When dealing with linear problems, we will typically consider (and this is definitely the case when applying the equivalence theorem) schemes which are linear themselves, that is, schemes in which the operator $S(\Delta; t_n, \cdot)$ is affine. Therefore, in the linear theory we examine schemes of the form

$$\begin{cases} V^{n+1} = S(\Delta; t_n, V^n) = B(\Delta; t_n)V^n + G^n(\Delta), \\ V^0 = U_0. \end{cases} \tag{4.6}$$

In the same spirit as for the general case, the dependence on Δ and/or t_n will possibly be omitted if redundant or unnecessary at all.

4.2 ▪ Convergence results for linear problems: The Lax–Richtmeyer theorem

We start the presentation of convergence theory for numerical methods with the more classical linear case, and in particular with time-invariant evolution operators. We assume therefore that the model to be approximated has the form

$$\begin{cases} u_t(x,t) + Au(x,t) = g(x,t), \quad (x,t) \in \Omega \times [0,T], \\ u(x,0) = u_0(x), \end{cases} \tag{4.7}$$

where A is a differential operator, and with suitable boundary conditions on $\partial\Omega$ or with $\Omega \equiv \mathbb{R}^d$. Equation (4.7) will be assumed to be well-posed in some space H, to be specified.

Accordingly, the scheme will be assumed to be in the form

$$\begin{cases} V^{n+1} = S(\Delta; V^n) = B(\Delta)V^n + G^n(\Delta), \\ V^0 = U_0. \end{cases} \tag{4.8}$$

It should be recalled from the very start that at the present state of the theory, the linear case is the only situation which allows for a general convergence theory of numerical schemes. This theory is basically summarized in the Lax–Richtmeyer equivalence theorem.

4.2.1 ▪ Consistency

The first requirement on a practical numerical scheme is that it should be a *small perturbation* of the original equation. This idea, which is common to both the linear and the nonlinear cases, is formalized in the concept of *consistency* (for a generic norm $\|\cdot\|$).

We define the *local truncation error*, or *consistency error*, as

$$L(\Delta; t, U(t)) = \frac{1}{\Delta t}\left[U(t+\Delta t) - S(\Delta; t, U(t))\right]. \tag{4.9}$$

Definition 4.1. *Let $U(t)$ be the set of nodal values of a solution $u(x,t)$, and let $T > 0$. The scheme S is said to be* consistent *if, for any $t \in [0,T]$,*

$$\|L(\Delta; t, U(t))\| \to 0 \tag{4.10}$$

as $\Delta \to 0$ for any u_0 in a dense set of initial conditions.

Remark 4.2. *In the usual practice, at least in the linear case, the dense set of solutions on which* (4.10) *is tested coincides with* C^∞*, which is dense in essentially all spaces of interest in the numerical approximation of PDEs. Moreover, once one chooses a smooth solution, the local truncation error typically satisfies a uniform bound on* $[0,T]$*, so that for a consistent scheme*

$$||L(\Delta;t,U(t))|| \le \tau(\Delta) \to 0. \tag{4.11}$$

This point will be implicitly assumed in what follows.

Remark 4.3. *The idea of consistency remains basically unchanged in the nonlinear case, but its formal statement may vary somewhat in the various formulations, as it will be shown in section* 4.4.

4.2.2 ▪ Stability

The second key assumption is that the small perturbations introduced by the discretization should not be amplified by the scheme to an uncontrolled magnitude. This leads to the definition of *stability*.

Definition 4.4. *The scheme* (4.6) *is said to be* stable *if, for any* n *such that* $n\Delta t \in [0,T]$,

$$||B(\Delta)^n|| \le M_s \tag{4.12}$$

for some constant $M_s > 0$ *independent of* Δ.

Note that the norm used in (4.12) must be an operator (matrix) norm compatible with the norm used in (4.10).

Remark 4.5. *While consistency is a standing assumption which also applies to nonlinear situations, stability as stated by* (4.12) *may be applied only to linear schemes. In fact, some form of stability is also necessary in nonlinear situations, but no such general concept exists, and each case is treated in a specific way.*

Assumption (4.12) is not easy to check in its greatest generality. Easier sufficient conditions for (4.12) to hold may be derived using the inequality (which holds for any matrix norm)

$$||B^n|| \le ||B||^n .$$

Then, a first trivial condition which implies stability is

$$||B|| \le 1,$$

so that we would also have $||B||^n \le 1$ for any n. This condition may be weakened as

$$||B|| \le 1 + C\Delta t \tag{4.13}$$

for some C independent of the discretization parameters. In fact, taking into account that $n \le T/\Delta t$, we have

$$\begin{aligned} ||B||^n &\le (1+C\Delta t)^n \\ &\le (1+C\Delta t)^{T/\Delta t} \\ &\le e^{CT}. \end{aligned}$$

More generally, if $||B|| \sim 1 + C\Delta t^\alpha$, then the scheme is stable if $\alpha \ge 1$.

4.2.3 ▪ Convergence

The definition of *convergent* scheme is related to the possibility of approximating not only a particular set of solutions (e.g., smooth solutions) but any solution in the space H. So we will define a *convergent* scheme as follows.

Definition 4.6. *The scheme S is said to be* convergent *if, for any n such that $n\Delta t \in [0,T]$,*

$$||V^n - U(t_n)|| \to 0 \tag{4.14}$$

for any initial condition $u_0 \in H$ as $\Delta \to 0$.

We have now all elements to state the main result, known as the Lax–Richtmeyer equivalence theorem.

Theorem 4.7. *Let the scheme* (4.8) *be consistent. Then, it is convergent if and only if it is stable. Moreover, if the solution is smooth enough to satisfy* (4.11)*, then, for any n such that $n\Delta t \in [0,T]$,*

$$||V^n - U(t_n)|| \le C\tau(\Delta) \tag{4.15}$$

for some positive constant C.

Proof. We give a sketch of the proof for the part related to sufficiency (stability implies convergence), and in particular to (4.15). First, the scheme and the consistency condition are rewritten at a generic step $k \in [0, n-1]$ as, respectively,

$$\begin{aligned} V^{k+1} &= B(\Delta)V^k + G^k(\Delta), \\ U(t_{k+1}) &= B(\Delta)U(t_k) + G^k(\Delta) + \Delta t L(\Delta; t_k, U(t_k)). \end{aligned}$$

Subtracting now both sides and defining the error at the step k as

$$e^k := U(t_k) - V^k,$$

we have

$$e^{k+1} = B(\Delta)e^k + \Delta t L(\Delta; t_k, U(t_k)).$$

Taking now into account that $e^0 = 0$ by the choice of the initial condition, we obtain at successive time steps

$$\begin{aligned} e^1 &= \Delta t L(\Delta; t_0, U(t_0)), \\ e^2 &= B(\Delta)e^1 + \Delta t L(\Delta; t_1, U(t_1)) \\ &= \Delta t \left(L(\Delta; t_1, U(t_1)) + B(\Delta)L(\Delta; t_0, U(t_0))\right), \\ &\vdots \\ e^n &= B(\Delta)e^{n-1} + \Delta t L(\Delta; t_{n-1}, U(t_{n-1})) \\ &= \Delta t \left(L(\Delta; t_{n-1}, U(t_{n-1})) + \cdots + B(\Delta)^{n-1} L(\Delta; t_0, U(t_0))\right) \\ &= \Delta t \sum_{i=0}^{n-1} B(\Delta)^i L(\Delta; t_{n-i-1}, U(t_{n-i-1})) \end{aligned}$$

so that, due to the stability assumption, we can bound the error e^n as

$$||e^n|| \le n\Delta t M_s \tau(\Delta) \le T M_s \tau(\Delta),$$

and using also the assumption of consistency, we obtain that $e^n \to 0$, and in particular (4.15), once we define $C = TM_s$. □

Remark 4.8. *Note that the computation assumes that the consistency error would vanish, so that in principle it holds only for smooth solutions. On the other hand, smooth solutions are a dense set, so it is possible to prove convergence to all solutions in H by a density argument.*

Remark 4.9. *The reverse implication (convergence implies stability) requires the use of the principle of uniform boundedness (Banach–Steinhaus theorem). We do not show this part of the proof, but remark that it allows us to interpret the Courant–Friedrichs–Lewy condition, which will be described later, as a necessary condition of stability.*

4.2.4 ▪ Time-dependent evolution operators

We turn now to the case of time-dependent evolution operators, that is,

$$\begin{cases} u_t(x,t) + A(t)u(x,t) = g(x,t), & (x,t) \in \Omega \times [0,T], \\ u(x,0) = u_0(x), \end{cases} \tag{4.16}$$

complemented again with suitable boundary conditions. Note that, in some respect, (4.16) might be considered as a model problem for a number of nonlinear models, in which linearization would lead to local evolution operators depending on the solution and therefore on time.

A scheme for (4.16) should be assumed to be in the more general form (4.6), and adapting the Lax–Richtmeyer equivalence theorem to this case is quite straightforward. First, the general definition of consistency (4.10) directly applies to a scheme in the form (4.6). Second, retracing the proof of the equivalence theorem, the expression of the error at the nth time step is replaced by

$$\begin{aligned} e^n &= B(\Delta; t_{n-1})e^{n-1} + \Delta t L(\Delta; t_{n-1}, U(t_{n-1})) \\ &= \Delta t\big(L(\Delta; t_{n-1}, U(t_{n-1})) + B(\Delta; t_{n-1})L(\Delta; t_{n-2}, U(t_{n-2}) + \cdots \\ &\quad + B(\Delta; t_{n-1})B(\Delta; t_{n-2}) \cdots B(\Delta; t_1)L(\Delta; t_0, U(t_0))\big), \end{aligned}$$

and it is immediate to see that the formulation (4.12) of stability should be replaced with the requirement that, for any n such that $n\Delta t \in [0,T]$,

$$\|B(\Delta; t_n)B(\Delta; t_{n-1}) \cdots B(\Delta; t_1)\| \le M_s$$

for some constant $M_s > 0$ independent of Δ. In practice, in what follows we will never apply the general stability condition (4.12), but rather the sufficient condition (4.13), which obviously works also in the time-dependent case once the constant C is independent of the time step considered.

4.2.5 ▪ The stationary case

To conclude this review on linear convergence theory, we sketch the basic ideas for treating stationary problems. In particular, we consider here problems of the linear form

$$Au(x) = g(x) \tag{4.17}$$

and, for our specific purposes, schemes in the fixed point form

$$V = S(\Delta; V) \tag{4.18}$$

in which a component v_j of the vector V is intended to approximate $u(x_j)$. In order to ensure that the discrete problem has a unique solution, we typically assume that S is a contraction, that is,

$$\|S(\Delta;V)-S(\Delta;W)\| \le L_S\|V-W\|, \quad L_S<1,$$

for V, W in some set $\mathcal{U}$, such that $S(\Delta;\mathcal{U}) \subseteq \mathcal{U}$ (note that in general the Lipschitz constant L_S should be assumed to depend on Δ). In this framework, the numerical solution may be computed as a limit of the fixed point iteration

$$\begin{cases} V^{(k+1)} = S\left(\Delta; V^{(k)}\right), \\ V^{(0)} \in \mathcal{U}. \end{cases} \tag{4.19}$$

The similarity between (4.19) and (4.5) is more than formal. An iteration like (4.19) is often the result of applying a scheme in the time-dependent form (4.5) to compute the stationary state (4.17) of the evolutive equation (4.7), provided such a stationary state exists. In this strategy, which produces the so-called *time-marching schemes*, the discrete solution is computed as a regime state of the scheme, just as the exact solution is a regime state of the equation. Also, for all these reasons, when considering a time-marching scheme we might find it useful to introduce a time discretization parameter even if the equation to be approximated is stationary.

As it has been remarked for the evolutive case, in the linear theory we consider schemes with $S(\cdot)$ affine, that is, schemes of the form

$$V = S(\Delta;V) = B(\Delta)V + G. \tag{4.20}$$

In this situation, the condition for S to be a contraction reads as

$$L_S = \|B(\Delta)\| < 1 \tag{4.21}$$

and the invariant set $\mathcal{U}$ coincides with the whole domain of S. Note that, in the proof of convergence, (4.21) will appear as a stability condition. Note also that if the fixed-point system (4.20) is recast as a linear system in the form

$$(I - B(\Delta))V = G,$$

and if condition (4.21) is enforced in the ∞-norm (which typically happens in monotone schemes), then by (4.21) the matrix $I - B(\Delta)$ is diagonally dominant and therefore nonsingular. Hence, the practical implementation of the scheme need not be carried out in iterative form.

Consistency

While the general idea of plugging the exact solution into the scheme also applies to schemes in the form (4.18) or (4.20), a correct scaling should be adopted to define the consistency error. We define the *local truncation error* for the scheme (4.18) as

$$L(\Delta;U) = \frac{1}{1-L_S}[U - S(\Delta;U)], \tag{4.22}$$

and the scheme is said to be *consistent* if

$$\|L(\Delta;U)\| \le \tau(\Delta) \to 0 \tag{4.23}$$

as $\Delta \to 0$ for any $u(x)$ in a dense set of solutions. When applying this definition to the particular case of (4.20), the local truncation error is given by

$$L(\Delta; U) = \frac{1}{1 - ||B(\Delta)||} \left[U - B(\Delta)U - G \right]. \tag{4.24}$$

Remark 4.10. *Typically, in time-marching schemes, the contraction coefficient satisfies $L_S \sim 1 - C\Delta t$, and therefore the definition of local truncation error* (4.22) *matches the definition* (4.9)*, when applied to a stationary solution.*

Convergence

We define convergence for the stationary scheme (4.18) in the obvious way, that is, requiring that

$$||V - U|| \to 0$$

as $\Delta \to 0$. We can now give a convergence result which adapts the Lax–Richtmeyer theorem to the framework of linear stationary schemes in the form (4.20).

Theorem 4.11. *Let the scheme* (4.20) *be consistent, and let it satisfy* (4.21)*. Then, it is convergent. Moreover, if the solution is smooth enough to satisfy* (4.23)*, then*

$$||V - U|| \leq \tau(\Delta). \tag{4.25}$$

Proof. Following the same guidelines of the equivalence theorem, the scheme and the consistency condition are rewritten as

$$\begin{aligned} V^{(k+1)} &= B(\Delta)V^{(k)} + G, \\ U &= B(\Delta)U + G + (1 - ||B(\Delta)||)L(\Delta; U). \end{aligned}$$

Subtracting both sides and defining the error at the iteration k as

$$e^{(k)} := U - V^{(k)},$$

we obtain

$$e^{(k+1)} = B(\Delta)e^{(k)} + (1 - ||B(\Delta)||)L(\Delta; U).$$

We are interested in giving a bound on the limit

$$\lim_{k\to\infty} e^{(k)} = \lim_{k\to\infty} \left[U - V^{(k)} \right]$$

which corresponds to the error on the fixed point solution V of (4.20). Since this solution is unique, it does not depend on $V^{(0)}$ and we can assume that $V^{(0)} = U$, so that $e^{(0)} = 0$, and we obtain for successive iterations

$$\begin{aligned} e^{(1)} &= (1 - ||B(\Delta)||)L(\Delta; U), \\ e^{(2)} &= B(\Delta)e^{(1)} + (1 - ||B(\Delta)||)L(\Delta; U) \\ &= (1 - ||B(\Delta)||)(I + B(\Delta))L(\Delta; U), \\ &\vdots \\ e^{(k)} &= B(\Delta)e^{(k-1)} + (1 - ||B(\Delta)||)L(\Delta; U) \\ &= (1 - ||B(\Delta)||)\left[I + B(\Delta) + \cdots + B(\Delta)^{k-1} \right] L(\Delta; U). \end{aligned}$$

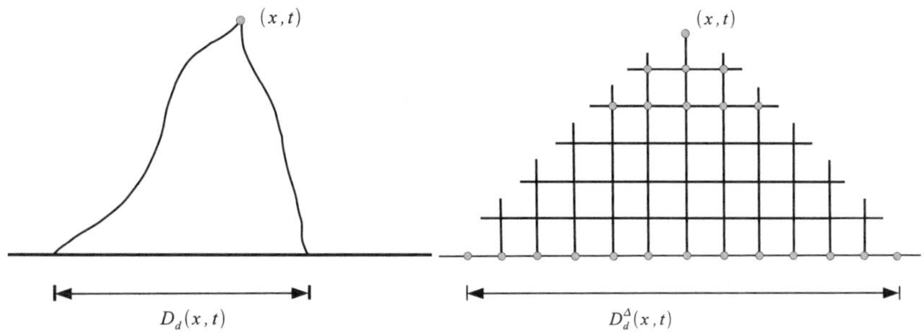

Figure 4.1. *Analytical and numerical domain of dependence.*

Now, giving an upper bound on the sum in square brackets as a geometric series, we can estimate the error $e^{(k)}$ as

$$\|e^{(k)}\| \le (1-\|B(\Delta)\|)\frac{1}{1-\|B(\Delta)\|}\|L(\Delta;U)\| \le \tau(\Delta),$$

and, by consistency, we obtain that $e^{(k)} \to 0$, and in particular (4.25) if the solution is smooth enough. □

4.3 ▪ More on stability

We turn back here to examine in greater detail the concept of stability, and in particular some special forms of this general assumption. Most of what will be said here can be applied to linear as well as nonlinear problems and schemes.

4.3.1 ▪ The CFL condition

In 1928, long before the theory of convergence of numerical schemes for PDEs had become an established matter, Courant, Friedrichs, and Lewy published a paper in which the concept of *domain of dependence* was singled out as a central point for the convergence of numerical schemes. To sketch the general idea, we first define the *analytical domain of dependence* $D_d(x,t)$ of the solution u at a point (x,t) as the smallest set such that if the initial condition u_0 of (4.7) is changed on the set $\mathbb{R}^d \setminus D_d(x,t)$, $u(x,t)$ remains unchanged. The discrete counterpart of this set is the *numerical domain of dependence* $D_d^\Delta(x,t)$ which, for a point $(x,t)=(x_i,t_n)$ in the space-time grid, is the set of all nodes x_j such that $u_0(x_j)$ affects the value v_i^n. Figure 4.1 illustrates these two concepts.

The Courant–Friedrichs–Lewy (CFL) necessary condition states that, for any point (x,t), the analytical domain of dependence must be included in the limsup of numerical domains of dependence, that is,

$$D_d(x,t) \subseteq D_d^0(x,t) = \limsup_{\Delta\to 0} D_d^\Delta(x,t). \tag{4.26}$$

More explicitly, the set $D_d^0(x,t)$ at the right-hand side of (4.26) is the set of all points for which any open neighborhood contains points of $D_d^\Delta(x,t)$ along some sequence $\Delta_m \to 0$.

The proof of the CFL condition is very simple: if (4.26) is not satisfied, then there exist subsets of $D_d(x,t)$ which have no intersection with $D_d^\Delta(x,t)$, at least as $\Delta \to 0$.

Therefore, since changing the value of v_0 in these subsets changes the value of $v(x,t)$ but not the value of the numerical approximation, the scheme cannot be convergent.

On the other hand, by the equivalence theorem (which was stated about thirty years later) we can interpret this situation in more general terms. Since a consistent, convergent scheme is necessarily stable, if a consistent scheme violates the CFL condition, then this scheme is *unstable*.

4.3.2 ▪ Monotonicity and L^∞ stability

For a number of differential problems (and in particular for most of the model problems which have been considered in the previous chapters), two important qualitative properties are satisfied:

- If u_0 is changed by adding a constant c, the corresponding solution is $u(x,t)+c$. In particular, in lack of source terms, constant solutions are preserved; that is, if $u_0 \equiv c$, then $u(x,t) \equiv c$ for any x and $t > 0$.
- Starting from two different initial conditions u_0 and w_0 such that $u_0(x) \geq w_0(x)$, this ordering is preserved also for $t > 0$, that is, $u(x,t) \geq w(x,t)$.

This makes it natural to require similar properties from the numerical scheme. In fact, we will see that this is a common, although very restrictive, notion of stability in the nonlinear case.

The first property is usually expressed in terms of *invariance with respect to the addition of constants* as follows.

Definition 4.12. *The scheme S is said to be* invariant with respect to the addition of constants *if*

$$S(V+c) = S(V)+c \tag{4.27}$$

for any vector V, where the sum of a vector with the constant c is to be intended component by component.

It is clear that, in order to preserve constants, (4.27) should be complemented with the assumption that $S(0)=0$.

In particular, when the scheme is linear, the matrix $B=(b_{ij})$ must satisfy

$$\begin{aligned} S_i(V+c) &= BV + Bc + G \\ &= S(V) + c\sum_j b_{ij}, \end{aligned}$$

and therefore (4.27) is satisfied if and only if $\sum_j b_{ij} = 1$ for any i (note that, here and in what follows, the dependence of the vector G on the time step is irrelevant).

The concept of monotonicity requires some more words. We first define a *monotone scheme*.

Definition 4.13. *The scheme S is said to be* monotone *if, for any Δ,*

$$S(\Delta; V) - S(\Delta; W) \geq 0 \tag{4.28}$$

for any couple of vectors V and W such that $V - W \geq 0$; this inequality to be intended component by component.

Usually, checking monotonicity of a scheme is relatively easy. First, if the scheme is linear, we may write for the ith component

$$\begin{aligned} S_i(V)-S_i(W) &= \sum_j b_{ij}v_j + g_i - \sum_j b_{ij}w_j - g_i \\ &= \sum_j b_{ij}(v_j - w_j) \end{aligned}$$

so that monotonicity is satisfied if and only if all entries b_{ij} of the matrix B are positive. If a linear scheme is invariant for the addition of constants and monotone, then it is stable in l^∞. In fact, under these assumptions the entries on a row of the matrix B must be positive and have unity sum, and hence the matrix has unity ∞–norm.

If the scheme is nonlinear, enforcing the condition

$$S_i(V)-S_i(W)\geq 0$$

for any $V \geq W$ requires that, for any i and j,

$$\frac{\partial}{\partial v_j}S_i(V)\geq 0.$$

Note that any (linear or nonlinear) scheme which is monotone and preserves constants is also L^∞ stable. In fact, defining two vectors $\overline{W}$ and $\underline{W}$ such that $\overline{w}_j \equiv \sup_x v_0(x)$, $\underline{w}_j \equiv \inf_x v_0(x)$, we obtain that

$$\underline{W} \leq V^0 \leq \overline{W}, \tag{4.29}$$

and due to the two properties mentioned above, successive steps satisfy

$$\underline{W} = S^n(\underline{W}) \leq V^n \leq S^n(\overline{W}) = \overline{W} \tag{4.30}$$

(where S^n denotes the n-iterated composition $S\circ S\circ\cdots\circ S$) so that the numerical solution is uniformly bounded in L^∞.

More generally, it can be proved that if a scheme is invariant with respect to the addition of constants and is monotone, then it is nonexpansive in L^∞, that is,

$$\|S(V)-S(W)\|_\infty \leq \|V-W\|_\infty. \tag{4.31}$$

In fact, set the constant $c = \|(V-W)^+\|_\infty$. Then, we also have that $V \leq W+c$, so that monotonicity of the scheme implies that

$$S(V)\leq S(W+c) = S(W)+c$$

and therefore

$$S(V)-S(W)\leq \|(V-W)^+\|_\infty,$$

which implies (4.31), once the roles of V and W are interchanged. Note that, in the linear case, this is directly implied by the fact that B has unity ∞–norm.

4.3.3 ▪ Monotonicity and Lipschitz stability

A further property of a monotone scheme, which is crucial in treating nonlinear problems, is nonexpansivity in the Lipschitz norm. This characteristic is not a trivial consequence of monotonicity and conservation of constants, since it not only requires that L^∞

norm be conserved but that Lipschitz constant be conserved too. This gap requires an additional assumption of *invariance by translation* which will be formally stated as follows.

Definition 4.14. *The scheme S is said to be* invariant by translation *if, defining the translation operator Θ_i such that*

$$(\Theta_i V)_j = V_{j+e_i},$$

we have, for any $i = 1, \dots, d$,

$$S(\Delta; \Theta_i V) = \Theta_i S(\Delta; V). \tag{4.32}$$

Clearly, this definition requires a uniform infinite space grid. Also, we do not expect that it could be satisfied when treating variable-coefficient equations. Now, define the ith partial right incremental ratio at the point x_j as

$$D_{i,j}[V] = \frac{v_{j+e_i} - v_j}{\Delta x_i} \qquad (i = 1, \dots, d). \tag{4.33}$$

Making use of the operator Θ_i we can write the maximum value over j of the right-hand side of (4.33) as

$$\|D_i[V]\|_\infty = \frac{\|\Theta_i V - V\|_\infty}{\Delta x_i}. \tag{4.34}$$

If this computation is performed on $S(V)$, we have

$$\|D_i[S(V)]\|_\infty = \frac{\|\Theta_i S(V) - S(V)\|_\infty}{\Delta x_i} = \frac{\|S(\Theta_i V) - S(V)\|_\infty}{\Delta x_i}, \tag{4.35}$$

so that, applying the property of L^∞-nonexpansivity of the scheme, we finally obtain

$$\|D_i[S(V)]\|_\infty \le \frac{\|\Theta_i V - V\|_\infty}{\Delta x_i} = \|D_i[V]\|_\infty. \tag{4.36}$$

Therefore, since none of the partial incremental ratios is expanded, this also holds for the Lipschitz constant.

4.3.4 ▪ Von Neumann analysis and L^2 stability

It is clear that condition (4.12) is strictly related to the eigenvalues of B. Since any matrix norm is bounded from below by the spectral radius, it would be natural to require that, for a generic eigenvalue ρ_m of B, the inequality

$$\rho_m(B) \le 1 + C\Delta t \tag{4.37}$$

should hold for a stable scheme. Condition (4.37) is usually referred to as a *Von Neumann stability condition*, and for the moment it will be considered a necessary condition. Indeed, if (4.37) is not satisfied, then successive powers of B cannot be uniformly bounded, and the scheme is clearly unstable. However, in general (4.37) is still far from being a sufficient condition for the reasons we will soon explain.

Let B be set in Jordan canonic form by a transformation T, so that

$$B(\Delta) = T(\Delta)^{-1} \Lambda(\Delta)\, T(\Delta) \tag{4.38}$$

(here, we have explicitly indicated that all these matrices depend on the discretization parameters). Now, successive powers of B have the form

$$B(\Delta)^n = T(\Delta)^{-1}\Lambda(\Delta)^n\, T(\Delta) \tag{4.39}$$

and therefore

$$||B(\Delta)^n|| \le \left\|T(\Delta)^{-1}\right\| \cdot ||\Lambda(\Delta)^n|| \cdot ||T(\Delta)||. \tag{4.40}$$

Giving a bound on successive powers of B via (4.40) may not be easy, either. Three main technical difficulties still hold:

(i) The transformation T itself depends on Δ, so that it is necessary to further give a uniform bound on the condition number $K(T) = ||T^{-1}|| \cdot ||T||$ as a function of Δ.

(ii) If Λ has multiple eigenvalues, then the fundamental solutions associated with such eigenvalues are of the kind $n^\nu \rho^n$. Even if $|\rho| < 1$, it is clear that these solutions are asymptotically stable, but may have extrema which are not easy to locate and bound.

(iii) The structure of eigenvectors is unknown in general, and this makes it difficult to look for eigenvalues.

Points (i) and (ii) can be overcome by assuming B to be a *normal matrix*, that is, a matrix which could be brought to diagonal form by means of an orthonormal transformation T. Actually, if this is the case, then all eigenvalues are simple (and, hence, the associated fundamental solutions are of the form ρ^n). Moreover, since T is orthonormal, working in the 2-norm we also have, for any Δ,

$$K_2(T(\Delta)) = \left\|T(\Delta)^{-1}\right\|_2 ||T(\Delta)||_2 = 1.$$

Taking also into account that the 2-norm of a diagonal matrix is given by its spectral radius, we get therefore

$$||B(\Delta)^n||_2 \le K_2(T(\Delta))||\Lambda(\Delta)^n||_2 = \sup_m |\rho_m(B(\Delta))^n|, \tag{4.41}$$

and condition (4.37) also becomes sufficient.

Concerning point (iii), if one restricts himself to *circulating matrices*, which are a subclass of normal matrices, then the structure of eigenvectors is also known. More precisely, if n_n is the number of nodes and $B \in \mathbb{R}^{n_n \times n_n}$ is a circulating matrix, any eigenvector V has components of the form $v_j = z^j$, with z an n_nth root of the unity. Writing more explicitly such a root of the unity in the form $z = e^{im\theta}$ with $\theta = 2\pi/n_n$ and $m = 0, 1, \ldots, n_n - 1$, we get

$$v_j = \exp\left(i\frac{2\pi mj}{n_n}\right), \tag{4.42}$$

and using this form in the relationship

$$\rho v_j = \sum_l b_{jl}\, v_l, \tag{4.43}$$

which defines the eigenvalues, we obtain

$$\rho_m \exp\left(i\frac{2\pi mj}{n_n}\right) = \sum_l b_{jl} \exp\left(i\frac{2\pi ml}{n_n}\right), \tag{4.44}$$

that is, for any j,

$$\rho_m = \sum_l b_{jl} \exp\left(i \frac{2\pi m(l-j)}{n_n} \right). \tag{4.45}$$

In the literature, the eigenvalue ρ_m is also referred to as the *amplification factor* associated with the harmonic component m. It depends on Δ via the coefficients b_{jl}.

Assuming that B is a circulating matrix is really restrictive, but still corresponds to a real (in a sense, the simpler) situation of application of a given scheme. In particular, most schemes can be brought to this form when applied to constant-coefficient equations in one space dimension, with periodic boundary conditions and with constant space discretization steps. This is precisely the framework in which Von Neumann condition (4.37) becomes sufficient for L^2 stability. Actually, this can be the only practical stability analysis for a number of schemes, and it is widely reputed to give at least an indication for more general cases.

In general, we may not really be interested in computing the eigenvalue ρ_m related to a specific harmonic component, but rather to prove the bound (4.37). Since (4.45) shows that there exists a mapping between the boundary of the unity disk of $\mathbb{C}$ and a curve (always in $\mathbb{C}$) containing the eigenvalues, it is possible to neglect the position of the specific eigenvalue on this curve and rather focus on the fact that this curve itself is contained in the unity disc. On the other hand, if one keeps the discretization steps constant and increases the number of nodes n_n (this means enlarging the interval over which the equation is posed), then it becomes clear that the roots of the unity become denser on the original curve, and therefore the eigenvalues become denser on the transformed curve. This makes it reasonable to treat the phase

$$\omega = \frac{2\pi m}{n_n} \in [0, 2\pi] \tag{4.46}$$

as a continuous parameter and to carry out the stability analysis on the basis of the condition

$$|\rho(\omega)| = \left| \sum_l b_{jl}(\Delta)\, e^{i\omega(l-j)} \right| \le 1 + C\Delta t. \tag{4.47}$$

In this form, the Von Neumann analysis is suitable for treating both the case of periodic conditions and the case posed on the whole of $\mathbb{R}$ (which is obtained as the limit case for $n_n \to \infty$).

Remark 4.15. *Clearly, the eigenvalues of the matrix $B(\Delta)$ could be estimated by means of Gershgorin's theorem, but, when $\rho_m(B)$ is estimated with, respectively, row or column sums, condition* (4.37) *becomes equivalent to* (4.13) *written with, respectively, the ∞- or the 1-norm of B.*

4.4 ▪ Convergence results for HJ equations

We move in this section to the study of the convergence theory for the convex HJ equation

$$\begin{cases} u_t(x,t) + H(Du(x,t)) = 0, & (x,t) \in \mathbb{R}^d \times [0,T], \\ u(x,0) = u_0(x). \end{cases} \tag{4.48}$$

Once out of the framework of linear equations (and linear schemes), no convergence result of the same generality of the Lax–Richtmeyer theorem exists. In practice, while consistency of a scheme can be defined in essentially the same way for a nonlinear scheme, stability becomes a more subtle topic, since uniform boundedness of numerical solutions does not suffice in general for convergence.

Therefore, different nonlinear stability concepts have been developed for nonlinear equations and schemes. In the framework of HJ equations two main concepts have been proposed: the more classical idea of monotone stability (which leads to the convergence theorems of Crandall and Lions and Barles and Souganidis) and the idea of uniform semi-concavity used in the Lin–Tadmor convergence theorem. We also mention that, in the specific field of SL schemes, Lipschitz stability can be an equally useful concept.

4.4.1 ▪ Crandall–Lions and Barles–Souganidis theorems

We collect together these two results which make use of monotonicity as a stability assumption.

The Crandall–Lions theorem is inspired by the result of convergence of monotone conservative schemes for conservation laws; therefore it assumes the scheme to have a structure which parallels the structure of conservative schemes. On the contrary, the Barles–Souganidis theorem does not assume any particular structure for the scheme, and is suitable for more general situations (including second-order HJ equations), provided a comparison principle holds for the exact equation. It also requires a more technical definition of consistency.

Schemes in differenced form: The Crandall–Lions theorem

We present this result referring to the case of two space dimensions, the extension to an arbitrary number of dimensions being straightforward. With a small abuse of notation, we rewrite (4.48) as

$$u_t + H(u_{x_1}, u_{x_2}) = 0. \tag{4.49}$$

The Crandall–Lions theorem works in the framework of difference schemes, so we assume that the space grid is orthogonal and uniform, Δx_i being the space step along the ith direction. We define an approximation of the partial derivative u_{x_i} at the point x_j by the right (partial) incremental ratio (4.33), that is,

$$D_{i,j}[V] = \frac{v_{j+e_i} - v_j}{\Delta x_i} \qquad (i = 1, 2).$$

In parallel with the definition of schemes in *conservative form* for conservation laws, we define here the class of schemes in *differenced form*.

Definition 4.16. *A scheme S is said to be* in differenced form *if it has the form*

$$v_j^{n+1} = v_j^n - \Delta t \mathscr{H}\left(D_{1,j-p}[V^n], \dots, D_{1,j+q}[V^n]; D_{2,j-p}[V^n], \dots, D_{2,j+q}[V^n]\right) \tag{4.50}$$

for two multi-indices p and q with positive components and for a Lipschitz continuous function $\mathscr{H}$ (called the numerical Hamiltonian*).*

In practice, (4.50) defines schemes in which the dependence on V^n appears only through its finite differences, computed on a rectangular stencil of points around x_j. The differenced form of a scheme lends itself to an easier formulation of the consistency condition, which is given in the following.

Definition 4.17. *A scheme in differenced form is* consistent *if, for any $a, b \in \mathbb{R}$,*

$$\mathcal{H}(a,\ldots,a;b,\ldots,b) = H(a,b). \tag{4.51}$$

We point out that the previous definition matches the usual one for schemes in differenced form. To show this fact, we rewrite the scheme in the form

$$\frac{v_j^{n+1} - v_j^n}{\Delta t} + \mathcal{H}\left(D_{1,j-p}[V^n],\ldots,D_{1,j+q}[V^n];D_{2,j-p}[V^n],\ldots,D_{2,j+q}[V^n]\right) = 0.$$

For the first term, it is clear that, once we replace V^n by $U(t_n)$,

$$\frac{u(x_j,t_{n+1}) - u(x_j,t_n)}{\Delta t} \to u_t(x_j,t_n),$$

whereas for the second, it is necessary to notice that, for $i = 1,2$ and for a fixed multi-index k,

$$D_{i,j+k}[U(t_n)] = u_{x_i}(x_j,t_n) + O(\Delta x).$$

Therefore, using this estimate in the scheme, we obtain

$$\mathcal{H}\left(D_{1,j-p}[U(t_n)],\ldots,D_{1,j+q}[U(t_n)];D_{2,j-p}[U(t_n)],\ldots,D_{2,j+q}[U(t_n)]\right)$$

$$= \mathcal{H}\left(u_{x_1}(x_j,t_n),\ldots,u_{x_1}(x_j,t_n);u_{x_2}(x_j,t_n),\ldots,u_{x_2}(x_j,t_n)\right) + O(\Delta x),$$

where all the error terms $O(\Delta x)$ have been collected due to the Lipschitz continuity of $\mathcal{H}$. On the other hand, applying (4.51), we obtain at the point (x_j,t_n)

$$\mathcal{H}(u_{x_1},\ldots,u_{x_1};u_{x_2},\ldots,u_{x_2}) + O(\Delta x) = H(u_{x_1},u_{x_2}) + O(\Delta x),$$

which implies the usual notion of consistency.

Note also that, in the nonlinear case, we expect that monotonicity may or may not hold depending on the speed of propagation of the solution, this speed being related to the Lipschitz constant of u_0 (in fact, in general, monotonicity does depend on the speed of propagation, but in the linear case this speed is given and unrelated to u_0). We will say that the scheme is monotone on $[-R,R]$ if (4.13) is satisfied for any V and W such that $|D_{i,j}[V]|, |D_{i,j}[W]| \le R$.

We have now all elements to state the Crandall–Lions convergence theorem.

Theorem 4.18. *Let $H : \mathbb{R}^2 \to \mathbb{R}$ be continuous, let u_0 in* (4.48) *be bounded and Lipschitz continuous (with Lipschitz constant L) on $\mathbb{R}^2$, and let $v_j^0 = u_0(x_j)$. Let the scheme* (4.50) *be monotone on $[-(L+1), L+1]$ and consistent for a locally Lipschitz continuous numerical Hamiltonian $\mathcal{H}$. Then, there exists a constant C such that, for any $n \le T/\Delta t$ and $j \in \mathbb{Z}$,*

$$\left|v_j^n - u(x_j,t_n)\right| \le C\Delta t^{1/2} \tag{4.52}$$

for $\Delta t \to 0$, $\Delta x_i = \lambda_i \Delta t$ $(i = 1,2)$.

We omit the proof, which is very technical. As a pure convergence result, the Crandall–Lions theory is generalized by the Barles–Souganidis theorem, although in the latter result no convergence estimate is obtained.

Remark 4.19. *In what follows, this convergence result will be applied to first-order, three-point stencil schemes in one space dimension. In this particular setting, a differenced scheme takes the simpler form (in which the index i of the variable is clearly omitted)*

$$v_j^{n+1} = v_j^n - \Delta t \mathscr{H}\left(D_{j-1}[V^n], D_j[V^n]\right), \tag{4.53}$$

involving only the points v_{j-1}^n, v_j^n, *and* v_{j+1}^n *in the computation of* v_j^{n+1}.

Exploiting the comparison principle: The Barles–Souganidis theorem

While it still requires monotonicity, the Barles–Souganidis convergence theory [BaS91] gives a more abstract and general framework for convergence of schemes, including the possibility of treating second-order, degenerate, and singular equations. Roughly speaking, this theory states that any monotone, stable, and consistent scheme converges to the exact solution, *provided there exists a comparison principle for the limiting equation.*

The Cauchy problem under consideration is

$$\begin{cases} u_t + H(x,u,Du,D^2u) = 0 & \text{on } \mathbb{R}^d \times (0,T), \\ u = u_0 & \text{on } \mathbb{R}^d \times \{t = 0\}. \end{cases} \tag{4.54}$$

The function $H : \mathbb{R}^d \times \mathbb{R} \times \mathbb{R}^d \times M^d \to \mathbb{R}$ (where M^d is the space of $d \times d$ symmetric matrices) is a continuous Hamiltonian, possibly depending on the Hessian matrix of second derivatives, and is assumed to be elliptic, i.e.,

$$H(x,w,p,A) \le H(x,w,p,B) \tag{4.55}$$

for all $w \in \mathbb{R}$, $x, p \in \mathbb{R}^d$, $A, B \in M^d$ such that $A \ge B$, that is, such that $A - B$ is a positive semidefinite matrix. In particular, this form includes the first-order HJ equation (4.48).

We will assume that a *comparison principle* holds true for (4.54); i.e., we assume that if u and v are, respectively, a super- and a subsolution of (4.54) on $\mathbb{R}^d \times (0,T) \to \mathbb{R}$, and if $u(\cdot,0) \le v(\cdot,0)$, then $u \le v$.

Let us consider a scheme in the general form (4.5). First, we require the property (4.27) of *invariance with respect to the addition of constants.* Then, a *generalized consistency condition* is assumed as follows.

Definition 4.20. *Let* $\Delta_m = (\Delta x_m, \Delta t_m)$ *be a generic sequence of discretization parameters, and let* (x_{j_m}, t_{n_m}) *be a generic sequence of nodes in the space-time grid such that, for* $m \to \infty$,

$$(\Delta x_m, \Delta t_m) \to 0 \quad \textit{and} \quad (x_{j_m}, t_{n_m}) \to (x,t). \tag{4.56}$$

Let $\phi \in C^\infty(\mathbb{R}^d \times (0,T])$. *Then, the scheme S is said to be* consistent *if*

$$\begin{aligned} \liminf_{m\to\infty} \frac{\phi(x_{j_m}, t_{n_m}) - S_{j_m}(\Delta_m; \Phi(t_{n_m - 1}))}{\Delta t_m} &\ge \phi_t(x,t) \\ &\quad + \underline{H}(x, \phi(x,t), D\phi(x,t), D^2\phi(x,t)), \\ \limsup_{m\to\infty} \frac{\phi(x_{j_m}, t_{n_m}) - S_{j_m}(\Delta_m; \Phi(t_{n_m - 1}))}{\Delta t_m} &\le \phi_t(x,t) \\ &\quad + \overline{H}(x, \phi(x,t), D\phi(x,t), D^2\phi(x,t)). \end{aligned} \tag{4.57}$$

Here, the index of the sequence is m, j_m, and n_m denoting the corresponding indices of a node with respect to the mth space-time grid, and we recall that by Φ or $\Phi(t)$ we denote the vector of node values for, respectively, $\phi(x)$ and $\phi(x,t)$. Moreover, $\overline{H}$ and $\underline{H}$ denote here lower and upper semicontinuous envelopes of H:

$$\underline{H}(x,\phi(x,t),D\phi(x,t),D^2\phi(x,t)) = \liminf_{(y,s)\to(x,t)} H(y,\phi(y,s),D\phi(y,s),D^2\phi(y,s)),$$

$$\overline{H}(x,\phi(x,t),D\phi(x,t),D^2\phi(x,t)) = \limsup_{(y,s)\to(x,t)} H(y,\phi(y,s),D\phi(y,s),D^2\phi(y,s)).$$

Note that if H is continuous (and this is definitely the case for the first-order HJ equation (4.48)), then in (4.57) the lim inf and the lim sup must coincide, and the definition reduces to the usual definition of consistency.

The standard definition of monotonicity is also replaced by a *generalized monotonicity* assumption stated as follows.

Definition 4.21. *Let $(\Delta x_m, \Delta t_m)$ and (x_{j_m}, t_{n_m}) be generic sequences satisfying* (4.56). *Then, the scheme S is said to be* monotone *(in the generalized sense) if it satisfies the following conditions:*

$$\text{if} \quad v_{j_m} \le \phi_{j_m}, \quad \text{then} \quad S_{j_m}(\Delta_m; V) \le S_{j_m}(\Delta_m; \Phi) + o(\Delta t_m); \tag{4.58}$$

$$\text{if} \quad \phi_{j_m} \le v_{j_m}, \quad \text{then} \quad S_{j_m}(\Delta_m; \Phi) \le S_{j_m}(\Delta_m; V) + o(\Delta t_m) \tag{4.59}$$

for any smooth function $\phi(x)$.

Also in this case, we have that if a scheme is monotone in the usual form (4.13), then it also satisfies (4.58)–(4.59).

Now, consider a numerical solution V^n (with v_j^n and its piecewise constant (in time) interpolation $v^{\Delta t}$) defined as

$$v^{\Delta t}(x,t) = \begin{cases} I[V^n](x) & \text{if } t \in [t_n, t_{n+1}), \\ v_0(x) & \text{if } t \in [0, \Delta t). \end{cases}$$

Here, $I[V^n]$ is assumed to be a general interpolation operator

$$I[V^n](x) = \sum_{x_l \in \mathscr{S}(x)} \psi_l(x) v_l^n, \tag{4.60}$$

where $\{\psi_l\}$ is a basis of cardinal functions in $\mathbb{R}^d$ (which in particular satisfy the property $\sum_l \psi_l(x) \equiv 1$), and $\mathscr{S}(x)$ is the stencil of nodes involved for interpolating at the point x. We assume here that it is contained in a ball of radius $O(\Delta x)$ around x and refer the reader to Chapter 3 for a detailed treatment of the various interpolation techniques.

The interpolation operator also has to verify a relaxed monotonicity property:

$$\text{if } v_j \le \phi_j \text{ for any } j \text{ such that } x_j \in \mathscr{S}(x), \quad \text{then} \quad I[V](x) \le I[\Phi](x) + o(\Delta t); \tag{4.61}$$

$$\text{if } \phi_j \le v_j \text{ for any } j \text{ such that } x_j \in \mathscr{S}(x), \quad \text{then} \quad I[\Phi](x) \le I[V](x) + o(\Delta t), \tag{4.62}$$

where V and ϕ denote vectors of node values of, respectively, a generic numerical solution and a smooth function $\phi(x)$. Moreover, $I[\cdot]$ satisfies

$$|I[\Phi](x) - \Phi(x)| = o(\Delta t). \tag{4.63}$$

Note that, once Δt and Δx are related to one another, bounds (4.61)–(4.63) (which are usually written in terms of the space discretization parameter) may also be understood in terms of Δt.

We can now state the extended version of the convergence result given in [BaS91].

Theorem 4.22. *Assume* (4.27), (4.57) *and* (4.58)–(4.63). *Let $u(x,t)$ be the unique viscosity solution of* (4.54). *Then, $v^{\Delta t}(x,t) \to u(x,t)$ locally uniformly on $\mathbb{R}^d \times [0,T]$ as $\Delta \to 0$.*

Proof. Let the bounded functions $\overline{u}, \underline{u}$ be defined by

$$\overline{u}(x,t) = \limsup_{\substack{(y,s)\to(x,t)\\ \Delta t\to 0}} v^{\Delta t}(y,s), \qquad \underline{u}(x,t) = \liminf_{\substack{(y,s)\to(x,t)\\ \Delta t\to 0}} v^{\Delta t}(y,s). \tag{4.64}$$

We claim that $\overline{u}(x,t), \underline{u}(x,t)$ are, respectively, a sub- and a supersolution of (4.54). Assume for the moment that the claim is true; then by the comparison principle $\overline{u}(x,t) \le \underline{u}(x,t)$ on $\mathbb{R}^d \times (0,T]$. Since the opposite inequality is obvious by the definition of $\overline{u}(x,t)$ and $\underline{u}(x,t)$, we have

$$u = \overline{u} = \underline{u}$$

and u is the unique continuous viscosity solution of (4.54). This fact together with (4.64) also implies the locally uniform convergence of $v^{\Delta t}$ to u.

Let us prove the previous claim. Let (x,t) be a local maximum of $\overline{u}-\phi$ on $\mathbb{R}^d \times (0,T]$ for some $\phi \in C^\infty(\mathbb{R}^d \times (0,T])$. Without any loss of generality, we may assume that (x,t) is a strict global maximum for $\overline{u}-\phi$ and that $\overline{u}(x,t) = \phi(x,t)$. Then, by a standard result from viscosity theory, there exist two sequences $\Delta t_m \in \mathbb{R}^+$ and $(y_m, \tau_m) \in \mathbb{R}^d \times [0,T]$, which are global maximum points for $v^{\Delta t_m} - \phi$, and as $m \to \infty$,

$$\Delta t_m \to 0, \quad (y_m, \tau_m) \to (x,t), \quad v^{\Delta t_m}(y_m, \tau_m) \to \overline{u}(x,t).$$

Then, for any x and t we have

$$v^{\Delta t_m}(x,t) \le \phi(x,t) + \xi_m \tag{4.65}$$

with $\xi_m = \left(v^{\Delta t_m} - \phi\right)(y_m, \tau_m)$ (note that $\overline{u}(x,t) = \phi(x,t)$, and hence $\xi_m \to 0$).

Since, in general, (y_m, τ_m) is not a grid point, we need to reconstruct the value attained by $v^{\Delta t_m}$ at such points. By the definition of $v^{\Delta t}$, there exists a t_{n_m} such that $\tau_m \in [t_{n_m}, t_{n_m+1})$ and $v^{\Delta t_m}(y_m, \tau_m) = v^{\Delta t_m}(y_m, t_{n_m})$. Furthermore, by the definition of $I[\cdot]$ in (4.60), there exists a set of nodes $\mathcal{S}(y_m)$ such that

$$I[V^{n_m}](y_m) = \sum_{x_j \in \mathcal{S}(y_m)} \psi_j(y_m) v_j^{n_m}. \tag{4.66}$$

Next, we apply (4.65) at $t = t_{n_m-1}$, $x = x_j \in \mathcal{S}(y_m)$ and deduce, from (4.27) and the monotonicity property (4.58), that

$$S_j\left(\Delta_m; V^{n_m-1}\right) \le S_j(\Delta_m; \Phi(t_{n_m-1})) + \xi_m + o(\Delta t_m).$$

Recalling that the left-hand side is nothing but $v_j^{n_m}$, we have

$$v_j^{n_m} \le S_j(\Delta_m; \Phi(t_{n_m-1})) + \xi_m + o(\Delta t_m),$$

which yields, applying (4.60), (4.61),

$$v^{\Delta t_m}(y_m,\tau_m)\le \sum_{x_j\in\mathcal{S}(y_m)} \psi_j(y_m)S_j(\Delta_m;\Phi(t_{n_m-1}))+\xi_m+o(\Delta t_m).$$

Now, by the definition of ξ_m, we get

$$\phi(y_m,\tau_m)\le \sum_{x_j\in\mathcal{S}(y_m)} \psi_j(y_m)S_j(\Delta_m;\Phi(t_{n_m-1}))+o(\Delta t_m). \tag{4.67}$$

We claim now that $\phi(y_m,\tau_m)=\phi(y_m,t_{n_m})+O(\Delta t_m^2)$. In fact, either $\tau_m=t_{n_m}$ (and the claim obviously holds), or $\tau_m\in(t_{n_m-1},t_{n_m})$. In the latter case, since $(v^{\Delta t_m}-\phi)(y_m,\cdot)$ has a maximum in τ_m and $v^{\Delta t_m}$ is constant in (t_{n_m-1},t_{n_m}), then $\phi_t(y_m,\tau_m)=0$ and we have $\phi(y_m,\tau_m)=\phi(y_m,t_{n_m})+O(\Delta t_m^2)$.

Using the previous claim in (4.67), we have

$$\phi(y_m,t_{n_m})\le \sum_{x_j\in\mathcal{S}(y_m)} \psi_j(y_m)S_j(\Delta_m;\Phi(t_{n_m-1}))+o(\Delta t_m) \tag{4.68}$$

and, by (4.63),

$$\phi(y_m,t_{n_m})=I[\Phi(t_{n_m})](y_m)+o(\Delta t_m)=\sum_{x_j\in\mathcal{S}(y_m)} \psi_j(y_m)\phi(x_j,t_{n_m})+o(\Delta t_m). \tag{4.69}$$

Now, (4.68) and (4.69) imply

$$\liminf_{m\to\infty}\sum_{x_j\in\mathcal{S}(y_m)} \psi_j(y_m)\frac{\phi(x_j,t_{n_m})-S_j(\Delta_m;\Phi(t_{n_m-1}))}{\Delta t_m}+o(1)\le 0.$$

Finally, by the consistency property (4.57), we obtain the desired result:

$$\phi_t(x,t)+\underline{H}(x,\phi(x,t),D\phi(x,t),D^2\phi(x,t))\le 0.$$

The proof that $\underline{u}$ is a supersolution follows the same arguments, except for replacing (4.58) with (4.59). We leave this adaptation to the reader. □

Remark 4.23. *It is possible, as in the original paper* [BaS91], *to replace the assumption of invariance for the addition of constants* (4.27) *with the assumption of* L^∞ *stability.*

Note that the consistency condition (4.57) might be reformulated so as to avoid any dependence on the variable t. In fact, adding and subtracting $\phi(x_j,t_{n-1})$, we obtain

$$\frac{\phi(x_j,t_n)-S_j(\Phi(t_{n-1}))}{\Delta t}=\frac{\phi(x_j,t_n)-\phi(x_j,t_{n-1})}{\Delta t}+\frac{\phi(x_j,t_{n-1})-S_j(\Phi(t_{n-1}))}{\Delta t}. \tag{4.70}$$

In the right-hand side of (4.70), the first term necessarily converges to $\phi_t(x_j,t_n)$, so that (4.57) is equivalent to

$$\underline{H}(x,\phi(x),D\phi(x),D^2\phi(x))\le \liminf_{m\to\infty}\frac{\phi(x_{j_m})-S_{j_m}(\Delta_m;\Phi)}{\Delta t_m} \tag{4.71}$$

$$\le \limsup_{m\to\infty}\frac{\phi(x_{j_m})-S_{j_m}(\Delta_m;\Phi)}{\Delta t_m}\le \overline{H}(x,\phi(x),D\phi(x),D^2\phi(x))$$

for a function $\phi(x)$ depending on x alone, and $x_{j_m}\to x$.

Once in the form (4.71), the definition of consistency error parallels the corresponding definition (4.22) for time-marching schemes. In fact, the Barles–Souganidis theorem can also be recast to work for schemes in the time-marching form (4.18)–(4.19) as follows.

Theorem 4.24. *Let $u(x)$ be the unique viscosity solution in $\mathbb{R}^d$ of the equation*

$$H(x,u,Du,D^2u)=0.$$

Assume that the scheme is in the form (4.18), *with $S(\Delta,\cdot)$ a contraction, and that the solution V is uniformly bounded in l^∞. Assume, moreover, that* (4.71), (4.58)–(4.63) *hold and that $L_S=1-C\Delta t+o(\Delta t)$. Then, $I[V](x)\to u(x)$ locally uniformly on $\mathbb{R}^d$ as $\Delta\to 0$.*

Remark 4.25. *The Barles–Souganidis theorem could be set in an even more general framework by considering schemes in the form*

$$\Sigma_j(\Delta;x_j,v_j,V)=0, \tag{4.72}$$

in which monotonicity *requires that Σ_j be nondecreasing with respect to the third argument and nonincreasing with respect to the fourth, while* consistency *is formulated by replacing* (4.71) *with the condition*

$$\begin{aligned}\underline{H}(x,\phi(x),D\phi(x),D^2\phi(x)) &\le \liminf_{m\to\infty}\frac{\Sigma_{j_m}(\Delta_m;x_{j_m},\phi(x_{j_m}),\Phi)}{\rho(\Delta_m)}\\ &\le \limsup_{m\to\infty}\frac{\Sigma_{j_m}(\Delta_m;x_{j_m},\phi(x_{j_m}),\Phi)}{\rho(\Delta_m)} \le \overline{H}(x,\phi(x),D\phi(x),D^2\phi(x))\end{aligned}$$

for a strictly positive (possibly constant) function ρ. Note that there is no unique way of setting a scheme in the form (4.72)*. Once a specific expression is chosen, however, it is expected that the definition of ρ leading to consistency is unique. For schemes in the time-marching form* (4.18)*, it is required that $\rho(\Delta)=\Delta t$, in which case this definition is equivalent to* (4.71).

4.4.2 ▪ Lin–Tadmor theorem and semiconcave stability

In the Lin–Tadmor convergence theory, a different concept of stability is singled out. Once more, this appears to be an adaptation to HJ equations of a parallel result obtained for conservation laws: as the Crandall–Lions theorem derives from the classical convergence theory of monotone conservative schemes, the Lin–Tadmor theorem derives from the *Lip′*-stability theory.

We start by giving the main concept of *semiconcave stability*.

Definition 4.26. *A family of approximate solutions u^ϵ of* (4.48) *is said to be* semiconcave stable *if there exists a function $k(t)\in L^1([0,T])$ such that*

$$D^2u^\epsilon(x,t)\le k(t)I \tag{4.73}$$

for $t\in[0,T]$.

More explicitly, condition (4.73) means that the matrix

$$D^2u^\epsilon(x,t)-k(t)I$$

is negative semidefinite, that is (since we are dealing with symmetric matrices), that all eigenvalues of $D^2 u$ are bounded from above by $k(t)$. The core of the theory is a first abstract result of convergence for perturbed semiconcave solutions.

Theorem 4.27. *Consider problem (4.48) for a semiconcave initial condition u_0 with compact support, and assume the family u^ϵ is semiconcave stable and H is semiconcave. Define the truncation error associated with u^ϵ as*

$$F(x,t) := u_t^\epsilon(x,t) + H(Du^\epsilon(x,t)). \tag{4.74}$$

Then, for any $t \in [0,T]$,

$$\|u(t) - u^\epsilon(t)\|_{L^1(\mathbb{R}^d)} \le C_1 \|u_0 - u^\epsilon(0)\|_{L^1(\mathbb{R}^d)} + C_2 \|F\|_{L^1(\mathbb{R}^d \times [0,T])}. \tag{4.75}$$

Proof. Let the error be defined by

$$e(x,t) := u(x,t) - u^\epsilon(x,t).$$

By (4.48) and (4.74), the error satisfies the advection equation

$$\begin{cases} e_t(x,t) + G(x,t) \cdot De(x,t) = F(x,t), \\ e(x,0) = u_0(x) - u^\epsilon(x,0), \end{cases} \tag{4.76}$$

where the advecting speed G is defined as the average

$$G(x,t) = \int_0^1 \frac{\partial H}{\partial p}\big(\eta Du(x,t) + (1-\eta) Du^\epsilon(x,t)\big) d\eta. \tag{4.77}$$

In order to estimate the norm $\|e(T)\|$ of the error at a given time T, we consider now the dual equation,

$$\begin{cases} \psi_t(x,t) + \operatorname{div}\big(G(x,t)\psi(x,t)\big) = 0, \\ \psi(x,T) = \psi_T(x), \end{cases} \tag{4.78}$$

where $\psi(T)$ is smooth and has support in a compact set $\Omega(T)$. From (4.76) and (4.78), using Green's formula, the L^2 scalar product $\big(e(t), \psi(t)\big)$ satisfies

$$\frac{d}{dt}\big(e(t), \psi(t)\big) + \int_{\partial\Omega(t)} e(x,t)\psi(x,t) G(x,t) \cdot n \, ds = \big(F(t), \psi(t)\big)$$

with $\Omega(t)$ a set depending on t, including the support of $\psi(x,t)$ at $T = t$. Hence, the boundary integral vanishes at T, and we obtain

$$\begin{aligned} \big(e(T), \psi_T\big) &= \big(e(0), \psi(0)\big) + \int_0^T \big(F(t), \psi(t)\big) dt \\ &\le \|e(0)\|_{L^1(\mathbb{R}^d)} \|\psi(0)\|_{L^\infty(\mathbb{R}^d)} + \|F\|_{L^1(\mathbb{R}^d \times [0,T])} \sup_{t \in [0,T]} \|\psi(t)\|_{L^\infty(\mathbb{R}^d)}, \end{aligned}$$

which implies, via Hölder's inequality,

$$\begin{aligned} \|e(T)\|_{L^1(\mathbb{R}^d)} \le &\sup_{\psi_T} \frac{\|\psi(0)\|_{L^\infty(\mathbb{R}^d)}}{\|\psi_T\|_{L^\infty(\mathbb{R}^d)}} \|e(0)\|_{L^1(\mathbb{R}^d)} \\ &+ \sup_{\psi_T} \frac{\sup_{t \in [0,T]} \|\psi(t)\|_{L^\infty(\mathbb{R}^d)}}{\|\psi_T\|_{L^\infty(\mathbb{R}^d)}} \|F\|_{L^1(\mathbb{R}^d \times [0,T])}, \end{aligned} \tag{4.79}$$

and hence (4.75), as soon as we are able to bound on $\|\psi(t)\|_\infty$ over the interval $[0,T]$, in terms of the final condition ψ_T.

To this end, we consider again (4.78), which gives

$$\frac{d}{dt}\|\psi(t)\|_{L^\infty(\mathbb{R}^d)} + \left(\operatorname{div} G(x_0,t)\right)\|\psi(t)\|_{L^\infty(\mathbb{R}^d)} = 0. \tag{4.80}$$

To derive (4.80), we multiply (4.78) by the sign of $\psi(x,t)$ and denote by x_0 the maximum point of $|\psi|$ at the time t. By Gronwall's inequality, we obtain from (4.80), for $t \in [0,T]$,

$$\|\psi(t)\|_{L^\infty(\mathbb{R}^d)} \le \|\psi_T\|_{L^\infty(\mathbb{R}^d)} \exp\left(\int_t^T \sup_x \left(\operatorname{div} G(x,s)\right) ds\right), \tag{4.81}$$

where by the definition (4.77), the divergence of G reads as

$$\operatorname{div} G = \int_0^1 \left(\eta \sum_{i,j} \frac{\partial^2 H}{\partial p_i \partial p_j}\frac{\partial^2 u}{\partial x_i \partial x_j} + (1-\eta)\sum_{i,j} \frac{\partial^2 H}{\partial p_i \partial p_j}\frac{\partial^2 u^\epsilon}{\partial x_i \partial x_j}\right) d\eta \tag{4.82}$$

(note that, in (4.82), arguments have been omitted, since we are looking for a uniform bound). The first sum in (4.82) may be estimated as

$$\begin{aligned}\sum_{i,j} \frac{\partial^2 H}{\partial p_i \partial p_j}\frac{\partial^2 u}{\partial x_i \partial x_j} &= \operatorname{trace}\,(D^2 H\, D^2 u)\\ &= \operatorname{trace}\left((D^2H)^{1/2}\, D^2 u\, (D^2 H)^{1/2}\right)\\ &\le d \cdot k(s) \cdot \beta.\end{aligned}$$

Here, we have used the facts that eigenvalues of $D^2 u$ are bounded from above by $k(t)$ and eigenvalues of $D^2 H$ are bounded from above by β, and that the trace is the sum of eigenvalues. Since (except for replacing u with u^ϵ) a similar estimate holds for the second sum in (4.82), we finally obtain

$$\|\psi(t)\|_{L^\infty(\mathbb{R}^d)} \le \|\psi_T\|_{L^\infty(\mathbb{R}^d)} \exp\left(d\beta \int_0^T k(s) ds\right),$$

which proves (4.79) and therefore (4.75). □

In practice, this result needs some refinement in order to be applied to a numerical scheme. In particular, the concept of semiconcave stability must be weakened and the measure of the local truncation error should be made more explicit.

Discrete semiconcave stability The assumption of semiconcave stability, as stated by (4.73), is too strict for numerical solutions produced by a real scheme. In fact, a reconstruction of the solution, typically in piecewise polynomial form, may not satisfy the upper bound on second derivatives at the interface between two cells of the grid, even in a monotone scheme. To overcome this problem, a discrete counterpart of the semiconcave stability is defined for the numerical solution v^Δ by a bound on the second directional incremental ratios in the form

$$\frac{v^\Delta(x+\delta,t) - 2v^\Delta(x,t) + v^\Delta(x-\delta,t)}{\|\delta\|^2} \le k(t). \tag{4.83}$$

Here, the function $k(t) \in L^1([0,T])$ plays the same role as in the original definition, and δ is a vector whose norm should remain bounded away from zero, more precisely

$$\|\delta\| \geq C\Delta x. \tag{4.84}$$

It is possible to prove (see [LT01]) that (4.83)–(4.84) imply that each numerical solution is close to a function of a family satisfying (4.73), so that Theorem 4.27 may be applied.

Computation of the truncation error The second ingredient of the Lin–Tadmor convergence theorem, the L^1 estimation of the truncation error, may also be difficult in general. In particular, (4.74) shows a somewhat reversed approach in measuring the truncation error, that is, by plugging the numerical solution into the exact equation.

An easier expression can be derived for Godunov-type schemes (as SL schemes are), taking into account that in this case numerical errors are generated only by the projection step, and not by the evolution operator, which is in principle exact. This expression will be introduced later, in the analysis of SL schemes for HJ equations.

4.5 ▪ Numerical diffusion and dispersion

Convergence analysis is concerned with the asymptotic behavior of a scheme. In particular, its endpoint is the possibility of generating numerical solutions as close as necessary to the exact one, although, in practice, we might be more interested in examining the qualitative response of the scheme for a finite, practical value of the discretization parameters. This is the object of a different and complementary analysis, in which we regard the values v_j^n of the numerical solutions as samples of a function $v(x,t)$ defined on $\mathbb{R} \times [0,T]$, so that $v_j^n = v(x_j, t_n)$. The function v is characterized in turn as the solution of a PDE obtained by a perturbation of the exact PDE. This perturbed equation is known as a *modified equation*.

Typically, such an analysis is performed on the constant-coefficient advection equation, and as a result the modified equation usually takes the form

$$v_t + cv_x = \nu \frac{\partial^k}{\partial x^k} v + o(\nu). \tag{4.85}$$

The coefficient $\nu = \nu(\Delta)$ depends on the discretization parameters, and, for convergence reasons, one expects that $\nu(\Delta) \to 0$ with the same order of the consistency error. Note that, in (4.85), we have made explicit the relevant perturbation, whereas the $o(\nu(\Delta))$ accounts for all the higher-order terms. In fact, we do not expect in general that a finite number of perturbing terms could suffice to completely characterize the function v.

For $k = 2$ the modified equation appears as an advection-diffusion equation, and in this case we term the behavior of the scheme as *diffusive* and the coefficient ν as the *numerical viscosity* (a positive number in stable schemes). If $k > 2$, the scheme is termed as *dispersive*. We show in Figure 4.2 the typical response of, respectively, a diffusive and a dispersive scheme on the advection of a characteristic function. In the first case, since the fundamental solution is represented by the advection of a Gaussian kernel (which is positive), the modified equation satisfies a maximum principle—this behavior is in fact characteristic of monotone schemes. In the second case, the fundamental solution is no longer positive and this results in the occurrence of under- and overshoots.

Rather then outlining a general procedure for obtaining the modified equation, we will delay its practical construction to future examples. In difference schemes, the usual

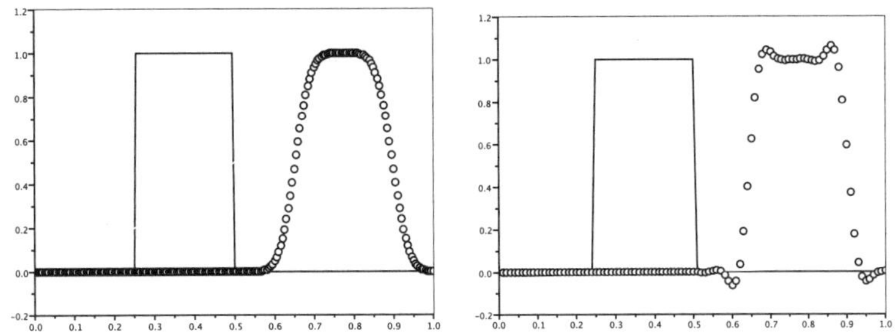

Figure 4.2. *Advection by a diffusive (left) or dispersive (right) scheme.*

technique is to express the numerical samples v_j^n as Taylor expansions on v, but this technique is unsuitable for SL schemes. To perform a diffusion/dispersion analysis on SL schemes, we derive therefore an analogue of the representation formula (1.3) for (4.85). Retracing the proof of (1.3), we can write

$$\frac{d}{dt}[v(\xi+ct,t)] = \nu \frac{\partial^k}{\partial x^k} v(\xi+ct,t) + o(\nu),$$

so that integrating between t_n and t_{n+1} we obtain

$$v(x_j,t_{n+1}) = v(x_j - c\Delta t, t_n) + \nu \int_{t_n}^{t_{n+1}} \frac{\partial^k}{\partial x^k} v(x_j - c(t_{n+1}-t), t)dt + o(\nu\Delta t),$$

and since, by a zeroth-order approximation,

$$\nu \int_{t_n}^{t_{n+1}} \frac{\partial^k}{\partial x^k} v(x_j - c(t_{n+1}-t), t)dt = \nu\Delta t \frac{\partial^k}{\partial x^k} v(x_j - c\Delta t, t_n) + O(\nu\Delta t^2),$$

we finally get

$$v(x_j,t_{n+1}) = v(x_j - c\Delta t, t_n) + \nu\Delta t \frac{\partial^k}{\partial x^k} v(x_j - c\Delta t, t_n) + o(\nu\Delta t). \tag{4.86}$$

In practice, in the operation of an SL scheme, the term associated with the kth space derivative of v is generated by the interpolation error, and this makes it possible to recover all terms of the modified equation, as it will be seen later on.

4.6 • Commented references

The Lax–Richtmeyer equivalence theorem was first published in [LR56], thus completing a theoretical study which had started from the work of Courant, Friedrichs, and Lewy (written in 1928, reprinted and translated into English in 1967 [CFL67]). By that time, the concept of stability had already been extensively studied and singled out as a crucial requirement for convergence of schemes, along with various stability criteria, including Von Neumann conditions [VR50]. Good reviews on the classical theory of convergence for difference schemes (including the study of modified equations) have been given in [RM67] and, more recently, in [Str89].

In the field of HJ equations, convergence results for monotone schemes have been formulated at the very start of the theory of viscosity solutions. In addition to the convergence theorem, the Crandall–Lions paper [CL84] also adapts to difference schemes the results of L^∞-nonexpansivity and of Lipschitz stability for monotone schemes which had been first proved in a simple and abstract way in [CT80].

Barles–Souganidis theorem appeared in its first version in [Sou85a], but has been refined in [BaS91] to include second-order equations, like the equation of Mean Curvature Motion. We have chosen to present a version of the theorem more suitable for real schemes [CFF10], since the original has a more abstract formulation. Explicit error bounds are obtained in [CL84] via a variable doubling technique (we will give an example in this direction in Chapter 8), whereas more recent technical tools for error estimation are discussed and applied, for example, in [BaJ07].

Finally, the Lin–Tadmor convergence theory has been proposed in [LT01] and is inspired by a similar result obtained for conservation laws [T91].

Chapter 5
First-order approximation schemes

This chapter is devoted to a comparison of the basic SL scheme (proposed by Courant, Isaacson, and Rees in 1952) with upwind and centered (Lax–Friedrichs) schemes. For each of the schemes, the presentation includes

- construction of the scheme,
- consistency,
- stability (CFL condition, monotonicity, Von Neumann analysis, uniform semiconcavity),
- convergence estimates, and
- numerical viscosity.

A comparison among the schemes is made, and simple numerical examples in one space dimension are presented to support our theoretical study. Note that, for the sake of clarity, most of the presentation is made on the model problems and in one space dimension, but the way to treat more general cases (including multiple dimensions) is sketched at the end of the sections, and further extensions of interest for applications will be considered in the following chapters. A subsection is also devoted to the treatment of boundary conditions.

5.1 ▪ Treating the advection equation

In this section, we introduce the schemes on the simpler case of the advection equation. We will start with the basic case of the constant-coefficient case,

$$\begin{cases} u_t(x,t)+cu_x(x,t)=0, & (x,t)\in\mathbb{R}\times[0,T], \quad c>0, \\ u(x,0)=u_0(x) \end{cases} \tag{5.1}$$

and generalize the ideas to the case with variable coefficients,

$$\begin{cases} u_t(x,t)+f(x,t)u_x(x,t)=g(x,t), & (x,t)\in\mathbb{R}\times[0,T], \\ u(x,0)=u_0(x). \end{cases} \tag{5.2}$$

Note that the simplified model (5.1) is also useful for performing L^2 (Von Neumann) stability analysis, as well as the study of numerical viscosity.

5.1.1 ▪ Upwind discretization: The first-order upwind scheme

As we have seen, in the physical and mathematical modeling of advection the advection speed (c or $f(x,t)$) is recognized as the vector field which makes the solution propagate. A first, natural strategy is therefore to let the numerical scheme mimic this behavior. More explicitly, *upwind schemes* are geometrically constructed so that the numerical domain of dependence follows the direction of propagation of the solution.

Construction of the scheme

Although the upwind scheme could be derived in a different framework, like the fully discrete structure of conservative schemes, we will sketch its construction here by using the method of lines, that is, introducing an intermediate semidiscrete version, which in our case will be understood as a formulation in which space derivatives (but not time derivatives) have been replaced by finite differences. In particular, we first focus on the constant-coefficient case (5.1) and then turn to the more general case (5.2).

The constant-coefficient equation Starting with the constant-coefficient equation (5.1) and approximating the space derivative $u_x(x_j,t)$ with the incremental ratio between x_{j-1} and x_j, the semidiscrete scheme computed at x_j reads as

$$\dot{v}_j(t)+c\,\frac{v_j(t)-v_{j-1}(t)}{\Delta x}=0. \tag{5.3}$$

A further discretization in time of (5.3) by means of the forward Euler method gives

$$\frac{v_j^{n+1}-v_j^n}{\Delta t}+c\,\frac{v_j^n-v_{j-1}^n}{\Delta x}=0, \tag{5.4}$$

that is,

$$v_j^{n+1}=v_j^n-\frac{c\Delta t}{\Delta x}\left[v_j^n-v_{j-1}^n\right]. \tag{5.5}$$

We define now the so-called *Courant number* $\lambda = c\Delta t/\Delta x$. Using this (adimensional) parameter, the scheme also takes the more explicit form

$$v_j^{n+1}=\lambda v_{j-1}^n+(1-\lambda)v_j^n. \tag{5.6}$$

A key point in the first-order upwind scheme (which also causes the presence of the term *upwind*) is that the incremental ratio used to approximate u_x should be performed on the correct side of x_j. While the importance of this point will be discussed in relationship with stability, we note that in the previous discussion, the choice of a left incremental ratio is related to the positive sign of c. A negative sign would rather require using the right incremental ratio.

The variable-coefficient equation When applying the upwind strategy to (5.2), we must therefore discriminate on the basis of the sign of $f(x,t)$: whenever $f(x_j,t_n)>0$, the space derivative is replaced by a left incremental ratio, and vice versa if $f(x_j,t_n)<0$ (in case $f(x_j,t_n)=0$, the equation simply says that $u_t(x_j,t_n)=g(x_j,t_n)$). It is convenient to consider the upwind scheme in the form (5.4), which immediately gives, for $f(x_j,t_n)>0$,

$$\frac{v_j^{n+1}-v_j^n}{\Delta t}+f(x_j,t_n)\,\frac{v_j^n-v_{j-1}^n}{\Delta x}=g(x_j,t_n), \tag{5.7}$$

and, for $f(x_j,t_n)<0$,

$$\frac{v_j^{n+1}-v_j^n}{\Delta t}+f(x_j,t_n)\frac{v_{j+1}^n-v_j^n}{\Delta x}=g(x_j,t_n). \tag{5.8}$$

To sum up, the form of the scheme for $f(x,t)$ changing sign is

$$v_j^{n+1}=\begin{cases} v_j^n-f(x_j,t_n)\dfrac{\Delta t}{\Delta x}\left[v_j^n-v_{j-1}^n\right]+\Delta t\, g(x_j,t_n) & \text{if } f(x_j,t_n)>0,\\ v_j^n+\Delta t\, g(x_j,t_n) & \text{if } f(x_j,t_n)=0,\\ v_j^n-f(x_j,t_n)\dfrac{\Delta t}{\Delta x}\left[v_{j+1}^n-v_j^n\right]+\Delta t\, g(x_j,t_n) & \text{if } f(x_j,t_n)<0. \end{cases} \tag{5.9}$$

Consistency

In order to analyze consistency, we apply the upwind scheme in the form (5.9) to a solution. For example, if $f(x_j,t_n)>0$, we estimate a single component of the consistency error (4.9) as

$$\begin{aligned} L_j^{Up}(\Delta;t,U(t))&=\frac{1}{\Delta t}\Big[u(x_j,t+\Delta t)-u(x_j,t)\\ &\quad+f(x_j,t)\frac{\Delta t}{\Delta x}(u(x_j,t)-u(x_{j-1},t))-\Delta t\, g(x_j,t)\Big]\\ &=\frac{u(x_j,t+\Delta t)-u(x_j,t)}{\Delta t}+f(x_j,t)\frac{u(x_j,t)-u(x_{j-1},t)}{\Delta x}\\ &\quad-g(x_j,t)\\ &=u_t(x_j,t)+O(\Delta t)+f(x_j,t)u_x(x_j,t)+O(\Delta x)-g(x_j,t)\\ &=O(\Delta t)+O(\Delta x), \end{aligned}$$

where the last two displays use the fact that the error in discretizing the space and time derivatives is of first order if the solution is smooth (or, at least, if it has bounded second derivatives in space and time) and that (5.2) is satisfied by a solution. Since an analogous computation can be carried out with $f(x_j,t_n)>0$, we can conclude that

$$\left|L_j^{Up}(\Delta;t,U(t))\right|\le C(\Delta t+\Delta x),$$

and therefore, passing to a generic normalized Hölder norm (in particular, the norms $\|\cdot\|_2$ and $\|\cdot\|_\infty$), we can state the consistency of the scheme by

$$\left\|L^{Up}(\Delta;t,U(t))\right\|\le C(\Delta t+\Delta x). \tag{5.10}$$

Remark 5.1. *Consistency of the upwind scheme would not require any particular relationship between the discretization steps, although a balance of the two terms suggests using steps of the same order of magnitude. In practice, this choice is also required for stability, as we will soon show.*

Stability

In the upwind scheme, all stability criteria lead to the same condition on the discretization steps, at least in the linear case. As we will see, this is no longer true in nonlinear situations.

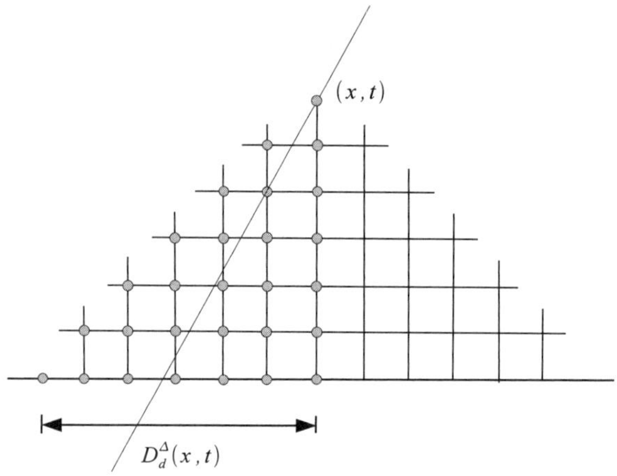

Figure 5.1. *Numerical domain of dependence for the upwind scheme.*

CFL condition For simplicity, let us refer to the constant-coefficient case with $c > 0$, that is, to the scheme in the version (5.6). The numerical solution v_j^n at $(x,t) = (x_j, t_n)$ depends on v_{j-1}^{n-1} and v_j^{n-1}, and in practice, stepping back in time, we construct a numerical domain of dependence in the form of a triangle, which is shown in Figure 5.1 along with the characteristic passing through (x_j, t_n). Due to the choice of approximating space derivatives with left finite differences, the triangle stands on the left of x_j—this agrees with the fact that, if $c > 0$, the characteristic passing through (x_j, t_n) is located precisely on this side. To fulfill the CFL condition, we must further impose that the foot of the characteristic $x_j - cn\Delta t$ (which represents the analytical domain of dependence $D_d(x_j, t_n)$) should be included in the numerical domain of dependence $[x_j - n\Delta x, x_j]$. This happens if $cn\Delta t < n\Delta x$, or more explicitly if

$$\lambda = \frac{c\Delta t}{\Delta x} \leq 1. \tag{5.11}$$

Clearly, the constant slope $1/c$ of characteristics in the x-t plane should be replaced, in the case of variable coefficients, by their minimum (positive or negative) slope $\pm 1/||f||_\infty$, so that in this case the CFL condition takes the form

$$\frac{||f||_\infty \Delta t}{\Delta x} \leq 1 \tag{5.12}$$

(note that here $||\cdot||_\infty$ stands for the norm in $L^\infty(\mathbb{R} \times [0, T])$).

Monotonicity We recall that the general condition for the monotonicity of a linear scheme in the form $V^{n+1} = B(\Delta; t_n)V^n + G^n(\Delta)$ (see Chapter 4) is that the entries of the matrix $B(\Delta; t_n)$ should be positive for any n. In the case of the upwind scheme, and assuming that $f(x_j, t_n) > 0$, the only nonzero entries on the jth row are

$$b_{j,j-1}^{Up}(\Delta; t_n) = f(x_j, t_n)\frac{\Delta t}{\Delta x},$$
$$b_{j,j}^{Up}(\Delta; t_n) = 1 - f(x_j, t_n)\frac{\Delta t}{\Delta x}$$

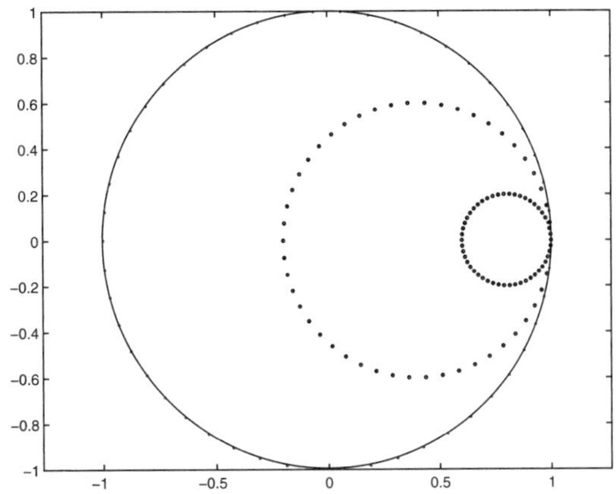

Figure 5.2. *Eigenvalues of the upwind scheme for $\lambda = 0.2$, 0.6, 1, with 50 nodes.*

(note that they have unity sum so that the scheme is also invariant with respect to the addition of constants). Since the first one is necessarily positive, the scheme is monotone, provided $1 - f(x_j, t_n)\frac{\Delta t}{\Delta x} > 0$ for any j and n, this being satisfied if

$$\frac{\|f\|_\infty \Delta t}{\Delta x} < 1. \tag{5.13}$$

Repeating the computations at points where $f(x_j, t_n) < 0$, we obtain that monotonicity is ensured under the same condition.

Von Neumann analysis We refer here to the scheme in the form (5.6). Given an eigenvector V of the scheme, in the form $v_j = e^{ij\omega}$, the condition which characterizes ρ as an eigenvalue reads as

$$\rho v_j = \lambda v_{j-1} + (1-\lambda)v_j,$$

that is,

$$\rho = (1-\lambda) + \lambda \cos\omega + i\lambda \sin\omega.$$

Summing up, eigenvalues are located on a circle of the complex plane centered at $1-\lambda$ and with radius λ (note that this is precisely the boundary of the Gershgorin discs of the matrix B^{Up}). Therefore, it results that for $\lambda \in [0,1]$ the eigenvalues are contained in the unity disc, and the Von Neumann condition is satisfied. Figure 5.2 shows the eigenvalues of B^{Up} for different values of $\lambda \in [0,1]$.

Convergence estimates

Finally, we summarize the convergence issues in a theorem. Taking into account the consistency estimate (5.10), as well as the monotonicity condition (5.13), which also ensures l^∞ stability, we have at last the following.

Theorem 5.2. *Let $f, g \in W^{1,\infty}(\mathbb{R})$, let u be the solution of* (5.2), *and let v_j^n be defined by* (5.9). *Then, for any $j \in \mathbb{Z}$ and $n \in [1, T/\Delta t]$,*

$$\left| v_j^n - u(x_j, t_n) \right| \to 0 \tag{5.14}$$

as $\Delta \to 0$, with the CFL condition (5.13).

Moreover, if u has bounded second derivatives in space and time, then

$$\|V^n - U(t_n)\|_\infty \leq C\Delta x. \tag{5.15}$$

Numerical viscosity

In order to derive the modified equation, we start from the Taylor expansions

$$v(x_j, t_{n+1}) = v + \Delta t v_t + \frac{\Delta t^2}{2} v_{tt} + O(\Delta t^3),$$
$$v(x_{j-1}, t_n) = v - \Delta x v_x + \frac{\Delta x^2}{2} v_{xx} + O(\Delta x^3),$$

where we have kept the convention of omitting the arguments for functions computed at (x_j, t_n). Plugging the expansions into the form (5.6) of the scheme, we obtain

$$v + \Delta t v_t + \frac{\Delta t^2}{2} v_{tt} + O(\Delta t^3) = \frac{c\Delta t}{\Delta x}\left[v - \Delta x v_x + \frac{\Delta x^2}{2} v_{xx} + O(\Delta x^3)\right] + \left(1 - \frac{c\Delta t}{\Delta x}\right) v$$

and hence, simplifying the terms in v, dividing by Δt, and taking into account that discretization steps vanish at the same rate,

$$v_t + c v_x = \frac{1}{2}(c\Delta x v_{xx} - \Delta t v_{tt}) + O(\Delta x^2). \tag{5.16}$$

On the other hand, deriving this equation first with respect to t and then with respect to x, and eliminating the mixed term v_{xt}, we obtain

$$v_{tt} = c^2 v_{xx} + O(\Delta x), \tag{5.17}$$

which, plugged into (5.16), gives at last

$$v_t + c v_x = \frac{c\Delta x}{2}(1 - \lambda) v_{xx} + O(\Delta x^2). \tag{5.18}$$

5.1.2 ▪ Central discretization: The Lax–Friedrichs scheme

The consistency rate of the upwind scheme is mostly limited by the fact that, using only two points in discretizing the space derivative, the approximation cannot go beyond the first order. An attempt to increase the consistency rate would therefore imply the use of a larger stencil of points. In the class of *central schemes*, this is typically accomplished by enlarging the stencil symmetrically. As we will see, usually this enlargement also causes a more dispersive behavior of the method.

Construction of the scheme

The simplest situation for a symmetric stencil is the use of x_{j-1}, x_j, and x_{j+1}. In this case, the space derivative $u_x(x_j, t)$ can be approximated as

$$u_x(x_j, t) = \frac{u(x_{j+1}, t) - u(x_{j-1}, t)}{2\Delta x} + O(\Delta x^2). \tag{5.19}$$

In principle, this larger stencil allows one to increase accuracy, although this result is not achieved in the Lax–Friedrichs (LF) scheme.

The constant-coefficient equation A first, naive way of plugging (5.19) into (5.1) would lead to the semidiscrete scheme

$$\dot{v}_j(t) + c\,\frac{v_{j+1}(t) - v_{j-1}(t)}{2\Delta x} = 0.$$

However, an explicit one-step time discretization of this scheme, and in particular that obtained by applying the forward Euler scheme,

$$v_j^{n+1} = v_j^n - \frac{c\Delta t}{2\Delta x}\left[v_{j+1}^n - v_{j-1}^n\right], \tag{5.20}$$

results in an unstable scheme. One classical way to stabilize the scheme is to turn to a midpoint two-step time approximation, thus obtaining to the so-called *leapfrog scheme*:

$$v_j^{n+1} = v_j^{n-1} - \frac{c\Delta t}{\Delta x}\left[v_{j+1}^n - v_{j-1}^n\right].$$

A second way, which will be studied in detail here, is to modify (5.20) in the LF form

$$v_j^{n+1} = \frac{v_{j-1}^n + v_{j+1}^n}{2} - \frac{c\Delta t}{2\Delta x}\left[v_{j+1}^n - v_{j-1}^n\right], \tag{5.21}$$

which clearly cannot be derived via the method of lines, although it also falls in the class of conservative schemes. In terms of the Courant number $\lambda = c\Delta t/\Delta x$, the LF scheme may also be rewritten as

$$v_j^{n+1} = \frac{1+\lambda}{2} v_{j-1}^n + \frac{1-\lambda}{2} v_{j+1}^n. \tag{5.22}$$

Note that, in this case, no assumption on the sign of c is necessary.

The variable-coefficient equation Passing to the variable-coefficient equation is easier in the LF scheme, since it is no longer required to take into account the sign of the advection term. By analogy with the upwind scheme, with a straightforward modification of the constant-coefficient case we finally obtain the form

$$v_j^{n+1} = \frac{v_{j-1}^n + v_{j+1}^n}{2} - f(x_j, t_n)\frac{\Delta t}{2\Delta x}\left[v_{j+1}^n - v_{j-1}^n\right] + \Delta t\, g(x_j, t_n). \tag{5.23}$$

Consistency

Before applying the definition of consistency, let us remark that, for a smooth solution $u(x,t)$, by elementary interpolation arguments we have

$$\frac{u(x_{j-1}, t_n) + u(x_{j+1}, t_n)}{2} = u(x_j, t_n) + O(\Delta x^2). \tag{5.24}$$

Then, applying the LF scheme (5.23) to a solution and using (5.24), we estimate a component of the consistency error (4.9) as

$$\begin{aligned} L_j^{LF}(\Delta;t,U(t)) &= \frac{1}{\Delta t}\Big[u(x_j,t+\Delta t)-u(x_j,t)+O(\Delta x^2) \\ &\quad +f(x_j,t)\frac{\Delta t}{2\Delta x}(u(x_{j+1},t)-u(x_{j-1},t))-\Delta t\, g(x_j,t)\Big] \\ &= \frac{u(x_j,t+\Delta t)-u(x_j,t)}{\Delta t}+O\left(\frac{\Delta x^2}{\Delta t}\right) \\ &\quad +f(x_j,t)\frac{u(x_{j+1},t)-u(x_{j-1},t)}{2\Delta x}-g(x_j,t) \\ &= u_t(x_j,t)+O(\Delta t)+O\left(\frac{\Delta x^2}{\Delta t}\right) \\ &\quad +f(x_j,t)u_x(x_j,t)+O(\Delta x^2)-g(x_j,t) \\ &= O(\Delta t)+O\left(\frac{\Delta x^2}{\Delta t}\right), \end{aligned}$$

where only the relevant terms have been considered. The consistency estimate in the Hölder norm for the LF scheme reads therefore as

$$\left\|L^{LF}(\Delta;t,U(t))\right\| \le C\left(\Delta t+\frac{\Delta x^2}{\Delta t}\right). \tag{5.25}$$

Remark 5.3. *Note that, as in the case of the upwind scheme, the balance of the two terms requires steps of the same order of magnitude, and this choice is also fine for stability reasons. However, in the case of the LF scheme, setting Δt of the order of Δx^2 or smaller would result in a nonconsistent scheme.*

Stability

Analysis of the LF scheme also leads to stability conditions which coincide among the various stability frameworks.

CFL condition Referring again to the constant-coefficient case and to the scheme in the version (5.22), the numerical solution v_j^n at $(x,t)=(x_j,t_n)$ depends on v_{j-1}^{n-1} and v_{j+1}^{n-1}. This causes a numerical domain of dependence in the form shown in Figure 5.3. Note that the triangle is symmetrical (so that no special care should be taken about the sign of the velocity c) and enlarges itself at the same rate as for the upwind scheme, so that we satisfy the CFL condition under the same relationship:

$$|\lambda| = \frac{|c|\Delta t}{\Delta x} \le 1. \tag{5.26}$$

As in the case of the upwind scheme, the condition may be extended to variable-coefficient equations by replacing the constant advection speed c with the norm $\|f\|_\infty$, so that the requirement becomes

$$\frac{\|f\|_\infty \Delta t}{\Delta x} \le 1. \tag{5.27}$$

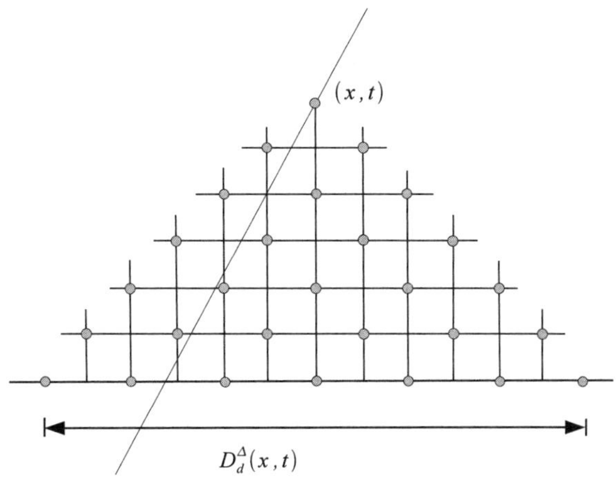

Figure 5.3. *Numerical domain of dependence for the LF scheme.*

Monotonicity We apply again the general condition for the monotonicity of a linear scheme. In the LF scheme, the only nonzero entries on the jth row are

$$b_{j,j-1}^{LF}(\Delta;t_n) = \frac{1}{2}\left(1 + f(x_j,t_n)\frac{\Delta t}{\Delta x}\right),$$
$$b_{j,j+1}^{LF}(\Delta;t_n) = \frac{1}{2}\left(1 - f(x_j,t_n)\frac{\Delta t}{\Delta x}\right).$$

Once again, they have unity sum, and one of the two is necessarily positive. Therefore, the scheme is invariant for the addition of constants and is monotone, provided the other term is positive for any j. This leads to the same condition obtained for the upwind scheme, that is,

$$\frac{\|f\|_\infty \Delta t}{\Delta x} \le 1. \tag{5.28}$$

Von Neumann analysis We refer here to the scheme in the form (5.22). Given an eigenvector V of the scheme, an eigenvalue ρ satisfies the condition

$$\rho v_j = \frac{1+\lambda}{2} v_{j-1} + \frac{1-\lambda}{2} v_{j+1};$$

that is, using the form $v_j = e^{ij\omega}$,

$$\rho e^{ij\omega} = \frac{1+\lambda}{2} e^{i(j-1)\omega} + \frac{1-\lambda}{2} e^{i(j+1)\omega},$$

which gives the eigenvalue ρ as

$$\begin{aligned}\rho &= \frac{1+\lambda}{2} e^{-i\omega} + \frac{1-\lambda}{2} e^{i\omega}\\ &= \cos\omega - i\lambda \sin\omega.\end{aligned}$$

In practice, eigenvalues are located on an ellipse centered at the origin, with a unity horizontal semi-axis and vertical semi-axis λ. The ellipse itself and the eigenvalues are therefore contained in the unity disc if $\lambda \in [0,1]$. Figure 5.4 shows the eigenvalues of B^{LF} for different values of λ.

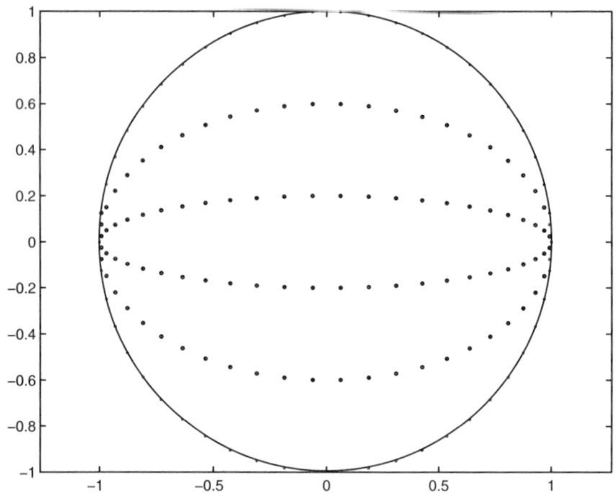

Figure 5.4. *Eigenvalues of the LF scheme for* $\lambda = 0.2, 0.6, 1$, *with* 50 *nodes.*

Convergence estimates

Again, we summarize the convergence study in a theorem. Consistency follows from (5.25), once we assume that $\Delta x = o(\Delta t^{1/2})$, and stability from the monotonicity condition (5.28). As for the upwind scheme, the smoothness required on the solution to achieve first-order convergence is to have bounded second derivatives in space and time, but also that $\Delta x \sim \Delta t$. We can therefore state the convergence result as follows.

Theorem 5.4. *Let* $f, g \in W^{1,\infty}(\mathbb{R})$, *let* u *be the solution of* (5.2), *and let* v_j^n *be defined by* (5.23). *Then, for any* $j \in \mathbb{Z}$ *and* $n \in [1, T/\Delta t]$,

$$\left| v_j^n - u(x_j, t_n) \right| \to 0 \tag{5.29}$$

as $\Delta \to 0$, *with* $\Delta x = o(\Delta t^{1/2})$ *and the CFL condition* (5.28).

Moreover, if u *has bounded second derivatives in space and time and* $\Delta x = c\Delta t$ *for some constant* c, *then*

$$\|V^n - U(t_n)\|_\infty \le C\Delta x. \tag{5.30}$$

Numerical viscosity

First, we compute the Taylor expansions

$$\begin{aligned}
v(x_j, t_{n+1}) &= v + \Delta t v_t + \frac{\Delta t^2}{2} v_{tt} + O(\Delta t^3), \\
v(x_{j-1}, t_n) &= v - \Delta x v_x + \frac{\Delta x^2}{2} v_{xx} + O(\Delta x^3), \\
v(x_{j+1}, t_n) &= v + \Delta x v_x + \frac{\Delta x^2}{2} v_{xx} + O(\Delta x^3)
\end{aligned}$$

(again, we have omitted the arguments for functions computed at (x_j, t_n)). Using such expansions in (5.21), we have

$$v + \Delta t v_t + \frac{\Delta t^2}{2} v_{tt} + O(\Delta t^3) = v + \frac{\Delta x^2}{2} v_{xx} + O(\Delta x^3) - \frac{c\Delta t}{2\Delta x}\left[2\Delta x v_x + O(\Delta x^3)\right]$$

so that by simplifying, dividing by Δt, and using again the fact that discretization steps vanish at the same rate, we get

$$v_t + c v_x = \frac{\Delta x^2}{2\Delta t} v_{xx} - \frac{\Delta t}{2} v_{tt} + O(\Delta x^2), \tag{5.31}$$

and using again (5.17) (which can be derived in the same way for (5.31)), we obtain at last

$$v_t + c v_x = \frac{1}{2}\left(\frac{\Delta x^2}{\Delta t} - c^2 \Delta t\right) v_{xx} + O(\Delta x^2). \tag{5.32}$$

Comparing the expression of the numerical viscosity with the corresponding expression for the upwind scheme, it is apparent that the LF scheme has a more viscous behavior. In fact, taking into account that $|\lambda| \le 1$,

$$\frac{1}{2}\left(\frac{\Delta x^2}{\Delta t} - c^2 \Delta t\right) = \frac{|c|\Delta x}{2}\left(\frac{1}{|\lambda|} - |\lambda|\right) \ge \frac{|c|\Delta x}{2}(1-|\lambda|)$$

so that the numerical viscosity is higher in the LF scheme than in the upwind scheme. For example, if $|\lambda| = 1/2$, then the viscosity of the upwind scheme is $|c|\Delta x/4$, whereas for the LF scheme we get a viscosity of $3|c|\Delta x/4$ (this is not surprising when considering that the numerical domain of dependence is twice as large). Note also that in the LF scheme the numerical viscosity is unbounded for vanishing Courant numbers. This is related to the loss of consistency pointed out in Remark 5.3.

5.1.3 ▪ SL discretization: The Courant–Isaacson–Rees scheme

In many respects, SL schemes are upwind schemes. They were first proposed in the field of hyperbolic systems by Courant, Isaacson, and Rees in [CIR52] in a form which precisely gives the first-order upwind scheme when applied to the advection equation. The possibility of making them work at large Courant numbers was recognized later, as soon as they were applied to problems with a relevant computational complexity.

Construction of the scheme

The construction of the Courant–Isaacson–Rees (CIR) scheme is performed by a discretization on the representation formula (1.6), (1.7), rather than on the equation itself. Following this strategy, an intermediate semidiscrete stage would be continuous in space and discrete in time. As in the previous examples, we start from the constant-coefficient case and then turn to the more general case.

The constant-coefficient equation In this case, the representation formula for the solution takes the simpler form (1.3). Note that we are interested in making the scheme evolve on a single time step, from the time t_n to t_{n+1}, and to assign a value to a given node x_j. Therefore, (1.3) should be rewritten as

$$u(x_j, t_{n+1}) = u(x_j - c\Delta t, t_n). \tag{5.33}$$

We point out that, except for computing the solution only at discrete points of the space-time grid, no numerical approximation has been introduced yet. However, since the point $x_j - c\Delta t$ is not in general a grid point, the unknown value $u(x_j - c\Delta t, t_n)$ has to be replaced by a numerical reconstruction obtained by interpolation between node values. In the original, first-order version of the CIR scheme, this is performed by a $\mathbb{P}_1$ interpolation.

We obtain therefore the numerical scheme

$$v_j^{n+1} = I_1[V^n](x_j - c\Delta t) \tag{5.34}$$

complemented with the initial condition $v_j^0 = u_0(x_j)$.

In order to write the right-hand side more explicitly, assume first that $c\Delta t < \Delta x$, so that

$$x_j - c\Delta t \in (x_{j-1}, x_j].$$

Setting again $\lambda = c\Delta t/\Delta x$, the $\mathbb{P}_1$ interpolation of the sequence V^n at this point may be written as

$$I_1[V^n](x_j - c\Delta t) = \lambda v_{j-1}^n + (1-\lambda)v_j^n,$$

so that the explicit form of (5.34) becomes

$$v_j^{n+1} = \lambda v_{j-1}^n + (1-\lambda)v_j^n. \tag{5.35}$$

Note that, despite being obtained through a different procedure, the result precisely coincides with the upwind discretization. But, although Courant, Isaacson, and Rees originally proposed their scheme in this version, since the representation formula (5.33) holds for any Δt, it is also possible to use Courant numbers beyond the unity, this meaning that the foot of the characteristic starting at x_j is no longer in an adjacent cell. Denoting by $\lfloor \cdot \rfloor$ the integer part, any $\lambda > 1$ is split as

$$\lambda = \lambda' + \lfloor \lambda \rfloor$$

and the cell in which the reconstruction is performed is $(x_{j-\lfloor \lambda \rfloor -1}, x_{j-\lfloor \lambda \rfloor}]$. Hence, for Courant numbers exceeding the unity, (5.34) takes the form

$$v_j^{n+1} = \lambda' v_{j-\lfloor \lambda \rfloor -1}^n + (1-\lambda')v_{j-\lfloor \lambda \rfloor}^n. \tag{5.36}$$

The variable-coefficient equation In the more general situation of (5.2), we start from the representation formula in the form (1.6), (1.7), written at the node x_j and on a single time step:

$$u(x_j, t_{n+1}) = \int_{t_n}^{t_{n+1}} g(y(x_j, t_{n+1}; s), s)ds + u(y(x_j, t_{n+1}; t_n), t_n), \tag{5.37}$$

where $y(x, t; s)$ solves (1.7).

In (5.37), in addition to the reconstruction of the value of $u(y(x_j, t_{n+1}; t_n), t_n)$, two more approximations are required. First, the point $y(x_j, t_{n+1}; t_n)$ itself should be approximated, since the exact solution of (1.7) is not known. Second, the integral should be evaluated by some quadrature formula. In the simplest situation, the point $y(x_j, t_{n+1}; t_n)$ is replaced by its Euler approximation,

$$x_j - \Delta t f(x_j, t_{n+1}) \approx y(x_j, t_{n+1}; t_n), \tag{5.38}$$

and the integral in (5.37) by the rectangle quadrature on $[t_n, t_{n+1}]$,

$$\Delta t\, g(x_j, t_{n+1}) \approx \int_{t_n}^{t_{n+1}} g(y(x_j, t_{n+1}; s), s)ds. \tag{5.39}$$

Plugging now (5.38), (5.39), and the $\mathbb{P}_1$ reconstruction of u into (5.37) and incorporating the initial condition, we obtain a scheme in the form

$$\begin{cases} v_j^{n+1} = \Delta t\, g(x_j, t_{n+1}) + I_1[V^n](x_j - \Delta t f(x_j, t_{n+1})), \\ v_j^0 = u_0(x_j). \end{cases} \tag{5.40}$$

Consistency

In analyzing the consistency of the CIR scheme, instead of using the equation, we will rather compare the scheme with the representation formula (1.6), (1.7). Let

$$S_j^{CIR}(\Delta; t, V) = \Delta t\, g(x_j, t+\Delta t) - I_1[V](x_j - \Delta t f(x_j, t+\Delta t)). \tag{5.41}$$

First, note that under smooth data, the elementary approximations introduced in the scheme satisfy the error estimates:

$$|x_j - \Delta t f(x_j, t+\Delta t) - y(x_j, t+\Delta t; t)| = O(\Delta t^2), \tag{5.42}$$

$$\left| \Delta t\, g(x_j, t+\Delta t) - \int_t^{t+\Delta t} g(y(x_j, t+\Delta t; s), s)\, ds \right| = O(\Delta t^2), \tag{5.43}$$

$$\|u - I_1[U]\|_\infty = O(\Delta x^2). \tag{5.44}$$

Bounds (5.42)–(5.44) are classical for, respectively, the Euler scheme, the rectangle quadrature, and the $\mathbb{P}_1$ interpolation. Writing now $u(x_j, t+\Delta t)$ by means of (5.37), we may now estimate the local truncation error at the node x_j as

$$\begin{aligned} \left| L_j^{CIR}(\Delta; t, U(t)) \right| &= \frac{1}{\Delta t} \left| u(x_j, t+\Delta t) - S_j^{CIR}(\Delta; t, U(t)) \right| \\ &\le \frac{1}{\Delta t} \left| \int_t^{t+\Delta t} g(y(x_j, t+\Delta t; s), s)\, ds + u(y(x_j, t+\Delta t; t), t) \right. \\ &\quad \left. - \Delta t\, g(x_j, t+\Delta t) - I_1[U(t)](x_j - \Delta t f(x_j, t+\Delta t)) \right| \\ &\le \frac{1}{\Delta t} \left[\left| \int_t^{t+\Delta t} g(y(x_j, t+\Delta t; s), s)\, ds - \Delta t\, g(x_j, t+\Delta t) \right| \right. \\ &\quad + \left| u(y(x_j, t+\Delta t; t), t) - I_1[U(t)](y(x_j, t+\Delta t; t)) \right| \\ &\quad \left. + \left| I_1[U(t)](y(x_j, t+\Delta t; t)) - I_1[U(t)](x_j - \Delta t f(x_j, t+\Delta t)) \right| \right]. \end{aligned} \tag{5.45}$$

The first two terms in the last display can be directly estimated by (5.43) and (5.44), whereas for the third we note that the Lipschitz constant of the reconstructed solution $I_1[U]$ is bounded by the Lipschitz constant L_u of u, so that applying again (5.42) we get

$$\left| L_j^{CIR}(\Delta; t, U(t)) \right| \le \frac{1}{\Delta t} \left[O(\Delta t^2) + O(\Delta x^2) + L_u O(\Delta t^2) \right],$$

which in turn implies the estimate

$$\left\| L^{CIR}(\Delta; t, U(t)) \right\| \le C \left(\Delta t + \frac{\Delta x^2}{\Delta t} \right). \tag{5.46}$$

At a closer look, however, this estimate is suboptimal. In fact, (5.46) would require the condition $\Delta x^2 = o(\Delta t)$ for consistency, while for small Courant numbers the CIR scheme reduces to an upwind scheme and should therefore retain consistency under any $\Delta x/\Delta t$ relationship. To include this case, we use the interpolation error in the more general form (3.18) and take into account that

$$\min_{m\in\mathbb{Z}} \frac{x-x_m}{\Delta x} \le \min\left(\frac{1}{2}, \frac{x-x_j}{\Delta x}\right),$$

where, for our purposes,

$$\begin{aligned} x &= y(x_j, t+\Delta t; t) \\ &= x_j + O(\Delta t) \end{aligned}$$

is the point at which interpolation is performed. Substituting into (3.18), we obtain therefore

$$|v(y(x_j, t+\Delta t; t)) - I_1[V](y(x_j, t+\Delta t; t))| \le C\Delta x^2 \min\left(1, \frac{\Delta t}{\Delta x}\right),$$

which gives, when used in (5.45), a consistency estimate in the form

$$\left\| L^{CIR}(\Delta; t, U(t)) \right\| \le C\left(\Delta t + \min\left(\Delta x, \frac{\Delta x^2}{\Delta t}\right)\right). \tag{5.47}$$

As in the upwind or LF scheme, the consistency rate is maximized when $\Delta t \sim \Delta x$. However, considering that the CIR scheme turns out to be unconditionally stable (and hence not restricted to operating with steps of the same order of magnitude), the interplay between the two discretization steps becomes less trivial. This effect will be analyzed in detail later in this section and again in Chapter 6.

Stability

The main issue in the stability analysis of the CIR scheme (and of SL schemes in general) is its unconditional stability, which allows for large Courant numbers. We analyze this feature of the scheme from different viewpoints.

CFL condition The fulfillment of the CFL condition is obtained in the CIR scheme by shifting the reconstruction stencil along characteristics in the neighborhood of the point $x_j - c\Delta t$. This causes some form of self-adaptation of the numerical domain of dependence, although at the additional cost of locating the foot of characteristics. Another advantage is the lower number of grid points involved, this generally implying a lower numerical viscosity. Figure 5.5 shows this point at a comparison with the corresponding situations for upwind and LF schemes (Figures 5.1 and 5.3). In particular, an enlargement of the numerical domain of dependence of $O(\Delta x)$ at each step is multiplied by a number of steps of $O(1/\Delta t)$, turning into a numerical domain of dependence of radius $O(\Delta x/\Delta t)$. If the refinement is made with diverging Courant numbers (and this can really be the case in SL schemes), then this domain collapses towards the foot of the characteristic.

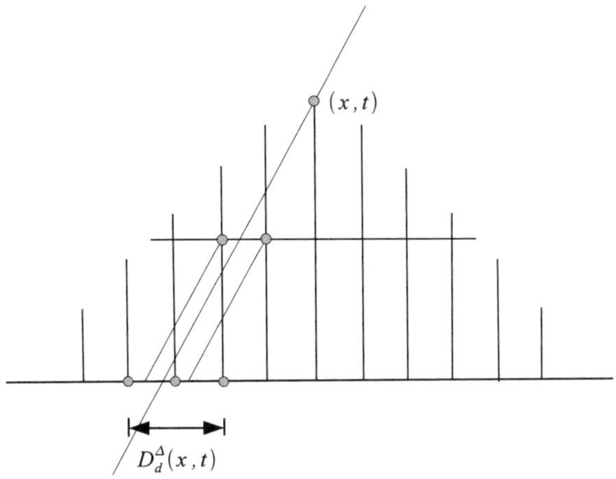

Figure 5.5. *Numerical domain of dependence for the CIR scheme.*

Monotonicity Rewriting the interpolation operator $I_1[V]$ in terms of the basis functions $\psi_i^{[1]}$, the CIR scheme takes the form

$$S_j^{CIR}(\Delta;t_n,V)=\Delta t\, g(x_j,t_{n+1})-\sum_i v_i \psi_i^{[1]}(x_j-\Delta t f(x_j,t_{n+1})) \tag{5.48}$$

so that

$$b_{ji}^{CIR}(\Delta;t_n)=\psi_i^{[1]}(x_j-\Delta t f(x_j,t_{n+1})). \tag{5.49}$$

Now, since $\psi_i^{[1]}(x)\geq 0$ and $\sum_i \psi_i^{[1]}(x)\equiv 1$ (in fact, the latter property is true for Lagrange interpolation of any order), the CIR scheme is invariant for the addition of constants and monotone for any Δx and Δt.

Von Neumann analysis As usual, we refer here to the scheme in the form (5.34). Given an eigenvector V of the scheme, in the form $v_j=e^{ij\omega}$, the condition which characterizes ρ as an eigenvalue reads as

$$\rho v_j=\lambda' v_{j-\lfloor\lambda\rfloor-1}+(1-\lambda')v_{j-\lfloor\lambda\rfloor},$$

that is,

$$\rho e^{ij\omega}=\lambda' e^{i(j-\lfloor\lambda\rfloor-1)\omega}+(1-\lambda')e^{i(j-\lfloor\lambda\rfloor)\omega},$$

which allows us to express the eigenvalue ρ as

$$\rho=e^{i\lfloor\lambda\rfloor\omega}\left[\lambda' e^{-i\omega}+(1-\lambda')\right]. \tag{5.50}$$

In the right-hand side of (5.50), the first is a pure phase term (depending only on $\lfloor\lambda\rfloor$) and is therefore irrelevant for stability, whereas the second, relevant term depends only on λ'. Restricting ourselves therefore to the second term and rearranging (5.50), we obtain (except for replacing λ with λ') the same conclusions as for the upwind scheme, that is,

$$|\rho|=|(1-\lambda')+\lambda'\cos\omega+i\lambda'\sin\omega|. \tag{5.51}$$

Therefore, if $\lambda\in[0,1]$, eigenvalues are located on a circle centered at $1-\lambda$ and with radius λ (here, the CIR scheme coincides with the upwind scheme). If $\lambda>1$, a further phase shift is superimposed on this curve. In both cases, $\lambda'\in[0,1]$ and the Von Neumann condition is satisfied. Figure 5.6 shows the eigenvalues of B^{CIR} for two different values of λ having the same fractional part.

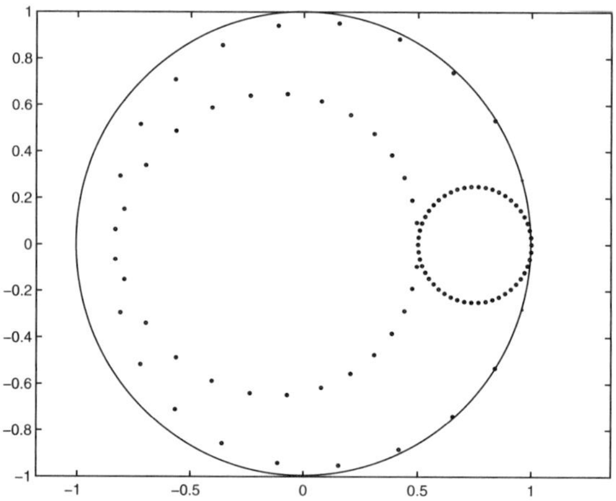

Figure 5.6. *Eigenvalues of the CIR scheme for* $\lambda = 0.25$, 2.25 *with* 50 *nodes.*

Convergence estimates

Finally, we summarize this convergence study. We point out that (5.42), (5.43) require the functions f and g to be in $W^{1,\infty}$, whereas (5.44) requires that the solution u be of $W^{2,\infty}$. Moreover, if $u \in W^{1,\infty}$, then (5.44) might be replaced by

$$\|u - I[U]\|_\infty \le C_u \Delta x$$

or, more in general, by (3.18) for $s = 1$. Taking into account the consistency of the scheme, we have at last the following.

Theorem 5.5. *Let $f, g \in W^{1,\infty}(\mathbb{R})$, let u be the solution of* (5.2), *and let v_j^n be defined by* (5.40). *Then, for any $j \in \mathbb{Z}$ and $n \in [1, T/\Delta t]$,*

$$\left| v_j^n - u(x_j, t_n) \right| \to 0 \tag{5.52}$$

as $\Delta \to 0$.

Moreover, if $u \in L^\infty([0,T], W^{s,\infty}(\mathbb{R}))$ $(s = 1, 2)$, then

$$\|V^n - U(t_n)\|_\infty \le C\left(\Delta t + \min\left(\Delta x^{s-1}, \frac{\Delta x^s}{\Delta t}\right)\right). \tag{5.53}$$

Numerical viscosity

To compute the numerical viscosity of the CIR scheme, we first plug a regular extension of the numerical solution into the scheme and use the estimate for the $\mathbb{P}_1$ Lagrange interpolation error to express the value $I_1[V^n](x_j - c\Delta t)$ as

$$I_1[V^n](x_j - c\Delta t) = v(x_j - c\Delta t, t_n) - \frac{\lambda'(\lambda' - 1)\Delta x^2}{2} v_{xx}(\xi_j),$$

with ξ_j an unknown point located in the same interval of $x_j - c\Delta t$. Next, taking into account that the computation of the second derivative in the interpolation error is performed at a point which differs from $x_j - c\Delta t$ by an $O(\Delta x)$, we get from (5.34)

$$v(x_j, t_{n+1}) = v(x_j - c\Delta t, t_n) + \frac{\lambda'(1 - \lambda')\Delta x^2}{2}\left[v_{xx}(x_j - c\Delta t) + O(\Delta x)\right],$$

which is in the form (4.86), with $k=2$, once we set the numerical viscosity as

$$\nu = \frac{\lambda'(1-\lambda')\Delta x^2}{2\Delta t}.$$

The viscosity coefficient is positive, depends on λ', and vanishes if $\lambda'=0$ (this is consistent with the fact that if the feet of characteristics coincide with grid nodes, the solution is advected without errors). In order to carry out a worst-case analysis, we note that its largest value is attained for $\lambda'=1/2$. We can then deduce that, in the worst case, the numerical solution corresponding to a local interpolation of first order solves the modified equation

$$v_t + c v_x = \frac{\Delta x^2}{8\Delta t} v_{xx} + o\left(\frac{\Delta x^2}{\Delta t}\right). \tag{5.54}$$

Note that the viscous term originates from the accumulation of interpolation errors and can be reduced by using large Courant numbers.

High-order characteristic tracking in the CIR scheme

The first-order discretization in (5.33) is suitable for the homogeneous, constant-coefficient case, where it provides the exact upwinding along characteristics. However, in nonhomogeneous problems, as well as in the variable-coefficient case, it could lead to an undesired error in time discretization. In practice, in order to balance the two error terms, the scheme could be forced again to work at small Courant numbers (hence, with an unnecessary computational complexity), in spite of its unconditional stability for large time steps. Another good reason for working at large time steps is the reduction of numerical viscosity, as it has just been noticed. Therefore, a more accurate approximation of characteristics and integral is desirable and in fact can be incorporated in the scheme without changing its stability properties.

Looking at the proof of consistency, it is clear that the consistency rates in (5.42) and (5.43) should be increased in parallel, the lower of the two being the relevant rate in the first term of the global estimate (5.46). Following [FF98], a general way to do this is to introduce the augmented system

$$\begin{pmatrix} \dot{y}(x,t;s) \\ \dot{\gamma}(x,t;s) \end{pmatrix} = \begin{pmatrix} f(y(x,t;s),s) \\ -g(y(x,t;s),s) \end{pmatrix} \tag{5.55}$$

with the initial conditions $y(x,t;t)=x$, $\gamma(x,t;t)=0$. In what follows, we denote by, respectively, $X^\Delta(x,t;s)$ and $G^\Delta(x,t;s)$ the approximations of $y(x,t;s)$ and $\gamma(x,t;s)$, where typically $x=x_j$, $t=t_{n+1}$, and $s=t_n$. For example, in the classical CIR scheme (5.40) the Euler/rectangle rule discretization used corresponds to the application of the Euler scheme to (5.55), so that

$$\begin{pmatrix} X^\Delta(x_j,t_{n+1};t_n) \\ G^\Delta(x_j,t_{n+1};t_n) \end{pmatrix} = \begin{pmatrix} x_j \\ 0 \end{pmatrix} - \Delta t \begin{pmatrix} f(x_j,t_{n+1}) \\ -g(x_j,t_{n+1}) \end{pmatrix}. \tag{5.56}$$

With this modification, the scheme (5.40) can be rewritten more generally as

$$\begin{cases} v_j^{n+1} = G^\Delta(x_j,t_{n+1};t_n) + I_1[V^n]\big(X^\Delta(x_j,t_{n+1};t_n)\big), \\ v_j^0 = u_0(x_j). \end{cases} \tag{5.57}$$

Assuming now that (5.42), (5.43) are replaced by the respective estimates of order p,

$$\left|X^{\Delta}(x_j,t_{n+1};t_n)-y(x_j,t_{n+1};t_n)\right|=O(\Delta t^{p+1}), \tag{5.58}$$

$$\left|G^{\Delta}(x_j,t_{n+1};t_n)-\int_{t_n}^{t_{n+1}} g(y(x_j,t_{n+1};s),s)ds\right|=O(\Delta t^{p+1}), \tag{5.59}$$

and retracing the proof of (5.46), we get a consistency error bounded by

$$\left\|L^{CIR}(\Delta;t,U(t))\right\|\le C\left(\Delta t^p+\frac{\Delta x^2}{\Delta t}\right) \tag{5.60}$$

(note that, in this case, we will make no use of the sharp estimate (3.18), since for $p>1$ the best performance of the scheme is obtained for large Courant numbers).

Usually, for (5.58), (5.59) to hold it is necessary to assume at least that $f,g\in W^{p,\infty}$. Since it is also immediate to check that monotonicity still holds for the scheme in the version (5.57), we can reformulate Theorem 5.5 in the following more general form.

Theorem 5.6. *Let $f,g\in W^{p,\infty}(\mathbb{R})$, let u be the solution of* (5.2), *and let v_j^n be defined by* (5.57). *Assume, moreover, that* (5.58), (5.59) *hold. Then, for any $j\in\mathbb{Z}$ and $n\in[1,T/\Delta t]$,*

$$\left|v_j^n-u(x_j,t_n)\right|\to 0 \tag{5.61}$$

as $\Delta\to 0$.

Moreover, if $u\in L^\infty([0,T],W^{s,\infty}(\mathbb{R}))$ $(s=1,2)$, then

$$\|V^n-U(t_n)\|_\infty\le C\left(\Delta t^p+\frac{\Delta x^s}{\Delta t}\right). \tag{5.62}$$

Note that, in comparison with the fully first-order CIR scheme, in this latter version the terms of the consistency error are balanced under increasing Courant numbers if $p>1$. Introducing the relationship $\Delta t=\Delta x^\alpha$, the consistency error is $O\left(\Delta x^{\alpha p}+\Delta x^{2-\alpha}\right)$ and its order is maximal with the choice $\alpha=2/(p+1)$ which balances the two terms. For example, using a second-order scheme in (5.63), we would have an optimal relationship for $\alpha=2/3$.

Examples: One-step schemes Applying a one-step scheme to (5.55) and restricting to a single backward time step, we have

$$\begin{pmatrix}X^{\Delta}(x,t;t-\Delta t)\\ G^{\Delta}(x,t;t-\Delta t)\end{pmatrix}=\begin{pmatrix}x\\ 0\end{pmatrix}-\Delta t\begin{pmatrix}\Phi_f\left(-\Delta t;x,t,X^{\Delta}(x,t;t-\Delta t)\right)\\ \Phi_{-g}\left(-\Delta t;x,t,X^{\Delta}(x,t;t-\Delta t)\right)\end{pmatrix}, \tag{5.63}$$

where the function Φ is split according to (5.55), and X^Δ is understood as the solution of a system in implicit schemes, in which the right-hand side of (5.63) depends genuinely on X^Δ itself. Table 5.1 puts in the form (5.63) the same examples of Table 3.1.

In these examples, we have taken for granted that the functions f and g can be computed for any generic argument, that is, that they have an explicit expression. In a number of situations, like the nonlinear advection in Fluid Dynamics models, it could rather happen that

Table 5.1. *First- and second-order one-step schemes for the augmented system* (5.55).

Scheme	Form	Order
FE	$\Phi_f\left(-\Delta t;x,t,X^\Delta\right)=f(x,t)$ $\Phi_{-g}\left(-\Delta t;x,t,X^\Delta\right)=-g(x,t)$	$p=1$
BE	$\Phi_f\left(-\Delta t;x,t,X^\Delta\right)=f\left(X^\Delta,t-\Delta t\right)$ $\Phi_{-g}\left(-\Delta t;x,t,X^\Delta\right)=-g\left(X^\Delta,t-\Delta t\right)$	$p=1$
H	$\Phi_f\left(-\Delta t;x,t,X^\Delta\right)=\frac{1}{2}[f(x,t)+f(x-\Delta t f(x,t),t-\Delta t)]$ $\Phi_{-g}\left(-\Delta t;x,t,X^\Delta\right)=-\frac{1}{2}[g(x,t)+g(x-\Delta t f(x,t),t-\Delta t)]$	$p=2$
CN	$\Phi_f\left(-\Delta t;x,t,X^\Delta\right)=\frac{1}{2}\left[f(x,t)+f\left(X^\Delta,t-\Delta t\right)\right]$ $\Phi_{-g}\left(-\Delta t;x,t,X^\Delta\right)=-\frac{1}{2}\left[g(x,t)+g\left(X^\Delta,t-\Delta t\right)\right]$	$p=2$

- both the advection and the source term result from previous computations and/or physical measures and are known only at grid nodes;
- the advection term is known at previous time steps, but not yet at the step t_{n+1}.

The first situation requires that f and g be interpolated to be computed at a generic point. Assuming that f is interpolated with degree q and therefore reconstructed with error $O(\Delta x^{q+1})$, the accuracy assumption (5.58) changes into

$$\left|X^\Delta(x,t;t-\Delta t)-y(x,t;t-\Delta t)\right|\le O\left(\Delta t^{p+1}\right)+O\left(\Delta t\Delta x^{q+1}\right)$$

and the consistency error into

$$\left\|L^{CIR}(\Delta;t,U(t))\right\|\le C\left(\Delta t^p+\Delta x^{q+1}+\frac{\Delta x^2}{\Delta t}\right)$$

(clearly, the same ideas also apply to the source term g). For example, if a second-order scheme is used to move along characteristics, the optimal relationship without interpolation for f would be, as we said above, $\Delta x=\Delta t^{3/2}$. Introducing the interpolation of f, we have in turn

$$\left\|L^{CIR}(\Delta;t,U(t))\right\|\le C\left(\Delta t^2+\Delta t^{\frac{3}{2}(q+1)}\right).$$

It suffices therefore to interpolate f with degree $q=1$ to preserve the consistency rate obtained without interpolation. Note also that in case a constant Courant number is kept

($\Delta x \sim \Delta t$), the consistency rate is preserved if $q+1 \geq p$, so a linear interpolation is still enough for a second-order time discretization.

The second situation, that is, f being unknown at the $(n+1)$th time step, is usually unsuitable to be treated with the one-step schemes of Table 5.1, except with a first-order approximation. Since under minimal smoothness assumptions we have

$$f(x,t) = f(x,t-\Delta t) + O(\Delta t),$$
$$g(x,t) = g(x,t-\Delta t) + O(\Delta t),$$

then, replacing the values of f and g computed at t in the scheme (5.40) with the same values computed at $t-\Delta t$, we would obtain for X^Δ

$$X^\Delta(x,t;t-\Delta t) = x - \Delta t f(x,t-\Delta t) + O\left(\Delta t^2\right)$$

(a similar error holds for G^Δ) so that X^Δ, G^Δ are computed using the values at the previous time step and without any loss in the consistency rate. An alternative choice is given by the backward Euler scheme, which requires only the knowledge of f at $t-\Delta t$. In this case, the foot of characteristic is located by solving the system

$$X^\Delta(x,t;t-\Delta t) = x - \Delta t f\left(X^\Delta(x,t;t-\Delta t), t-\Delta t\right),$$

whereas the source term is given by $G^\Delta(x,t;t-\Delta t) = g(X^\Delta(x,t;t-\Delta t), t-\Delta t)$. The system is already in fixed-point form, and could be solved for X^Δ by the iteration

$$X^\Delta_{k+1} = x - \Delta t f\left(X^\Delta_k, t-\Delta t\right)$$

provided the right-hand side is a contraction, that is, provided

$$\Delta t < \frac{1}{\|J_f\|},$$

J_f denoting the Jacobian matrix of the vector field f. Note that this introduces a bound on Δt, but this bound depends on the derivatives of the advection speed rather than on its magnitude. Note also that the method is first order, so that it gives no improvement in accuracy over the explicit Euler method used in the classical CIR scheme.

Examples: Multistep schemes The most classical application of multistep schemes for the approximation of the foot of characteristics involves the midpoint method (see Table 3.1). In this case, the equation of characteristics is integrated $2\Delta t$ back in time, and more precisely $X^\Delta(x_j, 2\Delta t)$ solves the fixed-point system

$$X^\Delta(x,t;t-2\Delta t) = x - 2\Delta t f\left(\frac{x + X^\Delta(x,t;t-2\Delta t)}{2}, t-\Delta t\right). \qquad (5.64)$$

Note that, according to the definition of the midpoint method, the vector field f should be computed at the point $X^\Delta(x,t;t-\Delta t)$. Along a smooth trajectory, however, we have

$$X^\Delta(x,t;t-\Delta t) = \frac{x + X^\Delta(x,t;t-2\Delta t)}{2} + O(\Delta t^2),$$

so that by the same arguments used for the explicit Euler method, the approximate midpoint method (5.64) still results in a second-order scheme. Accordingly, the source term is computed as

$$G^\Delta(x,t;t-2\Delta t) = 2\Delta t\, g\left(\frac{x + X^\Delta(x,t;t-2\Delta t)}{2}, t-\Delta t\right).$$

Thus, the scheme would compute the approximate solution at time t_{n+1} by advecting the approximate solution at t_{n-1} along the vector field computed at t_n (such a structure is often referred to as a *three-time-level scheme*). In general, the latter fixed-point system can also be solved iteratively under the same time-step limitation as for the backward Euler scheme, and we expect that in general the value of f at the right-hand side should be obtained by interpolation. Since the resulting scheme is second order in time, as remarked above, linear interpolation on f is enough to preserve the consistency rate.

Other multistep techniques have been proposed in the SL literature. Some of them (including the second-order Adams–Bashforth method) are studied, for example, in [TS87].

5.1.4 ▪ Multiple space dimensions

We briefly sketch here how the techniques introduced in the previous sections can be adapted to treat more general cases, in particular problems in higher dimensions and posed on bounded domains.

To treat the case of multiple dimensions, it is convenient to rewrite the advection equation,

$$u_t + f(x,t)\cdot Du = g(x,t), \qquad \mathbb{R}^d \times [0,T],$$

in the equivalent form

$$u_t + \sum_{i=1}^{d} f_i(x,t) u_{x_i} = g(x,t), \qquad \mathbb{R}^d \times [0,T]. \tag{5.65}$$

For simplicity, we review the various approaches referring to the two-dimensional case. Once given the general form, we briefly discuss consistency and monotonicity.

Upwind scheme The upwind scheme may be better explained referring to the homogeneous, constant-coefficient equation

$$u_t + c_1 u_{x_1} + c_2 u_{x_2} = 0.$$

Given that the consistency of the scheme is insensitive to the choice of taking right or left incremental ratios, the crucial point is to obtain a stable (in particular, monotone) scheme. Keeping the strategy of approximating space derivatives by left incremental ratios, we would obtain

$$\frac{v^{n+1}_{j_1,j_2} - v^n_{j_1,j_2}}{\Delta t} + c_1 \frac{v^n_{j_1,j_2} - v^n_{j_1-1,j_2}}{\Delta x_1} + c_2 \frac{v^n_{j_1,j_2} - v^n_{j_1,j_2-1}}{\Delta x_2} = 0,$$

which is clearly consistent. Solving this relationship for $v^{n+1}_{j_1,j_2}$, we obtain

$$\begin{aligned} v^{n+1}_{j_1,j_2} &= v^n_{j_1,j_2} - \frac{c_1\Delta t}{\Delta x_1}\left[v^n_{j_1,j_2} - v^n_{j_1-1,j_2}\right] - \frac{c_2\Delta t}{\Delta x_2}\left[v^n_{j_1,j_2} - v^n_{j_1,j_2-1}\right] \\ &= \left(1 - \frac{c_1\Delta t}{\Delta x_1} - \frac{c_2\Delta t}{\Delta x_2}\right) v^n_{j_1,j_2} + \frac{c_1\Delta t}{\Delta x_1} v^n_{j_1-1,j_2} + \frac{c_2\Delta t}{\Delta x_2} v^n_{j_1,j_2-1}, \end{aligned}$$

and it is immediate to see that all coefficients are nonnegative (for partial Courant numbers $c_i\Delta t/\Delta x_i$ small enough) if and only of $c_1, c_2 \geq 0$. This suggests that each of the terms $f_i(x,t)v_{x_i}$ should be discretized in an upwind form: with a left incremental ratio if $f_i(x_j,t_n) > 0$ and with a right incremental ratio if $f_i(x_j,t_n) < 0$. We give up the boring, but conceptually simple, description of the details.

LF scheme In extending the LF scheme to (5.65), it should be noted that the average which replaces $v^n_{j_1,j_2}$ should include all points involved in the finite difference approximations of space derivatives and therefore in this case ($d=2$) the four points $v^n_{j_1-1,j_2}$, $v^n_{j_1+1,j_2}$, $v^n_{j_1,j_2-1}$, and $v^n_{j_1,j_2+1}$. This gives the scheme

$$\frac{v^{n+1}_{j_1,j_2}-\frac{1}{4}\left(v^n_{j_1-1,j_2}+v^n_{j_1+1,j_2}+v^n_{j_1,j_2-1}+v^n_{j_1,j_2+1}\right)}{\Delta t}$$
$$+f_1(x_j,t_n)\frac{v^n_{j_1+1,j_2}-v^n_{j_1-1,j_2}}{\Delta x_1}+f_2(x_j,t_n)\frac{v^n_{j_1,j_2+1}-v^n_{j_1,j_2-1}}{\Delta x_2}=g(x_j,t_n)$$

(again, this discretization may be shown to be consistent using the same arguments of the one-dimensional case), and solving for $v^{n+1}_{j_1,j_2}$ we get

$$\begin{aligned}v^{n+1}_{j_1,j_2}=&\frac{1}{4}\left(v^n_{j_1-1,j_2}+v^n_{j_1+1,j_2}+v^n_{j_1,j_2-1}+v^n_{j_1,j_2+1}\right)\\&-f_1(x_j,t_n)\frac{\Delta t}{\Delta x_1}\left[v^n_{j_1+1,j_2}-v^n_{j_1-1,j_2}\right]\\&-f_2(x_j,t_n)\frac{\Delta t}{\Delta x_2}\left[v^n_{j_1,j_2+1}-v^n_{j_1,j_2-1}\right]+\Delta t\, g(x_j,t_n).\end{aligned}$$

In this way, the nonzero coefficients of the scheme are of the form

$$b^{LF}_{jk}(\Delta;t_n)=\frac{1}{4}\pm f_i(x_j,t_n)\frac{\Delta t}{\Delta x_i}$$

(for suitable values of the multi-index k and of the index i) and again can be made nonnegative by keeping the partial Courant numbers small enough.

CIR scheme In more than one space dimension, the CIR scheme can still be written in a formally unchanged way, that is,

$$v^{n+1}_j=G^\Delta(x_j,t_{n+1};t_n)+I_1[V^n]\big(X^\Delta(x_j,t_{n+1};t_n)\big).$$

In this case, however, the following should be understood:

- The augmented system (5.55) is now in dimension $d+1$: the approximation of the source term remains scalar, whereas characteristics are curves in $\mathbb{R}^d$. Thus, X^Δ is now a point of $\mathbb{R}^d$ itself.
- The reconstruction $I_1[V](\cdot)$ is a multidimensional $\mathbb{P}_1$ or $\mathbb{Q}_1$ interpolation.

The monotonicity of the scheme is proved following the same argument as in a single space dimension: we have in fact that the coefficients of the scheme are given by

$$b^{CIR}_{ji}(\Delta;t_n)=\psi^{[1]}_i\big(X^\Delta(x_j,t_{n+1};t_n)\big)$$

and are nonnegative if the basis functions for interpolation are nonnegative (in multiple dimensions, this is clearly true for tensorized linear bases and $\mathbb{P}_1$ finite elements).

5.1.5 ▪ Boundary conditions

The basic theory has been outlined so far without any reference to boundary conditions, which naturally arise when posing a problem on a bounded domain. A situation which allows one to keep essentially the same framework of unbounded grids is the case of periodic conditions given on a d-dimensional torus $[0,P_1]\times\cdots\times[0,P_d]$. Here, the periodicity is imposed in classical difference schemes by identifying the corresponding nodes (and values) on opposite sides of the torus, whereas in SL schemes a point outside the torus is identified with an internal point via the relationship

$$u(x_1 \pm P_1,\ldots,x_d \pm P_d) = u(x_1,\ldots,x_d).$$

The case of Dirichlet conditions is more interesting. Assume now that $\Omega \not\equiv \mathbb{R}^d$ and that Dirichlet data are imposed on the inflow part of the boundary $\partial\Omega$:

$$u(x,t) = b(x,t), \qquad (x,t) \in \Gamma_{in} \times [0,T]. \tag{5.66}$$

We review the main points in the implementation of this boundary condition in the upwind, LF, and CIR schemes, avoiding the most technical issues.

Upwind scheme To fix ideas, assume that the computational domain is a rectangle. Due to the definition of the scheme, nodes on each side the boundary are assigned a value based on internal nodes if the vector field f points outwards, that is, on the outflow boundary Γ_{out}, and this agrees with the fact that Dirichlet conditions cannot be enforced on this portion of the boundary. Nodes on Γ_{in} are assigned a value by (5.66), while nodes placed at the angles may be at the interface between Γ_{in} and Γ_{out}. In this situation, not all the partial incremental ratios may be computed using internal nodes, so on such a node we force the boundary condition (5.66). Clearly, more complex geometries require us to treat a larger number of possible situations.

LF scheme In the LF scheme, assigning a value to a node always requires that other nodes exist around it. So, while this strategy allows us to assign a boundary value to a node in Γ_{in} (since it need not be computed by the scheme), it does not allow us to compute the values at nodes of Γ_{out}, since it would require a further layer of nodes. On the other hand, if the boundary condition b is imposed on the whole of $\partial\Omega$, what typically happens is the development of a boundary layer on Γ_{out}. To get rid of this effect, a simple response would be to treat in upwind form the nodes on this part of the boundary.

CIR scheme The CIR scheme requires a more complex treatment of the Dirichlet condition. First, we rewrite the extended version (2.40) of the representation formula between t_n and t_{n+1}:

$$\begin{aligned} u(x_j,t_{n+1}) = &\int_{t_n \vee \theta(x_j,t_{n+1})}^{t_{n+1}} g(y(x_j,t_{n+1};s),s)\,ds \\ &+ u(y(x_j,t_{n+1};t_n \vee \theta(x_j,t_{n+1})),t_n \vee \theta(x_j,t_{n+1})). \end{aligned} \tag{5.67}$$

As we have discussed in Chapter 2, this amounts to singling out a case in which the solution is propagated from an internal point and a case in which it propagates from the boundary. The numerical counterpart of this procedure requires us to determine the point, if any exists, at which the discrete trajectory X^Δ crosses the boundary of Ω. If we assume for simplicity that the boundary $\partial\Omega$ is defined as the zero-level set of a continuous function $\varphi : \mathbb{R}^N \to \mathbb{R}$, so that $x \in \Omega$ if and only if $\varphi(x) < 0$, then we can define a variable time step $\delta(x_j, t_{n+1})$ as the minimum between Δt and the solution of the equation

$$\varphi\left(X^\Delta(x_j, t_{n+1}; t_{n+1} - \delta(x_j, t_{n+1}))\right) = 0.$$

Note that we expect at most one solution in $[t_n, t_{n+1}]$ for Δt small enough and "nonpathological" sets Ω. Note also that, for consistency reasons, $\delta(x_j, t_{n+1})$ should also replace Δt in the computation of X^Δ (e.g., in the construction of a one-step scheme).

Once we compute the variable step, we split the set of indices as $\mathscr{I} = \mathscr{I}_i \cup \mathscr{I}_b$, so that $\delta(x_j, t_{n+1}) = \Delta t$ for $j \in \mathscr{I}_i$ (*internal* nodes) and $\delta(x_j, t_{n+1}) \le \Delta t$ for $j \in \mathscr{I}_b$ (*near-boundary* nodes). Then, the CIR scheme is modified as

$$v_j^{n+1} = \begin{cases} G^\Delta(x_j, t_{n+1}; t_n) + I_1[V^n]\left(X^\Delta(x_j, t_{n+1}; t_n)\right) & \text{if } j \in \mathscr{I}_i, \\ G^\Delta(x_j, t_{n+1}; t_{n+1} - \delta(x_j, t_{n+1})) & \\ \quad + b\left(X^\Delta(x_j, t_{n+1}; t_{n+1} - \delta(x_j, t_{n+1})), t_{n+1} - \delta(x_j, t_{n+1})\right) & \text{if } j \in \mathscr{I}_b. \end{cases} \tag{5.68}$$

5.1.6 ▪ Examples

In order to evaluate the schemes which have been presented in this section, we use the simple one-dimensional example

$$\begin{cases} u_t(x,t) - (x - \overline{x})u_x(x,t) = 0, \quad (x,t) \in (0,1) \times (0,1), \\ u(x,0) = u_0(x) \end{cases} \tag{5.69}$$

with three initial conditions u_0 of different regularities and with bounded support. In particular, the first is a bounded discontinuous function:

$$u_0(x) = \mathbb{1}_{[0.25,0.5]}(x), \tag{5.70}$$

whereas the second is Lipschitz continuous:

$$u_0(x) = \max(1 - 16(x - 0.25)^2, 0), \tag{5.71}$$

and the third has a bounded second derivative:

$$u_0(x) = \max(1 - 16(x - 0.25)^2, 0)^2. \tag{5.72}$$

In (5.69), characteristics are curved lines and this makes time discretization significant in SL schemes. Also, characteristics converge towards the point $\overline{x} = 1.1$, so that the initial condition is advected rightwards and the support shrinks for increasing time.

The test has been carried out with the upwind, LF, and CIR schemes, the last being implemented both in its classical version (5.40) and in the version (5.57) with a second-order time discretization (denoted as CIR2). In the first three cases, grid refinement has been performed by keeping a constant $\Delta t/\Delta x$ relationship, and more precisely keeping the Courant number less than 0.6 in the upwind and LF schemes and less than 6 in the CIR scheme. In the case of the second-order time discretization for the CIR scheme, we have used the relationship $\Delta t \sim \Delta x^{2/3}$ (more precisely, $\Delta t \approx 2\Delta x^{2/3}$), which optimizes the consistency rate.

Table 5.2. *Errors in the 2-norm for the upwind, LF, and CIR (with first- and second-order time discretization) schemes.*

L^∞ solution				
n_n	Upwind	LF	CIR	CIR2
25	$1.93 \cdot 10^{-1}$	$2.45 \cdot 10^{-1}$	$1.06 \cdot 10^{-1}$	$1.1 \cdot 10^{-1}$
50	$1.66 \cdot 10^{-1}$	$2.3 \cdot 10^{-1}$	$1.16 \cdot 10^{-1}$	$7.35 \cdot 10^{-2}$
100	$1.35 \cdot 10^{-1}$	$2.2 \cdot 10^{-1}$	$7.59 \cdot 10^{-2}$	$6.31 \cdot 10^{-2}$
200	$1.1 \cdot 10^{-1}$	$1.9 \cdot 10^{-1}$	$6.64 \cdot 10^{-2}$	$4.71 \cdot 10^{-2}$
400	$9.2 \cdot 10^{-2}$	$1.56 \cdot 10^{-1}$	$5.38 \cdot 10^{-2}$	$4.01 \cdot 10^{-2}$
Rate	0.27	0.16	0.24	0.36

$W^{1,\infty}$ solution				
n_n	Upwind	LF	CIR	CIR2
25	$1.55 \cdot 10^{-1}$	$2.39 \cdot 10^{-1}$	$1.29 \cdot 10^{-1}$	$3.36 \cdot 10^{-2}$
50	$1.09 \cdot 10^{-1}$	$2.09 \cdot 10^{-1}$	$7.62 \cdot 10^{-2}$	$1.78 \cdot 10^{-2}$
100	$7.05 \cdot 10^{-2}$	$1.72 \cdot 10^{-1}$	$4.1 \cdot 10^{-2}$	$8.43 \cdot 10^{-3}$
200	$4.39 \cdot 10^{-2}$	$1.28 \cdot 10^{-1}$	$2.21 \cdot 10^{-2}$	$4.42 \cdot 10^{-3}$
400	$2.66 \cdot 10^{-2}$	$8.68 \cdot 10^{-2}$	$1.16 \cdot 10^{-2}$	$1.99 \cdot 10^{-3}$
Rate	0.64	0.37	0.87	1.02

$W^{2,\infty}$ solution				
n_n	Upwind	LF	CIR	CIR2
25	$1.52 \cdot 10^{-1}$	$2.17 \cdot 10^{-1}$	$1.32 \cdot 10^{-1}$	$2.97 \cdot 10^{-2}$
50	$1.14 \cdot 10^{-1}$	$1.96 \cdot 10^{-1}$	$7.52 \cdot 10^{-2}$	$1.44 \cdot 10^{-2}$
100	$7.61 \cdot 10^{-2}$	$1.68 \cdot 10^{-1}$	$4.01 \cdot 10^{-2}$	$6.09 \cdot 10^{-3}$
200	$4.63 \cdot 10^{-2}$	$1.32 \cdot 10^{-1}$	$2.08 \cdot 10^{-2}$	$2.6 \cdot 10^{-3}$
400	$2.63 \cdot 10^{-2}$	$9.27 \cdot 10^{-2}$	$1.06 \cdot 10^{-2}$	$1.05 \cdot 10^{-3}$
Rate	0.63	0.31	0.91	1.21

We list in Table 5.2 the numerical errors in the 2-norm for the various schemes, along with overall convergence rates computed on the extreme steps of the refinement. Figure 5.7 shows the behavior of the schemes on the advection of the characteristic function (5.70) (the exact solution is shown with a solid line).

The apparent point in the comparison of the various schemes is numerical diffusion. The accuracy of the LF scheme is reduced by its highly diffusive behavior, but ranging from the upwind to the CIR2 scheme the convergence rate has no dramatic change, although it is clear how the possibility of working at large Courant numbers reduces the numerical diffusion. In terms of absolute accuracy, the first-order CIR scheme is about

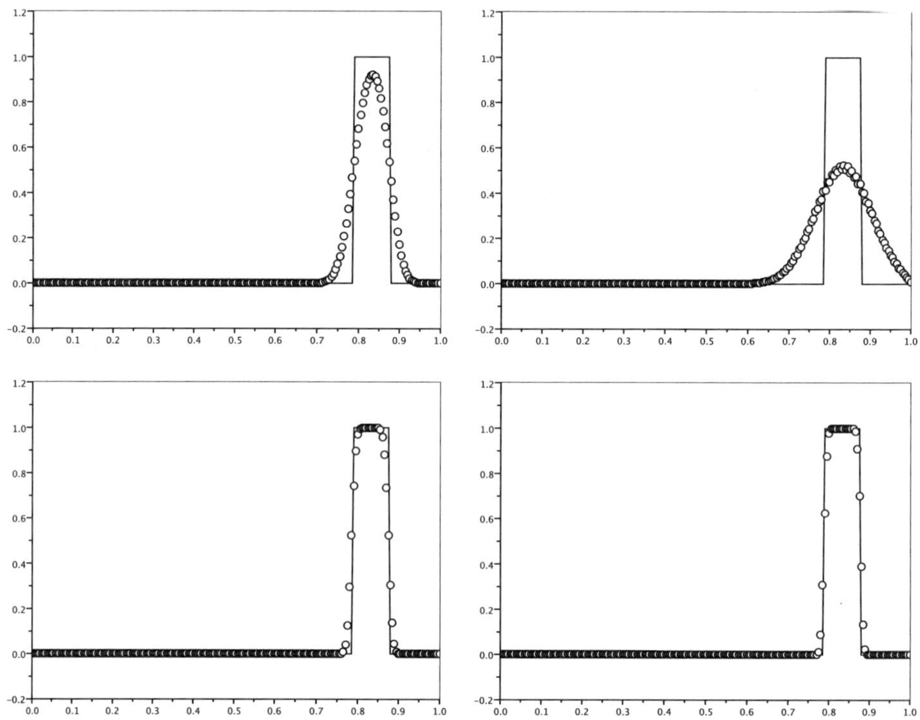

Figure 5.7. *Numerical results for the advection of a characteristic function obtained via the upwind (upper left), LF (upper right), CIR (lower left), and CIR with second-order time discretization (lower right) schemes, with* 200 *nodes.*

twice as accurate as the upwind scheme, and the increase in accuracy is even more apparent for the CIR2 scheme, especially as the solution gets smoother (here, the theoretical convergence rate would be $4/3 \approx 1.33$).

5.2 ▪ Treating the convex HJ equation

In this section, we show how the schemes introduced can be adapted to the one-dimensional, convex HJ equation

$$\begin{cases} u_t(x,t)+H(u_x(x,t))=0, & (x,t)\in\mathbb{R}\times[0,T], \\ u(x,0)=u_0(x). \end{cases} \tag{5.73}$$

We will make the standing assumption that H is convex and that there exists $\alpha_0 \in \mathbb{R}$ such that

$$\begin{cases} H'(\alpha)\leq 0 & \text{if } \alpha\leq\alpha_0, \\ H'(\alpha)\geq 0 & \text{if } \alpha\geq\alpha_0. \end{cases} \tag{5.74}$$

We also define

$$M_{H'}(L) = \max_{[-L,L]} |H'|, \tag{5.75}$$

which corresponds to the maximum speed of propagation of a solution with Lipschitz constant L.

5.2.1 ▪ Upwind discretization

In adapting the upwind scheme to the nonlinear case, it should be taken into consideration that the speed of propagation of the solution is $H'(v_x)$. While it is perfectly clear how to construct an upwind scheme for a speed of constant sign, care should be taken at points where the speed changes sign, in order to obtain a monotone scheme.

Construction of the scheme

The construction outlined will follow the guidelines of [CL84], in which monotone schemes for HJ equations are derived from monotone schemes for conservation laws, and the theory is carried out accordingly. The *differenced* form of the upwind scheme is (4.53),

$$v_j^{n+1} = v_j^n - \Delta t \mathcal{H}^{Up}\left(D_{j-1}[V^n], D_j[V^n]\right), \tag{5.76}$$

where the numerical Hamiltonian $\mathcal{H}^{Up}$ is defined by

$$\mathcal{H}^{Up}(\alpha,\beta) = \begin{cases} H(\alpha) & \text{if } \alpha,\beta \geq \alpha_0, \\ H(\beta) + H(\alpha) - H(\alpha_0) & \text{if } \alpha \geq \alpha_0, \beta \leq \alpha_0, \\ H(\alpha_0) & \text{if } \alpha \leq \alpha_0, \beta \geq \alpha_0, \\ H(\beta) & \text{if } \alpha,\beta \leq \alpha_0. \end{cases} \tag{5.77}$$

Note that the situation in which speed changes sign is subject to a different handling, depending on the fact that characteristics converge or diverge.

Consistency

Since the scheme is in differenced form, it actually suffices to apply Definition 4.17. If $\alpha = \beta = a$, then the numerical hamiltonian (5.77) satisfies

$$\mathcal{H}^{Up}(a,a) = H(a), \tag{5.78}$$

and the consistency condition (4.51) is satisfied. Note that, in (5.77), the second and third cases occur only if $a = \alpha_0$.

Stability

By construction, schemes in differenced form (in particular, upwind and LF in what follows) are necessarily invariant for the addition of constants. Therefore, the main stability issues in this context will be CFL condition and monotonicity.

CFL condition Since the maximum speed of propagation of the solution is $M_{H'}(L)$, the condition which keeps characteristics within the numerical domain of dependence is

$$\frac{M_{H'}(L)\Delta t}{\Delta x} \le 1. \tag{5.79}$$

In this case, this restriction is really only necessary—as we will see, monotonicity requires a stronger condition.

Monotonicity First, we write the partial derivative of the jth component of the scheme as

$$\frac{\partial}{\partial v_i} S_j^{Up}(\Delta; V) = \delta_{ij} - \Delta t \left[\frac{\partial \mathscr{H}^{Up}}{\partial \alpha} \frac{\partial D_{j-1}[V]}{\partial v_i} + \frac{\partial \mathscr{H}^{Up}}{\partial \beta} \frac{\partial D_j[V]}{\partial v_i} \right], \tag{5.80}$$

where α and β are the dummy variables used in the definition (5.77), and δ_{ij} is the Kronecker symbol. It is clear that

$$\frac{\partial D_{j-1}[V]}{\partial v_i} = \begin{cases} \frac{1}{\Delta x} & \text{if } i = j, \\ -\frac{1}{\Delta x} & \text{if } i = j-1, \\ 0 & \text{otherwise,} \end{cases}$$

$$\frac{\partial D_j[V]}{\partial v_i} = \begin{cases} \frac{1}{\Delta x} & \text{if } i = j+1, \\ -\frac{1}{\Delta x} & \text{if } i = j, \\ 0 & \text{otherwise,} \end{cases}$$

so that, substituting into (5.80), we obtain the more explicit form

$$\frac{\partial}{\partial v_i} S_j^{Up}(\Delta; V) = \begin{cases} -\Delta t \, \frac{\partial \mathscr{H}^{Up}}{\partial \alpha} \frac{\partial D_{j-1}[V]}{\partial v_{j-1}} & \text{if } i = j-1, \\ 1 - \Delta t \left[\frac{\partial \mathscr{H}^{Up}}{\partial \alpha} \frac{\partial D_{j-1}[V]}{\partial v_j} + \frac{\partial \mathscr{H}^{Up}}{\partial \beta} \frac{\partial D_j[V]}{\partial v_j} \right] & \text{if } i = j, \\ -\Delta t \, \frac{\partial \mathscr{H}^{Up}}{\partial \beta} \frac{\partial D_j[V]}{\partial v_{j+1}} & \text{if } i = j+1, \\ 0 & \text{otherwise.} \end{cases} \tag{5.81}$$

By the definition of $\mathscr{H}^{Up}$ we have

$$\frac{\partial \mathscr{H}^{Up}}{\partial \alpha}(\alpha, \beta) = \begin{cases} H'(\alpha) \ge 0 & \text{if } \alpha \ge \alpha_0, \\ 0 & \text{otherwise.} \end{cases} \tag{5.82}$$

$$\frac{\partial \mathscr{H}^{Up}}{\partial \beta}(\alpha, \beta) = \begin{cases} H'(\beta) \le 0 & \text{if } \beta \le \alpha_0, \\ 0 & \text{otherwise.} \end{cases} \tag{5.83}$$

Looking at the signs of the various terms, it is apparent that

$$\frac{\partial}{\partial v_i} S_j^{Up}(\Delta; V) \ge 0 \quad (i \ne j), \tag{5.84}$$

whereas, for $i = j$, we have

$$\left| \frac{\partial \mathscr{H}^{Up}}{\partial \alpha} \frac{\partial D_{j-1}[V]}{\partial v_j} + \frac{\partial \mathscr{H}^{Up}}{\partial \beta} \frac{\partial D_j[V]}{\partial v_j} \right| \leq \frac{2M_{H'}(L_V)}{\Delta x}, \tag{5.85}$$

where L_V denotes the Lipschitz constant of the sequence V. Therefore, using (5.85) in (5.81), we obtain that the scheme is monotone if

$$\frac{\Delta t}{\Delta x} \leq \frac{1}{2M_{H'}(L_V)}. \tag{5.86}$$

Note that, in contrast to the linear case, this condition is more stringent then the CFL condition.

Convergence estimates

Finally, we give the convergence result, which follows from consistency (5.78), monotonicity (5.86), and Theorem 4.18.

Theorem 5.7. *Let H satisfy the basic assumptions, let $u_0 \in W^{1,\infty}(\mathbb{R})$, let u be the solution of* (5.73) *with L as its Lipschitz constant, and let v_j^n be defined by* (5.76) *with $v_j^0 = u_0(x_j)$. Then, for any $j \in \mathbb{Z}$ and $n \in [1, T/\Delta t]$,*

$$\left| v_j^n - u(x_j, t_n) \right| \leq C\Delta t^{1/2} \tag{5.87}$$

as $\Delta \to 0$, with $2M_{H'}(L+1)\Delta t \leq \Delta x$.

5.2.2 ▪ Central discretization

In treating the LF scheme, we will follow again the guidelines of [CL84]. Rather than using more general forms of the scheme, we will restrict ourselves here to the particular form that directly generalizes the linear case.

Construction of the scheme

The simplest way to recast the LF scheme for the HJ equation is to define it in the form

$$v_j^{n+1} = \frac{v_{j-1}^n + v_{j+1}^n}{2} - \Delta t H\left(D_j^c[V^n]\right), \tag{5.88}$$

where $D_j^c[V^n]$ is the centered difference at x_j defined by

$$D_j^c[V^n] = \frac{v_{j+1}^n - v_{j-1}^n}{2\Delta x} = \frac{D_{j-1}[V^n] + D_j[V^n]}{2}. \tag{5.89}$$

This definition of the LF scheme completely parallels the linear case and is also suitable to be treated in the framework of the Crandall–Lions theorem. In fact, once we recall that

$$\frac{v_{j-1}^n + v_{j+1}^n}{2} = v_j^n + \frac{\Delta x}{2}\left(D_j[V^n] - D_{j-1}[V^n]\right),$$

(5.88) can be written in the differenced form

$$v_j^{n+1} = v_j^n - \Delta t \mathcal{H}^{LF}\left(D_{j-1}[V^n], D_j[V^n]\right)$$

by setting

$$\mathcal{H}^{LF}(\alpha,\beta) = H\left(\frac{\alpha+\beta}{2}\right) - \frac{\Delta x}{2\Delta t}(\beta-\alpha).$$

Note that, as for the advection equation, no special care is necessary to determine the direction of propagation for the solution (that is, to compare α and β with α_0), since the stencil is symmetric.

Consistency

The LF scheme (5.88) satisfies condition (4.51), and in fact

$$\mathcal{H}^{LF}(a,a) = H\left(\frac{a+a}{2}\right) = H(a). \tag{5.90}$$

Consistency is therefore satisfied.

Stability

We examine again the issues of CFL condition and monotonicity, which in this case give the same restriction on the discretization steps.

CFL condition Taking into account that the maximum speed of propagation is $M_{H'}(L)$, the CFL condition reads as

$$\frac{M_{H'}(L_V)\Delta t}{\Delta x} \le 1, \tag{5.91}$$

as for the upwind scheme. In this case, this condition is necessary and sufficient, since it also ensures monotonicity (as we will soon show).

Monotonicity In examining monotonicity, it is convenient to refer to the LF scheme in the form (5.88). Clearly, the jth component $S_j^{LF}(\Delta;V)$ depends only on the values $v_{j\pm1}$, so that

$$\frac{\partial}{\partial v_i} S_j^{LF}(\Delta;V) = 0 \qquad (i \ne j \pm 1).$$

On the other hand, if $i = j \pm 1$, we have

$$\begin{aligned}\frac{\partial}{\partial v_{j\pm1}} S_j^{LF}(\Delta;V) &= \frac{1}{2} - \Delta t H'\left(D_j^c[V]\right) \frac{\partial D_j^c[V]}{\partial v_{j\pm1}} \\ &= \frac{1}{2} \mp \frac{\Delta t}{2\Delta x} H'\left(D_j^c[V]\right),\end{aligned}$$

where we have used the fact that

$$\frac{\partial D_j^c[V]}{\partial v_{j\pm1}} = \pm \frac{1}{2\Delta x}.$$

Therefore, if L_V is the Lipschitz constant of the sequence V, the scheme is monotone, provided

$$\frac{\Delta t}{\Delta x} \le \frac{1}{M_{H'}(L_V)}. \tag{5.92}$$

Convergence estimates

Again, the convergence result is obtained from consistency (5.90) and monotonicity (5.92) and applying Theorem 4.18.

Theorem 5.8. *Let H satisfy the basic assumptions, let $u_0 \in W^{1,\infty}(\mathbb{R})$, let u be the solution of* (5.73) *with L as its Lipschitz constant, and let v_j^n be defined by* (5.88) *with $v_j^0 = u_0(x_j)$. Then, for any $j \in \mathbb{Z}$ and $n \in [1, T/\Delta t]$,*

$$\left| v_j^n - u(x_j, t_n) \right| \le C \Delta t^{1/2} \tag{5.93}$$

as $\Delta \to 0$, with $M_{H'}(L+1)\Delta t \le \Delta x$.

5.2.3 ▪ SL discretization

Except for replacing the formula of characteristics (1.6)–(1.7) with a suitable generalization, the extension of the CIR approach to HJ equations is relatively straightforward. In this chapter we analyze the monotone version of the SL scheme, that is, the version obtained with $\mathbb{P}_1$ interpolation.

Construction of the scheme

SL discretization of the HJ equation follows the same steps seen for the advection equation: namely, what is really discretized is the representation formula for the solution. In the case of convex HJ equations, the formula under consideration is the Hopf–Lax formula (2.28). Once rewritten in a single space dimension and at a point (x_j, t_{n+1}) of the space-time grid, it reads

$$\begin{aligned} u(x_j, t + \Delta t) &= \min_{a \in \mathbb{R}} \left\{ \Delta t H^*(a) + u(x_j - a\Delta t, t) \right\} \\ &= \Delta t H^*(\bar{a}_j) + u(x_j - \bar{a}_j \Delta t, t) \end{aligned} \tag{5.94}$$

(note that, now and then, we might find it useful to introduce the explicit notation $\bar{a}_j$ to denote the minimizer at the node x_j in (5.94)). In the special case of (5.73), characteristics are straight lines, so that no special care should be taken about the accuracy of time discretization (more complex situations will be considered in the next chapters). It is still necessary, however, to replace the value $u(x_j - a\Delta t, t)$ by a space reconstruction. At this stage, and in parallel with the CIR scheme, this will be chosen to be the $\mathbb{P}_1$ interpolation I_1. Once we set $t = t_n$, the resulting scheme is therefore

$$\begin{cases} v_j^{n+1} = \min_{\alpha \in \mathbb{R}} \left\{ \Delta t H^*(\alpha) + I_1[V^n](x_j - \alpha \Delta t) \right\}, \\ v_j^0 = u_0(x_j). \end{cases} \tag{5.95}$$

Since the SL scheme is not in differenced form, the convergence analysis will be carried out in the framework of, respectively, the Barles–Souganidis and Lin–Tadmor theorems.

Consistency

In order to develop a twofold convergence theory, consistency will be proved separately in two different frameworks. In the first, we use the usual notion (which is suitable for the use of the Barles–Souganidis theorem), which treats consistency as a pointwise concept. In the second, we adopt the L^1 notion of consistency used in the Lin–Tadmor theory, in the particular form suitable for Godunov-type schemes.

Pointwise theory As it has been done for the CIR scheme, the scheme will be compared with the representation formula for the solution, (2.28) in this case. We define

$$\begin{aligned} S_j^{SL}(\Delta; V) &= \min_{\alpha\in\mathbb{R}} \left\{ \Delta t H^*(\alpha) + I_1[V](x_j - \alpha\Delta t) \right\} \\ &= \Delta t H^*(\overline{\alpha}_j) + I_1[V](x_j - \overline{\alpha}_j \Delta t), \end{aligned} \tag{5.96}$$

where $\overline{\alpha}_j$ denotes the minimizer at the node x_j in (5.95).

Let u be a smooth solution of (5.73). First, we recall that for any smooth u the error estimate (5.44) is satisfied:

$$\|u - I_1[U]\|_\infty \leq C\Delta x^2.$$

Writing now $u(x_j, t + \Delta t)$ by means of (5.94), we give a first unilateral estimate on the left-hand term of (5.97) as

$$\begin{aligned} L_j^{SL}(\Delta; t, U(t)) &= \frac{1}{\Delta t}\left[u(x_j, t+\Delta t) - S_j^{SL}(\Delta; t, U(t)) \right] \\ &= \frac{1}{\Delta t}\Big[\Delta t H^*(\overline{a}_j) - u(x_j - \overline{a}_j \Delta t, t) \\ &\quad - \Delta t H^*(\overline{\alpha}_j) + I_1[U(t)](x_j - \overline{\alpha}_j \Delta t) \Big] \\ &\leq \frac{1}{\Delta t}\Big[\Delta t H^*(\overline{\alpha}_j) - u(x_j - \overline{\alpha}_j \Delta t, t) \\ &\quad - \Delta t H^*(\overline{\alpha}_j) + I_1[U(t)](x_j - \overline{\alpha}_j \Delta t) \Big] \\ &\leq \frac{1}{\Delta t} \|u(t) - I_1[U(t)]\|_\infty \leq C\frac{\Delta x^2}{\Delta t}. \end{aligned}$$

Note that the inequality follows from using $\overline{\alpha}_j$ instead of the exact minimizer $\overline{a}_j$ in (5.94). By reversing the roles of $\overline{\alpha}_j$ and $\overline{a}_j$ we get the opposite inequality,

$$-L_j^{SL}(\Delta; t, U(t)) = \frac{1}{\Delta t}\left[u(x_j, t+\Delta t) - S_j^{SL}(\Delta; t, U(t)) \right] \leq C\frac{\Delta x^2}{\Delta t},$$

and therefore

$$\left| L_j^{SL}(\Delta; t, U(t)) \right| \leq C\frac{\Delta x^2}{\Delta t}. \tag{5.97}$$

As it has been done in (5.47), it is also possible to use the sharp interpolation error bound (3.18) to obtain a consistency estimate in the form

$$\left| L_j^{SL}(\Delta; t, U(t)) \right| \leq C \min\left(\Delta x, \frac{\Delta x^2}{\Delta t} \right), \tag{5.98}$$

which can be applied to small Courant numbers, too.

Note that these estimates would suggest that the scheme achieves its best result when going to the final time in a single time step. In practice, this advantage is cut down by some side effects of very large time steps. First, in the situation in which characteristics are not straight lines, errors in characteristic tracking should be taken into consideration (see the consistency estimate for the linear case). Second, it might be useful to approximate the solution at intermediate times as well. Third, large time steps enlarge the numerical domain of dependence, thus requiring a more careful (and computationally more complex) procedure for the global minimum search. With respect to this latter argument, it should be noted that in order to obtain any computational advantage from the use of large time steps, it is necessary that the complexity of the minimum search would have a weak dependence on the size of the set in which this search is performed.

L^1 theory Before computing the consistency error in the sense (4.74) of Lin and Tadmor, we recall from [LT01] a result which simplifies the computation whenever the scheme is in *Godunov form*—which definitely occurs for SL schemes. For our purposes, we will consider a scheme to be in Godunov form when it is based on the repeated application of two operators:

- an exact evolution operator $E(\cdot)$ which make the numerical solution evolve (exactly) on a single time step, so that starting from $v^\Delta(t_n)$, we define the numerical solution on (t_n, t_{n+1}) as
$$v^\Delta(t) = E(t - t_n)v^\Delta(t_n), \quad t_n < t < t_{n+1}; \tag{5.99}$$
- an operator of projection $P^{\Delta x}$ which defines $v^\Delta(t_{n+1})$ by projecting the approximation defined above (computed at the end of the time step) on the specific space discretization:
$$v^\Delta(t_{n+1}) = P^{\Delta x} v^\Delta(t_{n+1}^-). \tag{5.100}$$
In our case, this operator can be defined as the $\mathbb{P}_1$ interpolation of v^Δ:
$$P^{\Delta x} v^\Delta(t_{n+1}^-) = I_1\left[V^\Delta(t_{n+1}^-)\right].$$

In the class of Godunov schemes, the local truncation error (4.74) can be expressed by means of the projection error as follows.

Theorem 5.9 (see [LT01]). *Let the approximation v^Δ be defined by* (5.99)–(5.100). *Then, the norm of the consistency error* (4.74) *may be bounded as*

$$\|F\|_{L^1(\mathbb{R}^d\times[0,T])} \le \frac{T}{\Delta t} \max_{0<t_k\le T} \left\| v^\Delta(t_k^-) - P^{\Delta x} v^\Delta(t_k^-)\right\|_{L^1(\mathbb{R}^d)}. \tag{5.101}$$

The simplest way to use this result is to note that both exact evolution and $\mathbb{P}_1$ interpolation are nonexpansive with respect to the Lipschitz constant of the numerical solution (in fact, the scheme is invariant for the sum of constants, and invariant by translation and monotone, and therefore Lipschitz stable). Then, the interpolation error satisfies

$$\left| v^\Delta(x,t) - I_1\left[V^\Delta(t)\right](x)\right| = O(\Delta x),$$

so that, considering that the solution is supposed to be compactly supported, and passing to the L^1 norm, we obtain

$$\|F\|_{L^1(\mathbb{R}^d\times[0,T])} \le CT\frac{\Delta x}{\Delta t}. \tag{5.102}$$

A deeper way to compute the projection error is to perform a genuinely integral estimation. To this end, consider a generic function $v(x)$ with an L^1 second derivative. On the interval $[x_j, x_{j+1}]$, of measure Δx, we can write $v(x)$ and its $\mathbb{P}_1$-interpolate as

$$\begin{aligned} v(x) &= v(x_j) + \int_{x_j}^{x} v'(\xi)d\xi \\ &= v(x_j) + \int_{x_j}^{x} \left[v'(x_j) + \int_{x_j}^{\xi} v''(\sigma)d\sigma \right] d\xi \\ &= v(x_j) + \int_{x_j}^{x_{j+1}} \mathbf{1}_{[x_j,x]}(\xi) \left[v'(x_j) + \int_{x_j}^{\xi} v''(\sigma)d\sigma \right] d\xi, \\ I_1[V](x) &= v(x_j) + \frac{x-x_j}{\Delta x} \int_{x_j}^{x_{j+1}} v'(\xi)d\xi \\ &= v(x_j) + \int_{x_j}^{x_{j+1}} \frac{x-x_j}{\Delta x} \left[v'(x_j) + \int_{x_j}^{\xi} v''(\sigma)d\sigma \right] d\xi \end{aligned}$$

(we denote by $\mathbf{1}_{[x_j,x]}$ the characteristic function of the interval $[x_j, x]$ and recall that the derivative of the $\mathbb{P}_1$-interpolate is the integral mean value of v' on $[x_j, x_{j+1}]$). We can therefore express the interpolation error as

$$\begin{aligned} |v(x) - I_1[V](x)| = \Bigg| & v'(x_j) \int_{x_j}^{x_{j+1}} \left(\mathbf{1}_{[x_j,x]}(\xi) - \frac{x-x_j}{\Delta x} \right) d\xi \\ & + \int_{x_j}^{x_{j+1}} \left(\mathbf{1}_{[x_j,x]}(\xi) - \frac{x-x_j}{\Delta x} \right) \int_{x_j}^{\xi} v''(\sigma)d\sigma d\xi \Bigg|. \end{aligned} \tag{5.103}$$

Note that, in (5.103), the first integral is identically zero, and hence, giving a bound on the second integral by Hölder's inequality,

$$\begin{aligned} |v(x) - I_1[V](x)| &\le \int_{x_j}^{x_{j+1}} \left| \mathbf{1}_{[x_j,x]}(\xi) - \frac{x-x_j}{\Delta x} \right| d\xi \cdot \sup_{\xi} \int_{x_j}^{\xi} |v''(\sigma)|d\sigma \\ &\le \Delta x \int_{x_j}^{x_{j+1}} |v''(\sigma)|d\sigma, \end{aligned} \tag{5.104}$$

where we have taken into account that the first term is the integral of a function taking values in $[0, 1]$ and that the second is maximized setting $\xi = x_{j+1}$. Integrating (5.104) on the interval $[x_j, x_{j+1}]$, we have

$$\int_{x_j}^{x_{j+1}} |v(x) - I_1[V](x)|dx \le \Delta x^2 \int_{x_j}^{x_{j+1}} |v''(\sigma)|d\sigma,$$

and last, summing over all elementary intervals, we obtain

$$\|v - I_1[V]\|_{L^1(\mathbb{R})} \le \Delta x^2 \|v''\|_{L^1(\mathbb{R})}. \tag{5.105}$$

Finally, in order to apply Theorem 5.9, we must recall that neither the evolution operator nor the $\mathbb{P}_1$ interpolation expands the L^1 norm of the second derivative, as it can be easily seen by interpreting the solution as the integral of the solution of the associated conservation law (2.42). We have therefore

$$||F||_{L^1(\mathbb{R}\times[0,T])} \le T||u_0''||_{L^1(\mathbb{R})}\frac{\Delta x^2}{\Delta t}. \tag{5.106}$$

Stability

Along with some remarks on the CFL condition, we will address here the two concepts of nonlinear stability used in the two different convergence theories, namely monotonicity and uniform semiconcavity.

CFL condition Looking at the basic definition (5.95) of the scheme, one could infer that, since α is allowed to vary over the whole of $\mathbb{R}$, the numerical domain of dependence also coincides with $\mathbb{R}$ and the CFL condition is always satisfied. In practice, whenever the Lipschitz constant of u_0 (hence of $u(t)$ for positive t) is known, it is possible to restrict the search for a minimum to a smaller subset of $\mathbb{R}$, still without violating the CFL condition.

Monotonicity First, note that the SL scheme is invariant for the addition of constants since, for a Lagrange interpolation of any order, $I[V+c](x)\equiv I[V](x)+c$.

To check that the SL scheme is monotone, consider two sequences V and W such that $V-W\ge 0$ componentwise. We have

$$\begin{aligned} S_j^{SL}(\Delta;V) &= \min_{\alpha\in\mathbb{R}}\left\{\Delta t H^*(\alpha)+I_1[V](x_j-\alpha\Delta t)\right\} \\ &= \Delta t H^*(\overline{\alpha}_j)+I_1[V](x_j-\overline{\alpha}_j\Delta t), \end{aligned}$$

$$\begin{aligned} S_j^{SL}(\Delta;W) &= \min_{\alpha\in\mathbb{R}}\left\{\Delta t H^*(\alpha)+I_1[W](x_j-\alpha\Delta t)\right\} \\ &= \Delta t H^*(\widetilde{\alpha}_j)+I_1[W](x_j-\widetilde{\alpha}_j\Delta t) \\ &\le \Delta t H^*(\overline{\alpha}_j)+I_1[W](x_j-\overline{\alpha}_j\Delta t) \end{aligned}$$

(as before, the last inequality is obtained by replacing the exact minimizer for W, $\widetilde{\alpha}_j$ with $\overline{\alpha}_j$). Therefore,

$$S_j^{SL}(\Delta;V)-S_j^{SL}(\Delta;W)\ge I_1[V](x_j-\overline{\alpha}_j\Delta t)-I_1[W](x_j-\overline{\alpha}_j\Delta t),$$

and since $\mathbb{P}_1$ interpolation is monotone itself (that is, $I_1[V]-I_1[W]\ge 0$), we get

$$S_j^{SL}(\Delta;V)-S_j^{SL}(\Delta;W)\ge 0. \tag{5.107}$$

Uniform discrete semiconcavity Assume that at the nth step the discrete semiconcavity assumption

$$v_{r-h}^n-2v_r^n+v_{r+h}^n\le C\Delta x^2 \tag{5.108}$$

holds for the discrete solution at any node x_r with an increment $\pm h\Delta x$ (h being an integer). Then, using the notation of (5.96), at the $(n+1)$th step we have

$$\begin{aligned}
v_{j-h}^{n+1} - 2v_j^{n+1} + v_{j+h}^{n+1} &= \Delta t H^*(\overline{\alpha}_{j-h}) + I_1[V^n](x_{j-h} + \overline{\alpha}_{j-h}\Delta t) \\
&\quad - 2\Delta t H^*(\overline{\alpha}_j) - 2I_1[V^n](x_j + \overline{\alpha}_j \Delta t)] \\
&\quad + \Delta t H^*(\overline{\alpha}_{j+h}) + I_1[V^n](x_{j+h} + \overline{\alpha}_{j+h}\Delta t) \\
&\le \Delta t H^*(\overline{\alpha}_j) + I_1[V^n](x_{j-h} + \overline{\alpha}_j \Delta t) \\
&\quad - 2\Delta t H^*(\overline{\alpha}_j) - 2I_1[V^n](x_j + \overline{\alpha}_j \Delta t) \\
&\quad + \Delta t H^*(\overline{\alpha}_j) + I_1[V^n](x_{j+h} + \overline{\alpha}_j \Delta t),
\end{aligned}$$

where we have bounded the sum from above by replacing the minimizers for the nodes $x_{j\pm h}$, i.e., $\overline{\alpha}_{j\pm h}$, with the minimizer for the node x_j. Now, for $I_1[V^n](x_j + \overline{\alpha}_j \Delta t) = \sum_i v_i^n \psi_i^{[1]}(x_j + \overline{\alpha}_j \Delta t)$, by periodicity of the grid we can set

$$\psi_i^{[1]}(x_j + \overline{\alpha}_j \Delta t) = \psi_{i\pm h}^{[1]}(x_{j\pm h} + \overline{\alpha}_j \Delta t) = \beta_i$$

so that $I_1[V^n](x_{j\pm h} + \overline{\alpha}_j \Delta t) = \sum_i \beta_i v_{i\pm h}^n$. Recalling that the $\mathbb{P}_1$ reconstruction consists of a convex combination of the values in neighboring nodes, we have that $\beta_i \ge 0$ and $\sum_i \beta_i \equiv 1$. We obtain therefore

$$\begin{aligned}
v_{j-h}^{n+1} - 2v_j^{n+1} + v_{j+h}^{n+1} &\le \sum_i \beta_i v_{i-h}^n - 2\sum_i \beta_i v_i^n + \sum_i \beta_i v_{i+h}^n \\
&= \sum_i \beta_i (v_{i-h}^n - 2v_i^n + v_{i+h}^n) \\
&\le \sum_i \beta_i C\Delta x^2 = C\Delta x^2,
\end{aligned} \tag{5.109}$$

and the solution satisfies the same semiconcavity estimate at the $(n+1)$th step.

Convergence

In view of the consistency estimate (5.98) and the monotonicity result (5.107), convergence of the first-order SL scheme follows from the Barles–Souganidis theorem.

Theorem 5.10. *Let H satisfy the basic assumptions, let $u_0 \in W^{1,\infty}(\mathbb{R})$, let u be the solution of* (5.73), *and let v_j^n be defined by* (5.95) *with $v_j^0 = u_0(x_j)$. Then, for any $j \in \mathbb{Z}$ and $n \in [1, T/\Delta t]$,*

$$\left| v_j^n - u(x_j, t_n) \right| \to 0 \tag{5.110}$$

locally uniformly as $\Delta \to 0$.

The parallel result, obtained by means of the Lin–Tadmor convergence theorem, uses the consistency estimate (5.106) and the uniform semiconcavity result (5.109), and may be stated as follows.

Theorem 5.11. *Let H satisfy the basic assumptions, let $u_0 \in W^{1,\infty}(\mathbb{R})$, let u be the solution of* (5.73), *and let v_j^n be defined by* (5.95) *with $v_j^0 = u_0(x_j)$. Assume, moreover, that u_0 is* *semiconcave and compactly supported. Then, for any* $j \in \mathbb{Z}$ *and* $n \in [1, T/\Delta t]$,

$$\|u(t_n) - I_1[V^n]\|_{L^1(\mathbb{R})} \leq C\frac{\Delta x^2}{\Delta t} \tag{5.111}$$

as $\Delta t \to 0$, *with* $\Delta x = o\left(\Delta t^{1/2}\right)$.

5.2.4 ▪ Multiple space dimensions

As it has been done for the advection equation, we briefly turn to the n-dimensional problem:

$$u_t + H(u_{x_1}, \ldots, u_{x_d}) = 0, \qquad \mathbb{R}^d \times [0, T]. \tag{5.112}$$

Most of the work of extending one-dimensional schemes for HJ equations to the multidimensional case follows the same principles of the linear case. In particular, schemes in differenced form have the general structure (4.50), so that, for example, the two-dimensional version of the LF scheme reads

$$\begin{aligned} v_{j_1,j_2}^{n+1} &= \frac{1}{4}\left(v_{j_1-1,j_2}^n + v_{j_1+1,j_2}^n + v_{j_1,j_2-1}^n + v_{j_1,j_2+1}^n\right) \\ &\quad + \Delta t H\left(\frac{v_{j_1+1,j_2}^n - v_{j_1-1,j_2}^n}{\Delta x_1}, \frac{v_{j_1,j_2+1}^n - v_{j_1,j_2-1}^n}{\Delta x_2}\right), \end{aligned}$$

and both consistency and monotonicity may be proved by the very same arguments used for the one-dimensional case. Also, the d-dimensional form of the SL scheme,

$$v_j^{n+1} = \min_{\alpha \in \mathbb{R}^d} \left\{\Delta t H^*(\alpha) + I_1[V^n](x_j - \alpha \Delta t)\right\},$$

retains the same formal structure of the one-dimensional form, although (as it has been noticed for the linear scheme) reconstruction and minimization are now performed in $\mathbb{R}^d$. Again, the proofs of consistency (which is obtained by comparison with the Hopf–Lax formula) and monotonicity follow the same criteria used in a single space dimension.

5.2.5 ▪ Boundary conditions

As we have seen in Chapter 2, boundary conditions for HJ equations must be interpreted in a weak sense, with the only exception of periodic conditions which can be numerically implemented, as it has been sketched for linear problems. Concerning boundary conditions of other kinds, we will limit ourselves here to the basic ideas for their implementation in SL schemes. In this case, the typical technique has been outlined in the treatment of Dirichlet conditions for the advection equation, which require a variable step algorithm to bring the foot of characteristics precisely on the boundary. In addition, it might happen that boundary conditions are applied at nodes belonging to *a neighborhood of the boundary* instead of just on the boundary nodes (as is usual in classical finite difference schemes). This point will be useful in treating Neumann conditions.

We consider three different boundary conditions: Dirichlet, Neumann, and state constraint.

Dirichlet boundary conditions The boundary operator for Dirichlet conditions is $B(x, u, D\varphi(x)) = u(x) - b(x)$, where b is the value imposed on the boundary $\partial\Omega$. Characteristics ending outside Ω should be brought to the boundary following the same idea used in (5.68). As a result, (5.95) is modified as

$$v_j^{n+1} = \min_{\substack{\alpha\in\mathbb{R}^d \\ \tau\le\Delta t \\ x_j-\alpha\tau\in\overline{\Omega}}} \left\{\Delta t H^*(\alpha) + I_1[V^n](x_j-\alpha\tau)\right\}, \tag{5.113}$$

where the value $I_1[V^n](x_j-\alpha\tau)$ should be understood as

$$I_1[V^n](x_j-\alpha\tau) = b(x_j-\alpha\tau)$$

if $x_j-\alpha\tau\in\partial\Omega$. The minimization in (5.113) should be interpreted as follows. The minimum is searched for with $\tau=\Delta t$. If the minimum is found at an internal point (i.e., for $x_j-\overline{\alpha}_j\Delta t\in\Omega$), then nothing else is necessary. If no minimum point is found in the interior of Ω, then among all boundary points which can be written as $x_j-\overline{\alpha}_j\tau$, it is necessary to select the one which achieves the minimum with respect to $\alpha\in\mathbb{R}^d$ and $\tau\le\Delta t$. In control applications, this boundary condition can be reinterpreted as a stopping cost.

Neumann boundary conditions Here, the boundary operator is

$$B(x,u,Du) = \frac{\partial u}{\partial \nu}(x) - m(x),$$

where m is a given function. In order to assign a value to points outside (but close to) Ω, we can use the directional derivative on the boundary to write

$$u(x+\delta\nu(x)) = u(x)+\delta m(x), \qquad x\in\partial\Omega. \tag{5.114}$$

Then, Neumann boundary conditions can be enforced by using directly (5.114) to replace the value $I_1[V^n](x_j-\alpha\Delta t)$ whenever $x_j-\alpha\Delta t\notin\Omega$. A different way is to extend the grid outside the domain Ω with the introduction of a neighborhood in which the values are assigned according to (5.114) and reconstructed as usual by interpolation in (5.95).

For example, if $m(x)\equiv 0$ and Ω is a rectangle of $\mathbb{R}^2$, then we just need to add an extra "frame" of nodes surrounding the boundary at a distance δ outside Ω, the corresponding values being defined as

$$u(x+\delta\nu(x)) = u(x), \qquad x\in\partial\Omega.$$

Clearly, the points $x_j-\alpha\Delta t$ must be kept within this enlarged domain, either by a suitable choice of δ or by a variable step technique.

State constraint boundary conditions This boundary condition is the easiest to implement. In fact, we recall that what is needed is just to extend the solution u outside Ω by setting

$$u(x) = \overline{C}, \qquad x\in\mathbb{R}^d\setminus\Omega,$$

where $\overline{C}$ is an upper bound for the $L^\infty(\Omega)$ norm of the solution. In the numerical implementation, in order to preserve the regularity of the function to be minimized in (5.95), it could be necessary to implement again a variable step algorithm, with $\overline{C}$ as a Dirichlet boundary datum. Note that, in a Dynamic Programming framework, this "large" value could be interpreted as a penalization (made through the stopping cost) of trajectories which violate the state constraints.

5.3 ▪ Examples

We present a new set of simple numerical examples, using the one-dimensional model problem

$$\begin{cases} u_t(x,t)+\frac{1}{2}|u_x(x,t)|^2=0, & (x,t)\in(0,1)\times(0,T), \\ u(x,0)=u_0(x) \end{cases} \tag{5.115}$$

with $T=0.05$ and two different initial conditions u_0 with bounded support. The first is a Lipschitz continuous function (the same used for the tests of subsection 5.1.6):

$$u_0(x)=\max(1-16(x-0.25)^2,0), \tag{5.116}$$

whereas the second, obtained by a simple change of sign, is also semiconcave:

$$u_0(x)=-\max(1-16(x-0.25)^2,0). \tag{5.117}$$

Using the initial condition (5.116), the solution eventually develops a singularity with a nonempty superdifferential. After the onset of the singularity, the exact solution reads as

$$u(x,t)=\begin{cases} \frac{\left(|x-\frac{1}{2}|-\frac{1}{4}\right)^2}{2t} & \text{if } \frac{1}{4}\le x\le\frac{3}{4}, \\ 0 & \text{else.} \end{cases}$$

On the other hand, using the initial condition (5.117), the solution has the expression

$$u(x,t)=\min\left(\frac{\left|x-\frac{1}{2}\right|^2}{2t+\frac{1}{16}}-1,0\right).$$

Again, the test is performed with the upwind, LF, and SL schemes. In this case, the refinement has been carried out with $\Delta t=\Delta x/40$ for the upwind scheme, $\Delta t=\Delta x/20$ for the LF scheme, and $\Delta t=0.01$ (fixed) for the SL scheme. Table 5.3 shows numerical errors in the ∞-norm, and Figure 5.8 compares exact with numerical solutions for the three schemes in both examples.

A first clear outcome of this example is that, although both initial conditions are Lipschitz continuous, having a semiconcave initial condition (and hence a *uniformly semiconcave solution*) makes a difference. In the first test the schemes basically respect the theory, by keeping a convergence rate of about $1/2$ for the upwind and LF schemes and of about 1 for SL scheme (note that this is the order of truncation error for a Lipschitz solution in the ∞-norm, once the time step is kept constant). On the other hand, in the second test convergence rates are almost doubled. A heuristic explanation of this effect is provided by Remark 2.25: the Hamiltonian and its conjugate being smooth and coercive, the foot of a characteristic must be a point of differentiability for u. Roughly speaking, this means that a semiconcave solution always propagates from regular points towards singularities, and this reduces numerical errors.

Note that, in spite of a similar convergence rate, the LF scheme gives a poor approximation in terms of absolute accuracy, even with respect to the upwind scheme. The SL scheme has the best performances in both convergence rate and absolute accuracy, still being monotone.

Table 5.3. *Errors in the ∞-norm for problem (5.115) for the upwind, LF, and SL schemes.*

$W^{1,\infty}$ initial condition			
n_n	Upwind	LF	SL
25	$1.13 \cdot 10^{-1}$	$2.84 \cdot 10^{-1}$	$8.82 \cdot 10^{-2}$
50	$1.01 \cdot 10^{-1}$	$2.51 \cdot 10^{-1}$	$3.53 \cdot 10^{-2}$
100	$6.62 \cdot 10^{-2}$	$1.89 \cdot 10^{-1}$	$1.81 \cdot 10^{-2}$
200	$4.08 \cdot 10^{-2}$	$1.27 \cdot 10^{-1}$	$8.26 \cdot 10^{-3}$
400	$2.42 \cdot 10^{-2}$	$8.0 \cdot 10^{-2}$	$3.94 \cdot 10^{-3}$
Rate	0.56	0.46	1.12

Semiconcave initial condition			
n_n	Upwind	LF	SL
25	$6.42 \cdot 10^{-2}$	$3.64 \cdot 10^{-1}$	$2.11 \cdot 10^{-2}$
50	$3.58 \cdot 10^{-2}$	$1.97 \cdot 10^{-1}$	$5.02 \cdot 10^{-3}$
100	$1.92 \cdot 10^{-2}$	$9.76 \cdot 10^{-2}$	$1.24 \cdot 10^{-3}$
200	$9.97 \cdot 10^{-3}$	$4.83 \cdot 10^{-2}$	$3.07 \cdot 10^{-4}$
400	$5.08 \cdot 10^{-3}$	$2.4 \cdot 10^{-2}$	$7.63 \cdot 10^{-5}$
Rate	0.91	0.98	2.03

5.4 ▪ Stationary problems

In this section, we will sketch how the theory introduced so far can be adapted to stationary problems of, respectively, linear and HJ types. As a general strategy, stationary equations will be regarded as limit states of evolutive equations, and as a consequence it will be natural to consider schemes in the time-marching form.

5.4.1 ▪ The linear case

When posed on the whole of $\mathbb{R}$, a general form of the stationary linear advection equation is

$$u(x)+f(x)u_x(x)=g(x), \qquad x\in\mathbb{R}, \tag{5.118}$$

whose solution $u(x)$ can be considered as the limit for $t\to\infty$ of a solution $u(x,t)$ of the equation

$$u_t(x,t)+u(x,t)+f(x)u_x(x,t)=g(x), \qquad x\in\mathbb{R}, \tag{5.119}$$

regardless of the initial condition used. Note that equations having a different coefficient of the zeroth-order term $u(x)$ may be brought to this form by a rescaling.

We will proceed to write the various schemes in the time-marching form (4.19) by applying them to (5.119). Note that the extension to more general situations (multiple space dimensions, boundary conditions) can be carried out in a fairly straightforward way by following the guidelines given for evolutive problems.

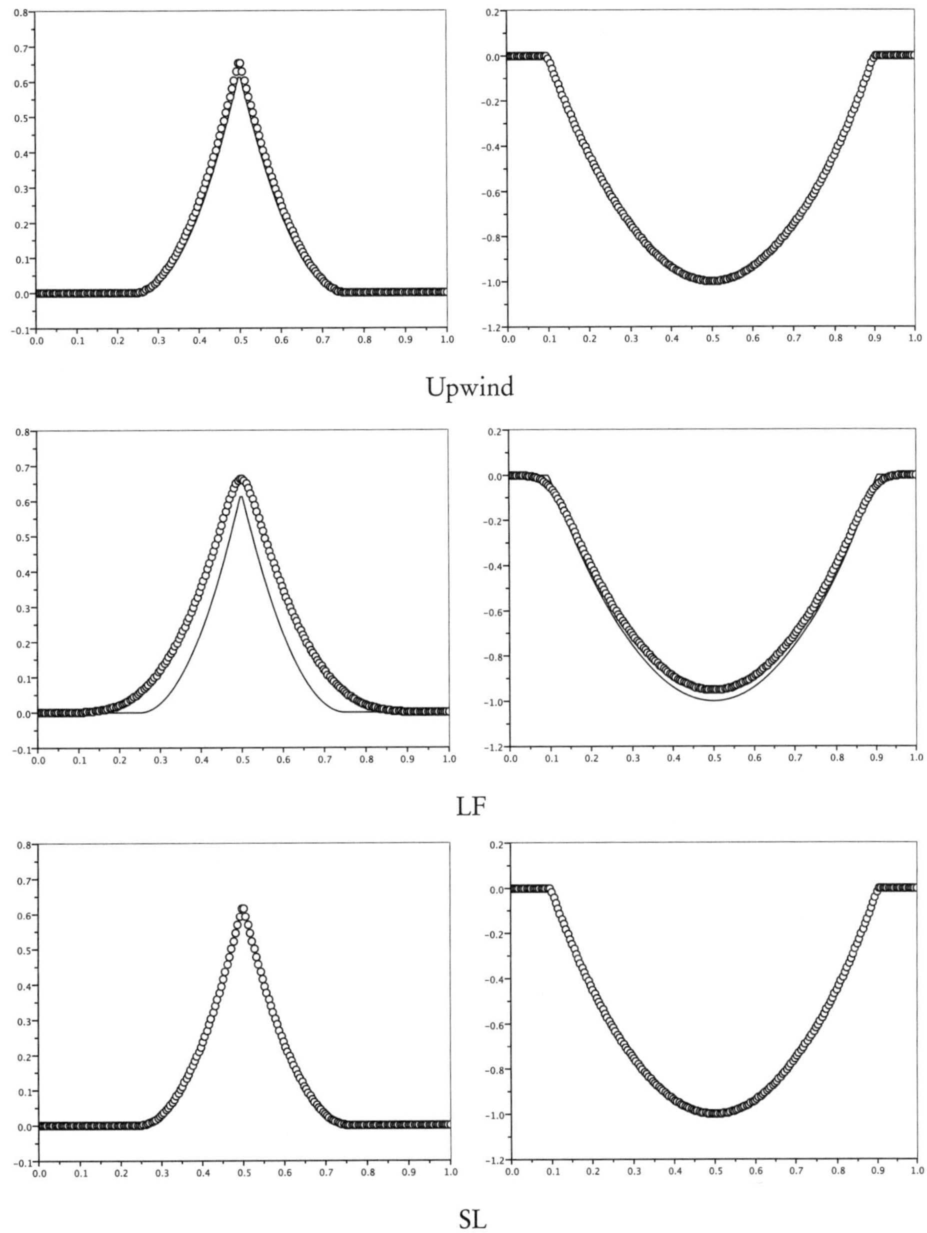

Figure 5.8. *Numerical results for problem* (5.115) *with Lipschitz (left) and semiconcave (right) initial conditions, obtained via the upwind (upper), LF (center), and SL (lower) schemes, with* 200 *nodes.*

Upwind discretization

Following the general strategy outlined in the first part of the chapter, (5.119) would be discretized, for $f(x_j) > 0$, as

$$\frac{v_j^{n+1} - v_j^n}{\Delta t} + v_j^n + f(x_j)\frac{v_j^n - v_{j-1}^n}{\Delta x} = g(x_j)$$

and, for $f(x_j, t_n) < 0$, as

$$\frac{v_j^{n+1} - v_j^n}{\Delta t} + v_j^n + f(x_j)\frac{v_{j+1}^n - v_j^n}{\Delta x} = g(x_j).$$

Once we replace the index of the time step by the index of iteration, the form of the scheme for $f(x)$ changing sign is therefore

$$v_j^{(k+1)} = \begin{cases} (1-\Delta t)v_j^{(k)} - f(x_j)\frac{\Delta t}{\Delta x}\left[v_j^{(k)} - v_{j-1}^{(k)}\right] + \Delta t\, g(x_j) & \text{if } f(x_j) > 0, \\ (1-\Delta t)v_j^{(k)} + \Delta t\, g(x_j) & \text{if } f(x_j) = 0, \\ (1-\Delta t)v_j^{(k)} - f(x_j)\frac{\Delta t}{\Delta x}\left[v_{j+1}^{(k)} - v_j^{(k)}\right] + \Delta t\, g(x_j) & \text{if } f(x_j) < 0. \end{cases} \tag{5.120}$$

Here and in the whole section, since we look for the limit solution, the choice of the initial vector $V^{(0)}$ is irrelevant.

Convergence Once we put the scheme in the form (4.20), it is easy to check that, under the slightly more restrictive stability condition

$$\frac{\|f\|_\infty \Delta t}{\Delta x} < 1 - \Delta t, \tag{5.121}$$

the scheme satisfies assumption (4.21), and more precisely $\|B^{Up}(\Delta)\|_\infty = 1 - \Delta t$. There also follows (see subsection 4.2.5) that it satisfies the consistency estimate (5.10) and converges by Theorem 4.11. We have therefore the following.

Theorem 5.12. *Let $f, g \in W^{1,\infty}(\mathbb{R})$, let u be the solution of* (5.118), *and let $v_j = \lim_k v_j^{(k)}$, with $v_j^{(k)}$ defined by* (5.120). *Then, for any $j \in \mathbb{Z}$,*

$$|v_j - u(x_j)| \to 0 \tag{5.122}$$

as $\Delta \to 0$, with the CFL condition (5.121).

Moreover, if u has a bounded second derivative, then

$$\|V - U\|_\infty \le C\Delta x. \tag{5.123}$$

Central discretization

For simplicity, the LF scheme will be adapted to (5.119) using once more the approximation

$$v(x_j) = \frac{v(x_{j-1}) + v(x_{j+1})}{2} + O(\Delta x^2)$$

in treating the zeroth-order term. With this choice, the time-marching form of the LF scheme is

$$v_j^{(k+1)} = (1-\Delta t)\frac{v_{j-1}^{(k)} + v_{j+1}^{(k)}}{2} - f(x_j)\frac{\Delta t}{2\Delta x}\left[v_{j+1}^{(k)} - v_{j-1}^{(k)}\right] + \Delta t\, g(x_j). \tag{5.124}$$

Convergence Comparing (5.124) with (4.20), it is easily seen that the LF scheme satisfies the stability condition $\|B^{LF}(\Delta)\|_\infty = 1-\Delta t$ under the CFL constraint (5.121). Moreover, as for the upwind scheme, the LF scheme in this form retains the consistency estimate of the evolutive scheme, i.e., (5.25). Therefore, applying Theorem 4.11, we obtain the following convergence result.

Theorem 5.13. *Let $f, g \in W^{1,\infty}(\mathbb{R})$, let u be the solution of* (5.118), *and let $v_j = \lim_k v_j^{(k)}$, with $v_j^{(k)}$ defined by* (5.124). *Then, for any $j \in \mathbb{Z}$,*

$$|v_j - u(x_j)| \to 0 \tag{5.125}$$

as $\Delta \to 0$, with $\Delta x = o\left(\Delta t^{1/2}\right)$ and the CFL condition (5.121).

Moreover, if u has a bounded second derivative and $\Delta x = c\Delta t$ for some constant c, then

$$\|V-U\|_\infty \le C\Delta x. \tag{5.126}$$

SL discretization

To write the time-marching version of the CIR scheme, we start from the representation formula (5.37) which is changed into its more general version

$$u(x,t) = \int_{t-\Delta t}^{t} e^{s-t} g(y(x,t;s))ds + e^{-\Delta t} u(y(x,t;t-\Delta t), t-\Delta t). \tag{5.127}$$

Here, the ODE satisfied by characteristics is driven by a vector field $f(x)$ which does not depend on t. We can therefore set conventionally $t=0$ and omit the initial time in the notation of y, thus obtaining the representation formula

$$\begin{aligned} u(x,t) &= \int_{-\Delta t}^{0} e^{s} g(y(x;s))ds + e^{-\Delta t} u(y(x;-\Delta t), t-\Delta t) \\ &= \int_{0}^{\Delta t} e^{-s} g(y(x;-s))ds + e^{-\Delta t} u(y(x;-\Delta t), t-\Delta t). \end{aligned}$$

In order to treat this case, the augmented system (5.55) is modified in the form

$$\begin{pmatrix} \dot y(x;s) \\ \dot \gamma(x;s) \end{pmatrix} = \begin{pmatrix} f(y(x;s)) \\ -e^{-s} g(y(x;s)) \end{pmatrix} \tag{5.128}$$

with the initial conditions $y(x;0) = x$, $\gamma(x;0) = 0$. Denoting again by $X^\Delta(x;s)$ and $G^\Delta(x;s)$ the approximations of $y(x;s)$ and $\gamma(x;s)$, and replacing the time index by the iteration index, we finally obtain the fixed-point form

$$v_j^{(k+1)} = G^\Delta(x_j;-\Delta t) + e^{-\Delta t} I_1\left[V^{(k)}\right]\left(X^\Delta(x_j;-\Delta t)\right). \tag{5.129}$$

Convergence Following the arguments used for the evolutive case, we can check that the scheme satisfies the same consistency estimate and that $\|B^{CIR}(\Delta)\|_\infty = e^{-\Delta t} = 1 - \Delta t + O\left(\Delta t^2\right)$. It is therefore possible to apply Theorem 4.11 to prove the following convergence result.

Theorem 5.14. *Let $f, g \in W^{p,\infty}(\mathbb{R})$, let u be the solution of* (5.118), *and let $v_j = \lim_k v_j^{(k)}$, with $v_j^{(k)}$ defined by* (5.129). *Assume, moreover, that* (5.58), (5.59) *hold. Then, for any $j \in \mathbb{Z}$,*

$$|v_j - u(x_j)| \to 0 \tag{5.130}$$

as $\Delta \to 0$.

Moreover, if $u \in W^{s,\infty}(\mathbb{R})$ $(s = 1,2)$, then

$$\|V - U\|_\infty \leq C\left(\Delta t^p + \min\left(\Delta x^{s-1}, \frac{\Delta x^s}{\Delta t}\right)\right). \tag{5.131}$$

5.4.2 ▪ The nonlinear case

In adapting the various schemes to stationary HJ equations, we refer to the stationary model which in some sense parallels (5.118), that is,

$$u(x) + H(u_x(x)) = g(x), \qquad x \in \mathbb{R}. \tag{5.132}$$

As before, we consider time-marching schemes, either in differenced form or of SL type.

Discretization in differenced form

In differenced time-marching schemes, the schemes are applied to the evolutive equation

$$u_t + u + H(u_x) = g(x)$$

whose solution converges to a regime state satisfying (5.132). Keeping the scheme in its most general form and adding the zeroth-order term, we have

$$\frac{v_j^{n+1} - v_j^n}{\Delta t} + v_j^n + \mathscr{H}\left(D_{j-p}[V^n], \ldots, D_{j+q}[V^n]\right) = g(x_j),$$

which is clearly a consistent scheme. Hence, replacing the time index with the iteration index, we obtain

$$v_j^{(k+1)} = (1 - \Delta t)v_j^{(k)} - \Delta t \mathscr{H}\left(D_{j-p}\left[V^{(k)}\right], \ldots, D_{j+q}\left[V^{(k)}\right]\right) + \Delta t\, g(x_j). \tag{5.133}$$

With some further work, it could be proved that the adaptation of monotone schemes in differenced form to (5.133) satisfies both the Barles–Souganidis and the Crandall–Lions theories. In the second case, [CL84] provides a short discussion of this point.

SL discretization

In the case of SL discretization, a generalized form of the Hopf–Lax formula, which applies to the solution of (5.132), will be proved in the chapter devoted to Dynamic Programming. For the moment, we simply mention that an SL-type discretization can be written in iterative form as

$$v_j^{(k+1)} = \min_{a \in \mathbb{R}} \left\{ (1 - e^{-\Delta t}) H^*(a) + e^{-\Delta t} I_1\left[V^{(k)}\right](x - a\Delta t) \right\} + \Delta t\, g(x_j). \tag{5.134}$$

Adapting the arguments used for the evolutive case, it can be shown again that the scheme is consistent, l^∞ stable, and monotone for any $\Delta t / \Delta x$ relationship and therefore convergent by Theorem 4.24. This study will be performed in detail in Chapter 8.

5.5 ▪ Commented references

The linear theory for classical monotone difference schemes like upwind or LF is widely treated in any textbook on difference schemes. Some general, relatively recent references on this subject are [Str89, L92, GR96]. However, the application to HJ equations is a much less established topic, and we are not aware of other monographs covering the subject. Except for some more explicit computation, the theory reported here is basically taken from [CL84].

The CIR has been proposed in [CIR52], and although the procedure of construction outlined in the paper clearly implies interpolation, its endpoint basically coincides with an upwind scheme. Moreover, the possibility of working at large Courant numbers was not recognized by that time. Most authors believe [W59] to be the first work proposing an SL technique. In fact, [W59], along with the later paper [Ro81], has originated a considerable streamline of literature concerning the application of SL techniques to Numerical Weather Prediction (NWP) and environmental Fluid Dynamics, for which [SC91] represents a classical (although no longer up-to-date) review paper. More details on this line of research will be given in Chapter 7.

In a completely independent way, SL schemes have been developed in the computational study of plasma physics. To our knowledge, the first paper presenting this approach is [CK76], whereas further developments may be found in [GBSJFF90, SBG99, BM08].

Once more independently, the Hopf–Lax representation formula (or its equivalent in optimal control, the Dynamic Programming Principle) has been used in order to construct numerical schemes for HJ equations. This approach originated in the 1970s with the probabilistic techniques of the so-called *Markov chain approximations*, for which a recent review can be found in the monograph [KuD01]. After the emergence of the theory of viscosity solutions, a reformulation of this approach has been given for deterministic problems. A review on this streamline of research is given in [F97]. We also mention the papers [FF94], in which the coupling between a monotone reconstruction and a high-order time discretization was first proposed, and [FF98, FF02], which analyze the SL scheme for, respectively, the advection and the HJ equations in a more numerical framework. Finally, the possibility of obtaining the improved consistency estimate (5.47) was first observed in [CDM13].

Chapter 6
High-order SL approximation schemes

It is well known that, despite the increasing number of accurate and efficient high-order schemes available in the literature, the related convergence theory is usually poorer. This chapter presents the state-of-the-art of theoretical analysis for high-order SL schemes in the fields of transport and HJ equations. The presentation is dedicated to SL schemes built by means of Lagrange, finite element, and nonoscillatory reconstructions. In its structure, the chapter parallels the previous one, except for the introduction of different and more complex tools at the level of stability analysis. Also, a subsection of one-dimensional examples is included for both the linear and the nonlinear models, and a qualitative study of the interplay between the two discretization steps is presented.

6.1 ▪ SL schemes for the advection equation

In turning our attention to high-order SL schemes, we get back to the case of the one-dimensional linear advection equation, which is rewritten here as

$$\begin{cases} u_t(x,t)+f(x,t)u_x(x,t)=g(x,t), & (x,t)\in\mathbb{R}\times[t_0,T], \\ u(x,0)=u_0(x), & x\in\mathbb{R}. \end{cases} \tag{6.1}$$

As it has been done in the previous chapter, we will also examine in detail the case with constant coefficients, namely,

$$u_t+cv_x=0 \tag{6.2}$$

(with $c>0$), whenever useful to illustrate the main ideas.

6.1.1 ▪ Construction of the scheme

A first step for obtaining a fully high-order discretization of the representation formula (5.33), (5.37) has already been presented concerning the CIR scheme and the possibility of tracking characteristics in a more accurate way. This increase in the accuracy of time discretization for the augmented system (5.63) is summarized in Theorem 5.6. The remaining step is to replace the $\mathbb{P}_1$ reconstruction used in the CIR scheme with an interpolation of order r, which will be denoted by I_r.

With this further improvement, the SL scheme can be rewritten as

$$\begin{cases} v_j^{n+1}=G^\Delta(x_j,t_{n+1};t_n)+I_r[V^n]\big(X^\Delta(x_j,t_{n+1};t_n)\big), \\ v_j^0=u_0(x_j). \end{cases} \tag{6.3}$$

Note that, as it has been remarked in the first-order case, this form also applies to a scheme in d space dimensions, as soon as the augmented system (5.55) is set in dimension $d+1$ and I_r is a d-dimensional reconstruction.

6.1.2 ▪ Consistency

Consistency analysis for the scheme (6.3) requires the basic estimates (5.58)–(5.59), which read as

$$\left|X^{\Delta}(x_j,t_{n+1};t_n)-y(x_j,t_{n+1};t_n)\right|=O(\Delta t^{p+1}), \tag{6.4}$$

$$\left|G^{\Delta}(x_j,t_{n+1};t_n)-\int_{t_n}^{t_{n+1}} g(y(x_j,t_{n+1};s),s)ds\right|=O(\Delta t^{p+1}),$$

as well as the corresponding estimate for interpolation:

$$\|u-I_r[U]\|_{\infty}=O(\Delta x^{r+1}). \tag{6.5}$$

Besides assuming that $f,g\in W^{p,\infty}$ for (5.58), (5.59) hold, (6.5) typically requires that $u\in W^{r+1,\infty}$. Then, retracing the proof of (5.46), (5.60) and incorporating the interpolation error (6.5), we obtain the consistency estimate

$$\left\|L^{SL}(\Delta;t,U(t))\right\|\leq C\left(\Delta t^p+\frac{\Delta x^{r+1}}{\Delta t}\right). \tag{6.6}$$

As it has been done in (5.47) for the CIR scheme, the estimate can be improved for small Courant numbers to give

$$\left\|L^{SL}(\Delta;t,U(t))\right\|\leq C\left(\Delta t^p+\min\left(\Delta x^r,\frac{\Delta x^{r+1}}{\Delta t}\right)\right). \tag{6.7}$$

6.1.3 ▪ Stability

While high-order characteristic tracking does not change the stability properties of the CIR scheme, the introduction of a space reconstruction of degree $r>1$ does. In this situation, the scheme is no longer monotone, and a completely different stability analysis must be performed.

A short digression: The Lagrange–Galerkin scheme

In this small detour, we introduce a class of schemes which will be useful in what follows for analyzing the stability of SL schemes under Lagrange reconstructions. Similar to SL schemes, *Lagrange-Galerkin* (LG) schemes are constructed by a discretization of the representation formula in the form (5.33) or (5.37). Unlike the SL schemes, however, LG schemes perform the space reconstruction by means of a Galerkin projection.

To give a more precise expression, assume for simplicity that $g\equiv 0$, so that the representation formula for the solution would be

$$v(x,t_{n+1})=v(y(x,t_{n+1};t_n),t_n). \tag{6.8}$$

The first step is to write the numerical solution at time t_n, v_h^n in the base $\{\phi_i\}$:

$$v_h^n(x)=\sum_i v_i^n\phi_i(x). \tag{6.9}$$

Note that we have used for the numerical solution the notation v_h^n, usual in Galerkin schemes (here, typically h denotes the space discretization step Δx), and that we have discriminated the Galerkin basis function ϕ_i from the basis functions ψ_i used for interpolation. In fact, the Galerkin basis functions have a conceptually different role, and in particular need not be cardinal functions. Rather, they are assumed to be in L^2, and to be asymptotically dense as $\Delta x \to 0$. The steps for turning the representation formula (6.8) into an LG scheme are

- replacing the exact solution by its approximation (6.9) and the exact displacement along characteristics y with its approximation X^Δ,
- multiplying both sides by a test function $w_h = \sum_i w_i^n \phi_i(x)$, and
- integrating with respect to the space variable.

This amounts to defining the scheme by

$$\int_{\mathbb{R}} v_h^{n+1}(\xi) w_h(\xi) d\xi = \int_{\mathbb{R}} v_h^n \left(X^\Delta(\xi, t^{n+1}; t^n)\right) w_h(\xi) d\xi, \tag{6.10}$$

where the equality has to hold for any w_h in the space generated by the base $\{\phi_i\}$. Hence, using (6.9) and using as test functions the basis functions ϕ_j, the resulting scheme is set in the form

$$\int_{\mathbb{R}} \sum_i v_i^{n+1} \phi_i(\xi) \phi_j(\xi) d\xi = \int_{\mathbb{R}} \sum_i v_i^n \phi_i \left(X^\Delta(\xi, t^{n+1}; t^n)\right) \phi_j(\xi) d\xi,$$

where j ranges over the set of admissible indices $\mathcal{I}$. This condition is actually enforced as

$$\sum_i v_i^{n+1} \int_{\mathbb{R}} \phi_i(\xi) \phi_j(\xi) d\xi = \sum_i v_i^n \int_{\mathbb{R}} \phi_i \left(X^\Delta(\xi, t^{n+1}; t^n)\right) \phi_j(\xi) d\xi. \tag{6.11}$$

Note now that the LG scheme, as it has been defined, is stable in L^2. For example, in the constant-coefficient case, we have that $v_h^n \left(X^\Delta(\xi, t^{n+1}; t^n)\right) = v_h^n(\xi - c\Delta t))$ is a pure translation, so that $\|v_h^n(X^\Delta(\cdot, t^{n+1}; t^n))\|_2 = \|v_h^n(\cdot))\|_2$. Therefore, using $w_h = v_h^{n+1}$ as a test function in (6.10) and applying Hölder's inequality, we get

$$\begin{aligned} \left\|v_h^{n+1}\right\|_2^2 &= \int_{\mathbb{R}} v_h^n \left(X^\Delta(\xi, t^{n+1}; t^n)\right) v_h^{n+1}(\xi) d\xi \\ &\le \|v_h^n\|_2 \left\|v_h^{n+1}\right\|_2, \end{aligned}$$

and this shows that the scheme is stable in the L^2 norm. More generally, the LG scheme is stable whenever the approximate evolution operator E^Δ defined by

$$E^\Delta(t - t_n) v_h^n(x) = v_h^n \left(X^\Delta(x, t; t^n)\right)$$

satisfies for $(t - t_n) \to 0^+$ the bound

$$\left\|E^\Delta(t - t_n)\right\| \le 1 + C(t - t_n),$$

in which the left-hand side is understood as the norm of an operator mapping L^2 into L^2.

It is worth pointing out that, in fact, the definition (6.10) of LG schemes does not allow in general for an exact implementation—this happens because the basis function are deformed by the approximate advection X^Δ, and the integrals appearing at the right-hand side of (6.10) cannot be exactly evaluated, except for constant advection speed. Thus, the real implementation of LG schemes requires some specific techniques, like approximate integration through quadrature formulae or area weighting (we will get back to this latter technique when treating variable-coefficient equations). Rather then giving details on the implementation of LG schemes, our interest here is to use the stability of their exact version (6.10) to prove the stability of SL schemes, via a result of equivalence between the two classes of schemes. This tool will be the used to treat the case of Lagrange reconstructions.

The constant-coefficient case

For the constant-coefficient case, the SL scheme (6.3) takes the form

$$\begin{cases} v_j^{n+1} = I[V^n](x_j - c\Delta t) = \sum_i v_i^n \psi_i(x_j - c\Delta t), \\ v_j^0 = u_0(x_j), \end{cases} \tag{6.12}$$

in which we have expressed the interpolation in terms of the basis functions. Depending on the strategy of reconstruction used, the scheme would require a different theory. Here, we will address two cases: the first is the case of odd-order Lagrange reconstruction, for which the theory is essentially complete; the second is the case of finite element reconstructions, for which a partial Von Neumann–type analysis will be presented.

Lagrange reconstructions In this situation, the reconstruction is invariant by translation and Von Neumann analysis (by means of Fourier methods) gives a necessary and sufficient stability condition. The Von Neumann condition (4.47) reads in this case as

$$|\rho(\omega)| = \left| \sum_l \psi_l(x_j - c\Delta t)\, e^{i\omega(l-j)} \right| \le 1 + C\Delta t.$$

Although an explicit proof of this inequality has been given in [BM08], such a proof is very technical and an easier way consists in showing that the SL scheme is equivalent to an LG scheme. We recast therefore (6.11), for the case under consideration, as

$$\sum_i v_i^{n+1} \int_{\mathbb{R}} \phi_i(\xi)\phi_j(\xi)d\xi = \sum_i v_i^n \int_{\mathbb{R}} \phi_i(\xi - c\Delta t)\phi_j(\xi)d\xi. \tag{6.13}$$

The equivalence between (6.12) and (6.13) is stated by the following.

Theorem 6.1. *Let the functions ψ_i be defined by*

$$\psi_i(\xi) = \psi\left(\frac{\xi}{\Delta x} - i\right) \tag{6.14}$$

for some reference function ψ such that

$$\psi \in W^{2,1}(\mathbb{R}), \tag{6.15}$$

$$\psi(y) = \psi(-y), \tag{6.16}$$

$$\psi(i) = \begin{cases} 1 & \text{if} \quad i = 0, \\ 0 & \text{if} \quad i \in \mathbb{Z},\, i \neq 0. \end{cases} \tag{6.17}$$

Then, there exists a basis $\{\phi_i\}$ *such that the SL scheme* (6.12) *is equivalent to the LG scheme* (6.13) *if and only the function* $\psi(y)$ *has a real nonnegative Fourier transform:*

$$\widehat{\psi}(\omega) \geq 0. \tag{6.18}$$

Proof. We look for a set of basis functions for the LG scheme in the form

$$\phi_i(\xi) = \frac{1}{\sqrt{\Delta x}}\,\phi\left(\frac{\xi}{\Delta x} - i\right) \tag{6.19}$$

for some reference function ϕ to be determined. Comparing (6.12) and (6.13), we obtain the set of conditions to be satisfied:

$$\int_{\mathbb{R}} \phi_i(\xi)\phi_j(\xi)d\xi = \delta_{ij}, \tag{6.20}$$

$$\int_{\mathbb{R}} \phi_i(\xi - c\Delta t)\phi_j(\xi)d\xi = \psi_i(x_j - c\Delta t). \tag{6.21}$$

Note that, by assumption (6.17), condition (6.20) is in fact included in (6.21), as can be seen setting $\Delta t = 0$. Using the definitions (6.14), (6.19) for ψ_k and ϕ_k, (6.21) can be rewritten as

$$\frac{1}{\Delta x}\int_{\mathbb{R}} \phi\left(\frac{\xi - c\Delta t}{\Delta x} - i\right)\phi\left(\frac{\xi}{\Delta x} - j\right)d\xi = \psi\left(\frac{x_j - c\Delta t}{\Delta x} - i\right); \tag{6.22}$$

that is, after defining $\eta = \xi/\Delta x - j$ and $\lambda = c\Delta t/\Delta x$,

$$\int_{\mathbb{R}} \phi(\eta - \lambda + j - i)\phi(\eta)d\eta = \psi(-\lambda + j - i).$$

This ultimately amounts to finding a function ϕ such that

$$\int_{\mathbb{R}} \phi(\eta + y)\phi(\eta)d\eta = \psi(y). \tag{6.23}$$

The left-hand side of (6.23) is the autocorrelation integral (see [P77]) of the unknown function ϕ. Working in the Fourier domain and transforming both sides of (6.23), we have

$$|\widehat{\phi}(\omega)|^2 = \widehat{\psi}(\omega). \tag{6.24}$$

Now, since ψ is a real and even function of y, its Fourier transform $\widehat{\psi}$ is also a real and even function of ω. Moreover, the assumption $\psi \in W^{2,1}(\mathbb{R})$ implies that $\widehat{\psi}(\omega)$ is bounded and

$$\widehat{\psi}(\omega) = O(\omega^{-2}) \quad (\omega \to \pm\infty).$$

Therefore, $\widehat{\psi}$ being also nonnegative, its square root $\widehat{\psi}(\omega)^{1/2}$ is real, even, and nonnegative. In addition, $\widehat{\psi}^{1/2}$ is also bounded and

$$\widehat{\psi}^{1/2} = O(\omega^{-1}) \quad (\omega \to \pm\infty),$$

and this implies that $\widehat{\psi}^{1/2} \in L^2(\mathbb{R})$. Finally, looking at the inverse Fourier transform $\mathcal{F}^{-1}$ as an operator mapping $L^2(\mathbb{R})$ into $L^2(\mathbb{R})$, we obtain that the solution ϕ defined by

$$\phi(y) = \mathcal{F}^{-1}\left\{\widehat{\psi}(\omega)^{1/2}\right\} \tag{6.25}$$

is a well-defined even real function of $L^2(\mathbb{R})$ solving (6.23). □

In conclusion, the SL scheme (6.12) is L^2 stable, being equivalent to a stable scheme. Note that the stability of LG schemes is intended in terms of the L^2 norm $||v_h^n||_2$, but in our case this coincides with the stability in the discrete norm $||V^n||_2$. In fact, we have

$$\begin{aligned}||v_h^n||_2^2 &= \int_{\mathbb{R}} v_h^n(x)^2 dx \\ &= \int_{\mathbb{R}} \left(\sum_i v_i^n \phi_i(x)\right)\left(\sum_j v_j^n \phi_j(x)\right) dx \\ &= \sum_{i,j} v_i^n v_j^n \int_{\mathbb{R}} \phi_i(x)\phi_j(x) dx \\ &= \sum_i \left(v_i^n\right)^2 = \frac{1}{\Delta x}||V^n||_2^2,\end{aligned}$$

where we have applied the orthogonality relationship (6.20). Thus, continuous and discrete norms coincide up to the scaling factor $1/\Delta x$.

Remark 6.2. *Due to a theorem of Bochner (see* [Sch99]*), the functions having nonnegative Fourier transforms could also be characterized as* positive semidefinite functions, *defined as follows.*

Definition 6.3. *A complex-valued function $g : \mathbb{R}^d \to \mathbb{R}$ is said to be positive semidefinite if*

$$\sum_{k=1}^{n}\sum_{j=1}^{n} a_k g(x_k - x_j)\overline{a}_j \geq 0 \tag{6.26}$$

for any $x_k \in \mathbb{R}^d$, $a_k \in \mathbb{C}$ $(k = 1, \ldots, n)$ and for all $n \in \mathbb{N}$.

Remark 6.4. *We explicitly note that* (6.25) *need not have (and, in fact, does not have in general) a unique solution. Equating $\widehat{\psi}$ and $|\widehat{\phi}|^2$, we neglect any information about the phase function of ϕ, and this in turn makes it possible to have multiple solutions. We will get back to this point when treating the case with variable coefficients.*

An example: $\mathbb{P}_1$ interpolation Although the case of $\mathbb{P}_1$ interpolation does not need this kind of stability analysis, we start by treating this simple case, for which explicit computations can be performed. This will allow us to illustrate some remarkable points, in particular the lack of uniqueness for the solution. We recall that in this case the reference function has the form

$$\psi^{[1]}(y) = \begin{cases} 1+y & \text{if } \; -1 \leq y \leq 0, \\ 1-y & \text{if } \; 0 \leq y \leq 1, \\ 0 & \text{elsewhere} \end{cases} \tag{6.27}$$

whose Fourier transform is

$$\widehat{\psi}^{[1]}(\omega) = \frac{2 - 2\cos\omega}{\omega^2} = \frac{\left(\sin\frac{\omega}{2}\right)^2}{\left(\frac{\omega}{2}\right)^2}. \tag{6.28}$$

Now, taking the square root of this transform as in (6.25), we get

$$\widehat{\phi}^{[1]}(\omega) = \frac{\left|\sin\frac{\omega}{2}\right|}{\left|\frac{\omega}{2}\right|} \tag{6.29}$$

and, accordingly,

$$\phi^{[1]}(y) = \mathscr{F}^{-1}\left\{ \frac{\left|\sin\frac{\omega}{2}\right|}{\left|\frac{\omega}{2}\right|} \right\}. \tag{6.30}$$

This solution, numerically computed by the FFT, is shown along with the reference function (6.27) in the upper row of Figure 6.1.

On the other hand, a different (and possibly more natural) solution can be picked up by noting that $\widehat{\psi}^{[1]}(\omega)$ is also the squared magnitude of

$$\widehat{\phi}^{[1]}(\omega) = \frac{\sin\frac{\omega}{2}}{\frac{\omega}{2}} \tag{6.31}$$

whose inverse Fourier transform is explicitly computable as

$$\phi^{[1]}(y) = \begin{cases} 1 & \text{if} \quad -1/2 \le y \le 1/2, \\ 0 & \text{elsewhere.} \end{cases} \tag{6.32}$$

This example shows more precisely the effect of losing uniqueness by neglecting the phase information—in fact, we can obtain an infinity of solutions to (6.23) which differ in the phase term. This is not a major problem here, since at this stage we need only existence of a solution. When dealing with variable-coefficient equations, however, we will need a more careful choice of the solution.

Lagrange interpolation of odd order As shown in Chapter 3, the general form of the reference basis function for Lagrange interpolation of odd degree is

$$\psi^{[r]}(y) = \begin{cases} \displaystyle\prod_{k\neq 0, k=-[r/2]}^{[r/2]+1} \frac{y-k}{-k} & \text{if } 0 \le y \le 1, \\ \vdots & \\ \displaystyle\prod_{k=1}^{r} \frac{y-k}{-k} & \text{if } [r/2] \le y \le [r/2]+1, \\ 0 & \text{if } y > [r/2]+1 \end{cases} \tag{6.33}$$

and extended by symmetry for $y < 0$. Since (6.33) is piecewise polynomial and compactly supported, its second derivative is the sum of a bounded compactly supported term, plus a finite number of Dirac masses, and hence $\psi \in W^{2,1}(\mathbb{R})$. The symbolic computation of the Fourier transforms (see [Fe10a, Fe13a]) shows that, for all odd orders $r \le 13$, they have the structure

$$\widehat{\psi}^{[r]}(\omega) = \frac{a_0 + a_2\omega^2 + \cdots + a_{r-1}\omega^{r-1}}{\omega^{r+1}} \left(\sin\frac{\omega}{2}\right)^{r+1}, \tag{6.34}$$

where the polynomial contains only positive terms of even degree ($a_{2m} > 0$, $a_{2m+1} = 0$). All Fourier transforms of the form (6.34) are therefore nonnegative. A natural conjecture would be that this structure holds for any odd value of r; however, rather than proving such a property, we simply note here that it holds for any order of practical interest.

The basis functions $\psi^{[r]}(y)$ and the solutions $\phi^{[r]}(y) = \mathscr{F}^{-1}\{\widehat{\psi}^{[r]}(\omega)^{1/2}\}$ for $r = 3$ and $r = 5$ are shown in, respectively, the middle and the lower rows of Figure 6.1. In this case, direct and inverse transforms have been computed numerically by the FFT.

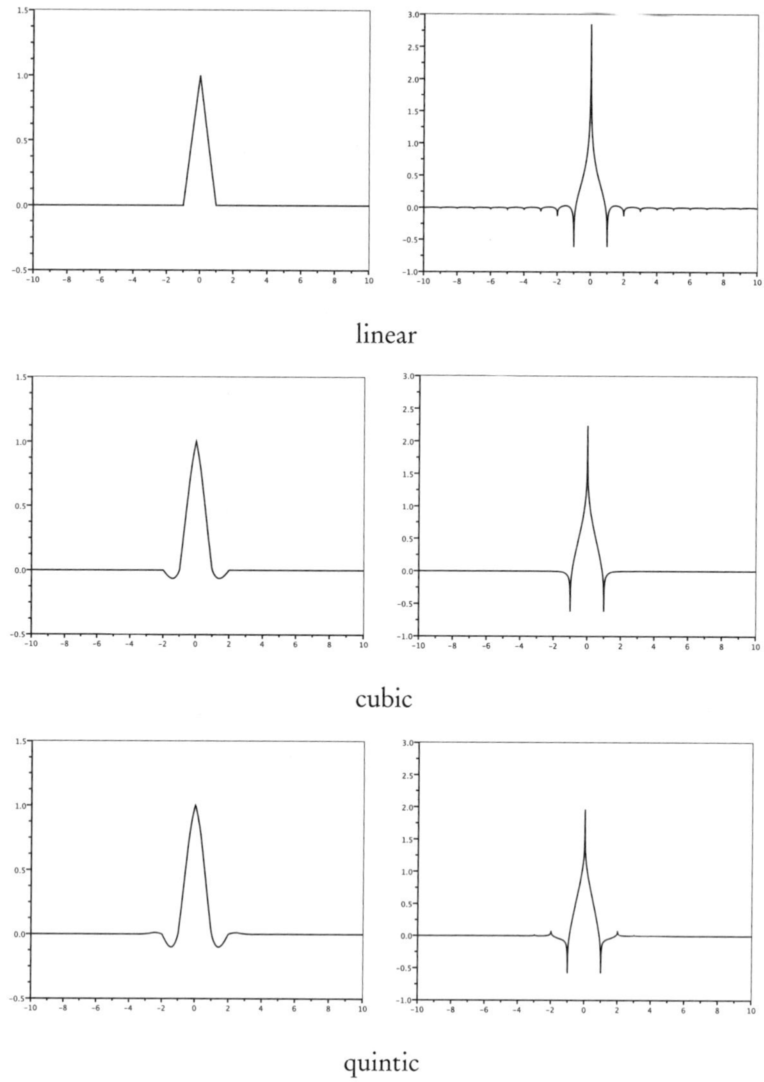

Figure 6.1. *Reference basis function for interpolation and equivalent LG reference basis functions obtained via* (6.25) *for the linear (upper), cubic (middle), and quintic (lower) reconstructions* [Fe13a].

Finite element reconstructions In this situation, since finite element reconstructions fail to be translation invariant (this point has already been remarked in Chapter 3), the technique used for Lagrange reconstructions cannot be applied and we will rather perform a Von Neumann analysis. Unfortunately, due to the lack of translation invariance, B is no longer a circulating matrix, and therefore such analysis gives only a necessary condition.

Assume that (6.2) is set on the interval $[0,1]$ with periodic conditions and that the interval is split into N subintervals, the interpolation degree being r on each subinterval. The total number of nodes is $n_n = Nr$, and $\Delta x = 1/(Nr)$ is the space step between nodes. We assume that the Courant number $\lambda = c\Delta t/\Delta x$ is in $(0,1)$, whereas the case of large Courant numbers can be recovered as a byproduct of this analysis, as it has been shown for the CIR scheme.

In the jth subinterval, the reconstruction depends on the values of the numerical solution at the nodes $x_{(j-1)r}, \ldots, x_{jr}$. Accordingly, the matrix B has the block circulating structure

$$B = \begin{pmatrix} B_1 & 0 & 0 & \cdots & 0 & 0 & B_2 \\ B_2 & B_1 & 0 & \cdots & 0 & 0 & 0 \\ 0 & B_2 & B_1 & \cdots & 0 & 0 & 0 \\ \cdots & \cdots & \cdots & \cdots & \cdots & \cdots & \cdots \\ 0 & 0 & 0 & \cdots & B_2 & B_1 & 0 \\ 0 & 0 & 0 & \cdots & 0 & B_2 & B_1 \end{pmatrix}$$

with $B_1, B_2 \in \mathbb{R}^{r\times r}$, and B_2 of the form

$$B_2 = \begin{pmatrix} 0 & | & b \end{pmatrix},$$

b being a column vector of $\mathbb{R}^r$. In more detail, setting $k = jr$ (that is, denoting by x_k the right extremum of the jth subinterval), we can write the jth block of the scheme as

$$\begin{cases} u_k^{n+1} = \ell_0(\lambda)u_k^n + \cdots + \ell_r(\lambda)u_{k-r}^n, \\ \vdots \\ u_{k-r+1}^{n+1} = \ell_0(\lambda+r-1)u_k^n + \cdots + \ell_r(\lambda+r-1)u_{k-r}^n, \end{cases} \tag{6.35}$$

where $\ell_0(\cdot), \ldots, \ell_r(\cdot)$ are the Lagrange basis functions expressed in a reference interval with unity step between nodes, and, by periodicity, $u_0^n \equiv u_{n_n}^n$ if $j = 1$. Note that, formally, the sign of the reference variable in the ℓ_i is reversed with respect to the physical variable because of the positive sign of λ, which corresponds to a negative shift along characteristics. Now, assuming that ρ is an eigenvalue for the eigenvector v, we have from (6.35)

$$\begin{cases} \left[\ell_0(\lambda) - \rho\right] v_k + \cdots + \ell_r(\lambda) v_{k-r} = 0, \\ \vdots \\ \ell_0(\lambda + r - 1) v_k + \cdots \\ \quad + \left[\ell_{r-1}(\lambda + r - 1) - \rho\right] v_{k-r+1} + \ell_r(\lambda + r - 1) v_{k-r} = 0. \end{cases} \tag{6.36}$$

Since (6.36) has r equations and $r+1$ unknowns, it is possible, for example via Gaussian elimination, to express v_{k-r} as a function of v_k in the form

$$v_{k-r} = h(\rho, \lambda)\, v_k \tag{6.37}$$

with $h(\cdot,\cdot)$ an iterated composition of rational functions. Then, repeating the procedure for N subintervals and imposing the periodicity condition we obtain

$$h(\rho, \lambda)^N = 1; \tag{6.38}$$

that is, $h(\rho, \lambda)$ must be an Nth root of the unity. As it has been shown in the classical case of circulating matrices, we can pass to the modulus, obtaining the equation of the curve containing the eigenvalues in the implicit form

$$|h(\rho, \lambda)| = 1. \tag{6.39}$$

Note that this latter condition suits the case of $N \to \infty$, in which the Nth roots of the unity become dense on the unit circle of $\mathbb{C}$. A different interpretation, when treating an unbounded grid, is that (6.39) states that an eigenvector v should be bounded in l^∞.

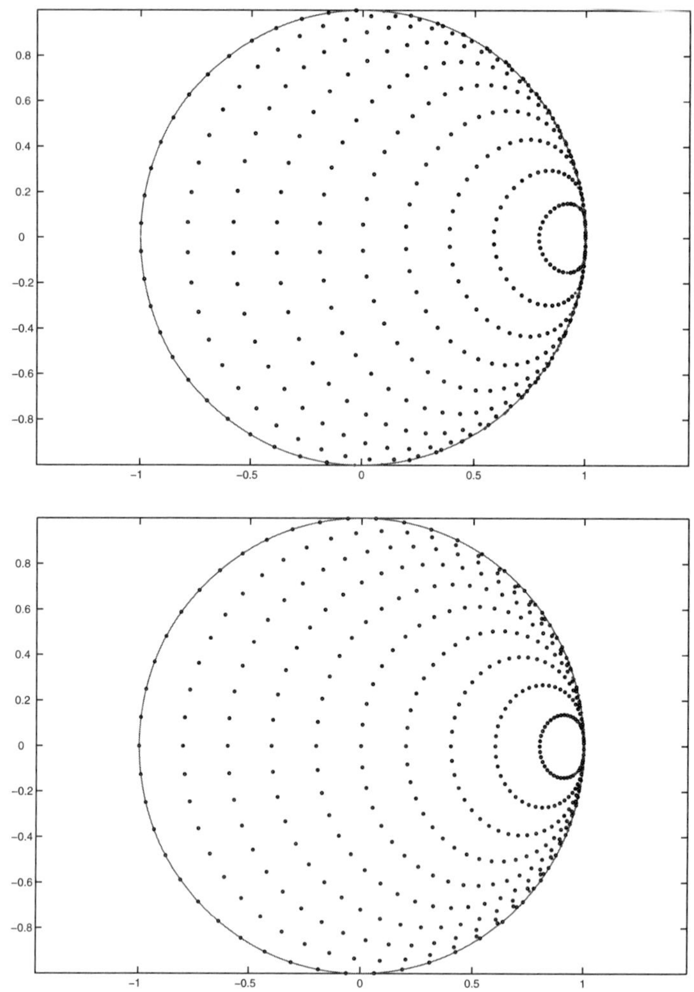

Figure 6.2. *Eigenvalues of the $\mathbb{P}_2$ (upper) and $\mathbb{P}_3$ (lower) schemes for* $\lambda = 0.1, 0.2, \ldots, 1$ [FP07].

Last, the Von Neumann condition is satisfied if

$$|h(\rho,\lambda)| = 1 \quad \text{implies that} \quad |\rho| \le 1. \tag{6.40}$$

We show in Figure 6.2 the eigenvalues of the $\mathbb{P}_2$ and $\mathbb{P}_3$ schemes for different values of λ, with, respectively, $N = 25$ (50 nodes) and $N = 17$ (51 nodes).

The $\mathbb{P}_2$ case We treat separately the $\mathbb{P}_2$ case, for which the basis functions in (6.35) have the form

$$\begin{aligned} \ell_0(y) &= \frac{1}{2}(y-1)(y-2), \\ \ell_1(y) &= y(2-y), \\ \ell_2(y) &= \frac{1}{2}y(y-1), \end{aligned} \tag{6.41}$$

and allow for an explicit computation. More precisely, we have the following.

Theorem 6.5. *Consider the problem* (6.2)*. Assume that the scheme S is in the block form* (6.35)*, with $r=2$ and ℓ_0, ℓ_1, ℓ_2 given by* (6.41)*, that the space step Δx is constant, and that $0<\lambda<1$. Then, the scheme S satisfies condition* (6.40).

Proof. We refer the reader to [P06] for the detailed computations, while sketching here only the main steps. Carrying out the computations outlined above, we obtain for (6.37) the expression

$$v_{k-2} = \left[\frac{\rho - \ell_0(\lambda) + \frac{\ell_1(\lambda)\ell_0(\lambda+1)}{\rho+\ell_1(\lambda+1)}}{\ell_2(\lambda) - \frac{\ell_1(\lambda)\ell_2(\lambda+1)}{\rho+\ell_1(\lambda+1)}} \right] v_k$$

which, via some algebra, gives the condition

$$\left[\frac{2\rho^2 + (\lambda+4)(\lambda-1)\rho + (\lambda-2)(\lambda-1)}{\lambda(\lambda-1)\rho + \lambda(\lambda+1)} \right]^N = 1.$$

Passing to the modulus and writing $\rho = z + iw \in \mathbb{C}$, we obtain the equation

$$\begin{aligned} w^4 &+ \left[2z^2 + (\lambda^2+3\lambda-4)z + (2\lambda^3 - \lambda^2 - 3\lambda + 2)\right]w^2 \\ &+ \left[z^4 + (\lambda^2+3\lambda-4)z^3 + (2\lambda^3 + \lambda^2 - 9\lambda + 6)z^2 \right. \\ &\quad \left. + (-5\lambda^2 + 9\lambda - 4)z - 2\lambda^3 + 3\lambda^2 - 3\lambda + 1\right] = 0. \end{aligned}$$

By working in the auxiliary variable w^2 we obtain, for $z \in [1-2\lambda, 1]$, w^2 as the following function of z and λ:

$$w^2 = \frac{-\left[2z^2 + (\lambda+4)(\lambda-1)z + (2\lambda^2 + \lambda - 2)(\lambda-1)\right] + \sqrt{D(z,\lambda)}}{2}$$

with

$$\begin{aligned} D(z,\lambda) = {} & \lambda^2(\lambda-1)(\lambda+7)z^2 + 2\lambda^2(\lambda-1)(2\lambda^2+7\lambda-7)z \\ & + \lambda^2(4\lambda^4 - 4\lambda^3 - 11\lambda^2 + 22\lambda - 7). \end{aligned}$$

Finally, for $z \in [1-2\lambda, 1]$, it is easy to verify that $w^2 < 1 - z^2$, which amounts to proving that eigenvalues are in the unit disc of $\mathbb{C}$. □

Nonuniform space grids In this situation, if $H_j = r\Delta x_j$ is the measure of the jth subinterval, we can define *local* Courant numbers

$$\lambda_j = \frac{c\Delta t}{\Delta x_j}.$$

We set again $k = jr$ with $j = 1, \ldots, N$, so that the jth block of the scheme reads now as

$$\begin{cases} u_k^{n+1} = \ell_0(\lambda_j)u_k^n + \cdots + \ell_r(\lambda_j)u_{k-r}^n, \\ \vdots \\ u_{k-r+1}^{n+1} = \ell_0(\lambda_j + r - 1)u_k^n + \cdots + \ell_r(\lambda_j + r - 1)u_{k-r}^n. \end{cases} \tag{6.42}$$

Theorem 6.6. *Assume that, for any j, $0 < \lambda_{\min} \le \lambda_j \le \lambda_{\max} < 1$. Assume, moreover, that the constant step scheme* (6.35) *satisfies condition* (6.40) *for any $\lambda \in [\lambda_{\min}, \lambda_{\max}]$. Then condition* (6.40) *is also satisfied by the variable step scheme* (6.42).

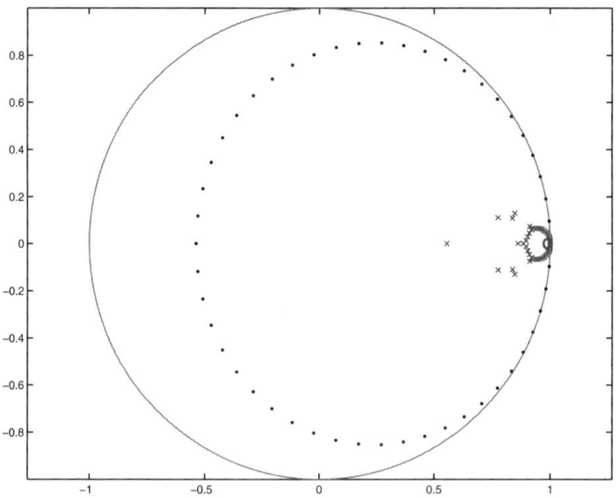

Figure 6.3. *Eigenvalues of the* $\mathbb{P}_2$ *scheme for* $0.014 \le \lambda_j \le 0.77$, *with* 50 *nodes* [FP07].

Proof. Following the same procedure as above, condition (6.38) is now replaced by

$$\prod_{j=1}^{N} h(\rho,\lambda_j) = 1. \tag{6.43}$$

Keeping the analogy with the uniform mesh case, the condition which characterizes the curve containing the eigenvalues may be rewritten as

$$\prod_{j=1}^{N} \left| h(\rho,\lambda_j) \right| = 1. \tag{6.44}$$

Now, all the terms in the product satisfy the bounds

$$\min_j \left| h(\rho,\lambda_j) \right| \le \left| h(\rho,\lambda_j) \right| \le \max_j \left| h(\rho,\lambda_j) \right| \tag{6.45}$$

so that

$$\min_j \left| h(\rho,\lambda_j) \right|^N \le \prod_{j=1}^{N} \left| h(\rho,\lambda_j) \right| \le \max_j \left| h(\rho,\lambda_j) \right|^N, \tag{6.46}$$

and we can conclude that if condition (6.40) is satisfied with a uniform grid for any $\lambda \in [\lambda_{\min}, \lambda_{\max}]$, then it is also satisfied in the nonuniform case. □

We show in Figure 6.3 (in crosses) the eigenvalues of the $\mathbb{P}_2$ scheme, obtained with 25 subintervals of random measure and $0.014 \le \lambda_j \le 0.77$, compared with (in dots) the eigenvalues associated with the extreme values of the local Courant numbers.

Remark 6.7. *The stability analysis for the case of a nonuniform grid also covers the case of variable-coefficient equations, except in the situation of low Courant numbers. The Von Neumann condition* $|\rho| \le 1$ *should be regarded in general as a necessary condition, since we have no informations on the normality of the matrix B. However, there is strong evidence that this matrix is* quasi-normal, *this meaning that in its Jordan decomposition* (4.40), *the condition number* $K(T) = \|T^{-1}\| \cdot \|T\|$, *although not unity, still remains uniformly bounded with respect to* Δ.

The general case for translation invariant reconstructions

When trying to extend to variable-coefficient equations the stability analysis based on the equivalence with LG schemes, it is clear that the relationship (6.23) can justify replacing Galerkin projection with interpolation, but cannot account for the deformation of the basis functions. On the other hand, if in (6.11) the approximate advection $X^\Delta(\xi, t^{n+1}; t^n)$ is replaced with a rigid translation $\xi - x_i + X^\Delta(x_i, t^{n+1}; t^n)$ along the approximate characteristic passing through x_i, we obtain an approximate LG scheme of the form

$$\sum_i v_i^{n+1} \int_{\mathbb{R}} \phi_i(\xi)\phi_j(\xi)d\xi = \sum_i v_i^n \int_{\mathbb{R}} \phi_i\left(\xi - x_i + X^\Delta(x_i, t^{n+1}; t^n)\right)\phi_j(\xi)d\xi. \quad (6.47)$$

With some abuse of notation, we will refer to a scheme in the form (6.47) as an *Area-weighted Lagrange-Galerkin (ALG) scheme*. Schemes of this kind are used to circumvent the problem of computing the integrals of deformed basis functions in exact LG schemes, and their stability analysis is relatively easy if the basis functions ϕ_k are piecewise polynomial and compactly supported, as usual in the finite element setting. Here, on the contrary, the ϕ_k are obtained through a reference function ϕ solving (6.23), which is characterized in a somewhat weaker way by means of its Fourier transform.

Thus, while the equivalence between (6.47) and (6.3) can be proved by the same ideas of the constant-coefficient case, an ad hoc proof is necessary to show that (6.47) represents a "small" perturbation of an exact LG scheme, that is,

$$\left\|B^{SL}(\Delta; t_n)\right\|_2 = \left\|B^{ALG}(\Delta; t_n)\right\|_2 \le \left\|B^{LG}(\Delta; t_n)\right\|_2 + C\Delta t. \quad (6.48)$$

In turn, this will be proved in the form

$$\left\|B^{ALG} - B^{LG}\right\|_2 \le \left(\left\|B^{ALG} - B^{LG}\right\|_1 \cdot \left\|B^{ALG} - B^{LG}\right\|_\infty\right)^{1/2} \le C\Delta t, \quad (6.49)$$

which is satisfied if both the 1-norm and the ∞-norm of $B^{ALG} - B^{LG}$ are bounded by an $O(\Delta t)$.

In order to suit the main situation of interest, i.e., Lagrange interpolation of odd degree, the LG reference basis function will be assumed to be piecewise smooth, with at most a countable set of isolated discontinuities y_k. More precisely, we will assume the following:

- The function $\phi(y)$ satisfies the decay condition

$$|\phi(y)| \le \frac{C_\phi}{1 + |y|^3}. \quad (6.50)$$

- Its derivative is in the form

$$\phi'(y) = \phi_s'(y) + \sum_{k=-\infty}^{\infty} w_k \delta(y - y_k), \quad (6.51)$$

where the regular and the singular parts satisfy the further bounds

$$|\phi_s'(y)| \le \frac{C_s}{1 + y^2}, \quad (6.52)$$

$$|w_k| \le \frac{C_w}{1 + k^2}, \quad (6.53)$$

and the singularities y_k have the expression

$$y_k = \alpha k + \beta. \tag{6.54}$$

- The vector field $f(\cdot,t)$ is uniformly Lipschitz continuous, that is,

$$|f(x^1,t)-f(x^2,t)| \le L_f |x^1 - x^2| \tag{6.55}$$

for any $t \in \mathbb{R}$.

- The approximation $X^\Delta(\cdot)$ is consistent, that is,

$$z_i = X^\Delta(x_i, t^{n+1}; t^n) = x_i - \Delta t\, f(x_i, t^{n+1}) + O(\Delta t^2). \tag{6.56}$$

We give a couple of preliminary lemmas, which will be useful in order to estimate the perturbation appearing in (6.49).

Lemma 6.8. *For any $a \in \mathbb{R}^+$, $b \in \mathbb{R}$,*

$$\int_{\mathbb{R}} \frac{dy}{(1+y^2)(1+(ay+b)^2)} = \frac{\pi(1+a)}{(1+a)^2+b^2}, \tag{6.57}$$

$$\sum_{k=-\infty}^{\infty} \frac{1}{(1+k^2)(1+(ak+b)^2)} \le \frac{C\pi(1+a)}{(1+a)^2+b^2} \tag{6.58}$$

for some positive constant C.

Proof. The integral (6.57) can be computed explicitly (the result has been obtained by symbolic integration with *Mathematica*; see [Fe13a]). The inequality (6.58) can be obtained by observing that all terms of the series are bounded, and therefore its sum can be given an upper bound, up to a multiplicative constant, by comparison with the corresponding integral (6.57). □

Lemma 6.9. *Let assumptions* (6.55), (6.56) *hold. Then, for any $c \in \mathbb{R}^+$, $d \in \mathbb{R}$, both series*

$$\sum_{j=-\infty}^{\infty} \frac{1}{c+(j-\frac{z_i}{\Delta x}+d)^2}, \tag{6.59}$$

$$\sum_{i=-\infty}^{\infty} \frac{1}{c+(j-\frac{z_i}{\Delta x}+d)^2} \tag{6.60}$$

are uniformly bounded as functions of, respectively, the index i and the index j.

Proof. Assume first that the smallest value attained by the term $(j-\frac{z_i}{\Delta x}+d)^2$ is exactly zero.

Consider the series (6.59). Let i (and therefore z_i) be fixed, and denote by j_0 the index for which $(j_0-\frac{z_i}{\Delta x}+d)^2 = 0$ (that is, $j_0 = \frac{z_i}{\Delta x} - d$). Accordingly, we have

$$\sum_{j=-\infty}^{\infty} \frac{1}{c+(j-\frac{z_i}{\Delta x}+d)^2} = \sum_{j=-\infty}^{\infty} \frac{1}{c+(j-j_0)^2}$$
$$= \sum_{j=-\infty}^{\infty} \frac{1}{c+j^2},$$

where the last form is obtained by an obvious shift in the summation index and is independent of the index i.

Consider now (6.60), and let j be fixed. Denote by i_0 the index for which $(j - \frac{z_{i_0}}{\Delta x} + d)^2 = 0$ (that is, $\frac{z_{i_0}}{\Delta x} = j + d$). By the consistency assumption (6.56), we have

$$z_i = x_i - \Delta t f(x_i) + O(\Delta t^2), \tag{6.61}$$

$$z_{i_0} = x_{i_0} - \Delta t f(x_{i_0}) + O(\Delta t^2) \tag{6.62}$$

so that using the Lipschitz continuity of f and the triangular inequality (in the form of a difference), we obtain

$$\begin{aligned} |z_i - z_{i_0}| &= \left| x_i - x_{i_0} - \Delta t (f(x_i) - f(x_{i_0})) + O(\Delta t^2) \right| \\ &\geq |x_i - x_{i_0}| - \Delta t |f(x_i) - f(x_{i_0})| (1 + O(\Delta t)) \\ &\geq (1 - \Delta t L_f) |x_i - x_{i_0}| (1 + O(\Delta t)) \end{aligned} \tag{6.63}$$

and, for Δt small enough,

$$|z_i - z_{i_0}| \geq \frac{1}{2} |x_i - x_{i_0}|. \tag{6.64}$$

Turning back to (6.60), we have, using (6.64) and again a shift in the summation index,

$$\begin{aligned} \sum_{i=-\infty}^{\infty} \frac{1}{c + (j - \frac{z_i}{\Delta x} + d)^2} &= \sum_{i=-\infty}^{\infty} \frac{1}{c + (\frac{z_{i_0}}{\Delta x} - \frac{z_i}{\Delta x})^2} \\ &\leq \sum_{i=-\infty}^{\infty} \frac{1}{c + \frac{1}{4}(\frac{x_{i_0}}{\Delta x} - \frac{x_i}{\Delta x})^2} \\ &= \sum_{i=-\infty}^{\infty} \frac{1}{c + \frac{1}{4}(i_0 - i)^2} \\ &= \sum_{i=-\infty}^{\infty} \frac{1}{c + \frac{1}{4} i^2}, \end{aligned} \tag{6.65}$$

which is independent of j.

Last, if the smallest value for the term $(j - \frac{z_i}{\Delta x} + d)^2$ is nonzero, then the same arguments may be applied, except for a fractional perturbation of the constant d. We note here that the sum of both series is continuous with respect to such a perturbation, so that it is possible to pass to the supremum, obtaining a uniform bound in this case as well. The technical details are left to the reader. □

We present now the main result. It states the stability of the ALG scheme under assumptions which generalize the parallel result in [MPS88]. This will allow us to apply the theory to LG basis functions obtained by solving (6.23).

Theorem 6.10. *Let the basic assumptions* (6.19) *and* (6.50)–(6.56) *hold, and assume* $x_j = j\Delta x$. *Then, there exists a positive constant* C *independent of* n, Δx, *and* Δt *such that, at a generic step* $n \in \mathbb{N}$ *and for* Δt *small enough,*

$$\left\| B^{ALG}(\Delta; t_n) - B^{LG}(\Delta; t_n) \right\|_2 \leq C \Delta t.$$

Proof. The proof uses the bound (6.49) and is split into several steps.

Step 1. We first need to estimate the difference $|X^\Delta(\xi)-(\xi-x_i+X^\Delta(x_i,t^{n+1};t^n))|$. Denoting for shortness $X^\Delta(\xi,t^{n+1};t^n)$ as $X^\Delta(\xi)$ and again $X^\Delta(x_i,t^{n+1};t^n)$ as z_i, by (6.55) and (6.56) we obtain

$$|X^\Delta(\xi)-(\xi-x_i+z_i)|\le C_X|\xi-x_i|\Delta t. \tag{6.66}$$

At a later time, we will use again the change of variables $\eta=\xi/\Delta x-i$. In this new variable, (6.66) is rewritten as

$$|X^\Delta((\eta+i)\Delta x)-(\eta\Delta x+z_i)|\le C_X|\eta|\Delta x\Delta t. \tag{6.67}$$

Step 2. We proceed to estimate $|b_{ij}^{ALG}(\Delta;t_n)-b_{ij}^{LG}(\Delta;t_n)|$. By definition, we have

$$\begin{aligned}\left|b_{ij}^{ALG}-b_{ij}^{LG}\right| &= \left|\int_{\mathbb{R}}\phi_j(\xi-x_i+z_i))\phi_i(\xi)d\xi-\int_{\mathbb{R}}\phi_j\left(X^\Delta(\xi)\right)\phi_i(\xi)d\xi\right|\\ &\le \frac{1}{\Delta x}\int_{\mathbb{R}}\left|\phi\left(\frac{\xi-x_i+z_i}{\Delta x}-j\right)-\phi\left(\frac{X^\Delta(\xi)}{\Delta x}-j\right)\right|\cdot\left|\phi\left(\frac{\xi}{\Delta x}-i\right)\right|d\xi.\end{aligned}$$

Now, introducing the variable $\eta=\xi/\Delta x-i$, we obtain

$$\begin{aligned}\left|b_{ij}^{ALG}-b_{ij}^{LG}\right| &\le \int_{\mathbb{R}}\left|\phi\left(\frac{\eta\Delta x+z_i}{\Delta x}-j\right)-\phi\left(\frac{X^\Delta((\eta+i)\Delta x)}{\Delta x}-j\right)\right|\cdot|\phi(\eta)|d\eta\\ &\le \int_{\mathbb{R}}\left|\int_{\frac{\eta\Delta x+z_i}{\Delta x}-j}^{\frac{X^\Delta((\eta+i)\Delta x)}{\Delta x}-j}\phi'(y)dy\right|\cdot|\phi(\eta)|d\eta\\ &\le \int_D|\phi'(y)|\cdot|\phi(\eta)|dyd\eta,\end{aligned}$$

where D is defined by

$$D=\left\{(\eta,y):\eta\in\mathbb{R},y\in\left[\underline{y}(\eta),\overline{y}(\eta)\right]\right\}$$

and $\underline{y}(\eta),\overline{y}(\eta)$ by

$$\begin{aligned}\underline{y}(\eta) &= \min\left(\frac{\eta\Delta x+z_i}{\Delta x},\frac{X^\Delta((\eta+i)\Delta x)}{\Delta x}\right)-j,\\ \overline{y}(\eta) &= \max\left(\frac{\eta\Delta x+z_i}{\Delta x},\frac{X^\Delta((\eta+i)\Delta x)}{\Delta x}\right)-j.\end{aligned}$$

Note that the integration domain is modified so as to have a y-projection of positive measure for any η.

Since the integrand is positive, we can give a further bound on this integral by enclosing the integration domain in a larger domain on which the integrand could be bounded in a more straightforward way. By (6.67), the domain D can be included in a set of the form

$$E=\left\{(\eta,y):\eta\in\mathbb{R},y\in\left[\eta-C_X\Delta t|\eta|+\frac{z_i}{\Delta x}-j,\eta+C_X\Delta t|\eta|+\frac{z_i}{\Delta x}-j\right]\right\}.$$

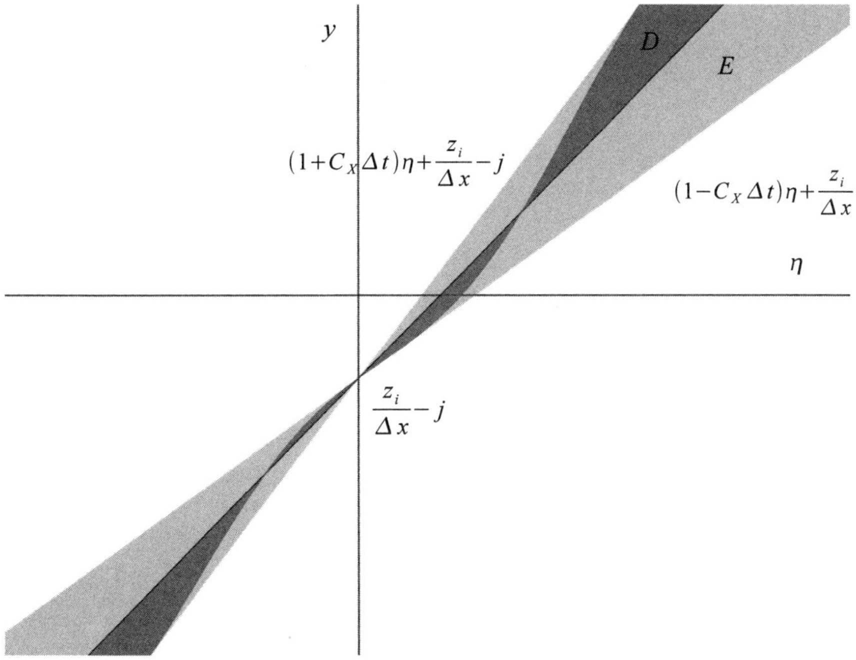

Figure 6.4. *Inclusion of the integration domain in the set E* [Fe13a].

Figure 6.4 shows the inclusion of the integration domain D in the set E defined above. In the figure, the line

$$y=\frac{\eta\Delta x+z_i}{\Delta x}-j$$

and the curve

$$y=\frac{X^{\Delta}((\eta+i)\Delta x)}{\Delta x}-j$$

of the η-y plane enclose the dark-shaded domain D, which in turn appears included in the light-shaded set E.

We have therefore

$$\begin{aligned}\left|b_{ij}^{ALG}-b_{ij}^{LG}\right| &\le \int_D |\phi'(y)|\cdot|\phi(\eta)|\,dy\,d\eta\\
&\le \int_{\mathbb{R}}\int_{\eta-C_X\Delta t|\eta|+\frac{z_i}{\Delta x}-j}^{\eta+C_X\Delta t|\eta|+\frac{z_i}{\Delta x}-j} |\phi'(y)|\cdot|\phi(\eta)|\,dy\,d\eta\\
&= \int_{\mathbb{R}}\int_{\eta-C_X\Delta t|\eta|+\frac{z_i}{\Delta x}-j}^{\eta+C_X\Delta t|\eta|+\frac{z_i}{\Delta x}-j} |\phi_s'(y)|\cdot|\phi(\eta)|\,dy\,d\eta\\
&\quad+\int_{\mathbb{R}}\int_{\eta-C_X\Delta t|\eta|+\frac{z_i}{\Delta x}-j}^{\eta+C_X\Delta t|\eta|+\frac{z_i}{\Delta x}-j}\left[\sum_{k=-\infty}^{\infty}|w_k|\delta(y-y_k)\right]\cdot|\phi(\eta)|\,dy\,d\eta\\
&= A_{ij}+B_{ij}.\end{aligned}\qquad(6.68)$$

Step 3. Estimation of A_{ij}. Using the decay informations on ϕ and ϕ_s', we can estimate A_{ij} as

$$\begin{aligned}A_{ij} &= \int_{\mathbb{R}} \int_{\eta - C_X\Delta t|\eta| + \frac{z_i}{\Delta x} - j}^{\eta + C_X\Delta t|\eta| + \frac{z_i}{\Delta x} - j} |\phi_s'(y)| dy \; |\phi(\eta)| d\eta \\ &\le C_s C_\phi \int_{\mathbb{R}} \int_{\eta - C_X\Delta t|\eta| + \frac{z_i}{\Delta x} - j}^{\eta + C_X\Delta t|\eta| + \frac{z_i}{\Delta x} - j} \frac{1}{1+y^2} dy \; \frac{1}{1+|\eta|^3} d\eta \\ &\le 2 C_s C_\phi C_X \Delta t \int_{\mathbb{R}} \frac{1}{1+\left((1-C_X\Delta t)\eta + \frac{z_i}{\Delta x} - j\right)^2} \frac{|\eta|}{1+|\eta|^3} d\eta, \end{aligned} \tag{6.69}$$

where the inner integral has been estimated by multiplying the sup of the function to be integrated by the measure of the interval. Now, it is easy to see that we can bound the second term of the integrand in (6.69) as

$$\frac{|\eta|}{1+|\eta|^3} \le \frac{2}{1+\eta^2},$$

so that, applying Lemma 6.8 and collecting all multiplicative constants in a single constant C_A, we obtain

$$\begin{aligned} A_{ij} &\le 4 C_s C_\phi C_X \Delta t \int_{\mathbb{R}} \frac{1}{1+\left((1-C_X\Delta t)\eta + \frac{z_i}{\Delta x} - j\right)^2} \frac{1}{1+\eta^2} d\eta \\ &= 4 C_s C_\phi C_X \Delta t \frac{\pi(2-C_X\Delta t)}{(2-C_X\Delta t)^2 + \left(\frac{z_i}{\Delta x} - j\right)^2} \\ &\le \frac{C_A \Delta t}{1+\left(\frac{z_i}{\Delta x} - j\right)^2}. \end{aligned} \tag{6.70}$$

where the last inequality has been obtained assuming also that Δt is small enough.

Step 4. Estimation of B_{ij}. The integral B_{ij} may be split as

$$\begin{aligned} B_{ij} &= \int_{\mathbb{R}} \int_{\eta - C_X\Delta t|\eta| + \frac{z_i}{\Delta x} - j}^{\eta + C_X\Delta t|\eta| + \frac{z_i}{\Delta x} - j} \left[\sum_{k=-\infty}^{\infty} |w_k| \delta(y - y_k) \right] \cdot |\phi(\eta)| \, dy \, d\eta \\ &= \sum_{k=-\infty}^{\infty} |w_k| \int_{\mathbb{R}} \int_{\eta - C_X\Delta t|\eta| + \frac{z_i}{\Delta x} - j}^{\eta + C_X\Delta t|\eta| + \frac{z_i}{\Delta x} - j} \delta(y - y_k) \cdot |\phi(\eta)| \, dy \, d\eta. \end{aligned} \tag{6.71}$$

Note that the measure to be integrated has its support on the segment of intersection between the line $y = y_k$ and the set E. Denoting by η_k^- the abscissa of the intersection of the line $y = y_k$ with the boundary $y = (1 + C_X\Delta t)\eta + \frac{z_i}{\Delta x} - j$ (this represents the extremum of smallest magnitude for this segment), and vice versa η_k^+ for the boundary $y = (1 - C_X\Delta t)\eta + \frac{z_i}{\Delta x} - j$ (the extremum of largest magnitude), we have (see Figure 6.5)

$$\begin{aligned} |\eta_k^+ - \eta_k^-| &= \frac{2 C_X \Delta t}{1 - C_X^2 \Delta t^2} \left| y_k - \frac{z_i}{\Delta x} + j \right| \\ &= \frac{2 C_X \Delta t}{1 - C_X^2 \Delta t^2} \left| \alpha k + \beta - \frac{z_i}{\Delta x} + j \right| \end{aligned}$$

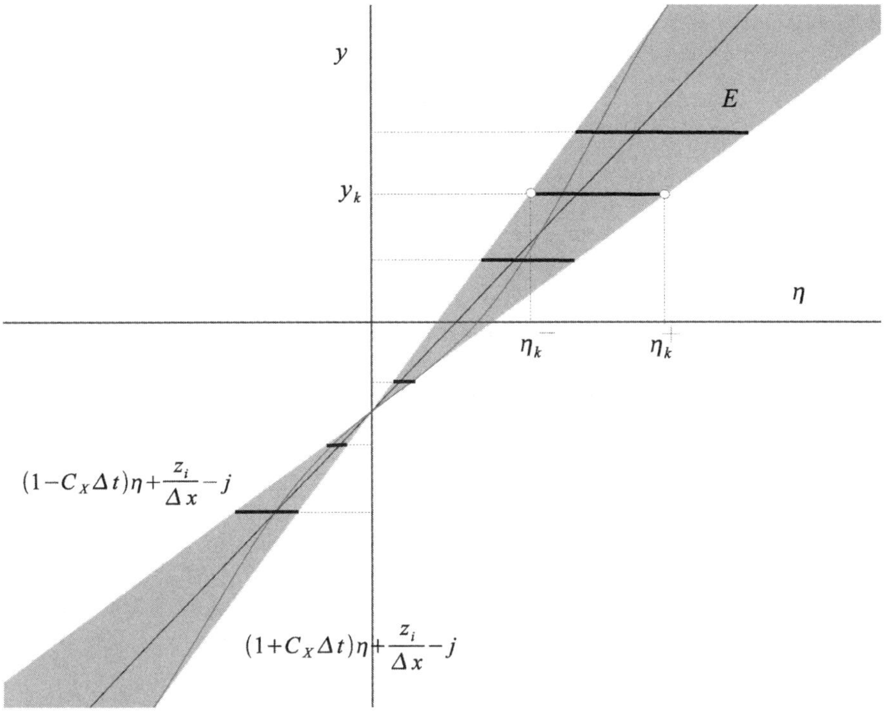

Figure 6.5. *Integration of the singular part of ϕ' on the set E* [Fe13a].

and

$$|\eta_k^-| = \frac{1}{1+C_X\Delta t}\left|y_k - \frac{z_i}{\Delta x}+j\right|$$
$$= \frac{1}{1+C_X\Delta t}\left|\alpha k+\beta - \frac{z_i}{\Delta x}+j\right|$$

so that for Δt small enough we also have

$$|\eta_k^+ - \eta_k^-| \le 4C_X\Delta t\left|\alpha k+\beta - \frac{z_i}{\Delta x}+j\right|,$$
$$|\eta_k^-| \ge \frac{1}{2}\left|\alpha k+\beta - \frac{z_i}{\Delta x}+j\right|.$$

Now, using such bounds and again the decay estimate for ϕ as in Step 3, we obtain

$$\int_{\mathbb{R}}\int_{\eta-C_X\Delta t|\eta|+\frac{z_i}{\Delta x}-j}^{\eta+C_X\Delta t|\eta|+\frac{z_i}{\Delta x}-j} \delta(y-y_k)\cdot|\phi(\eta)|\,dy\,d\eta \le 2C_X\Delta t\frac{\left|\alpha k+\beta-\frac{z_i}{\Delta x}+j\right|}{1+\left|\alpha k+\beta-\frac{z_i}{\Delta x}+j\right|^3}$$
$$\le 4C_X\Delta t\frac{1}{1+\left(\alpha k+\beta-\frac{z_i}{\Delta x}+j\right)^2}$$

and therefore

$$\begin{aligned} B_{ij} &\leq 4C_X\Delta t \sum_{k=-\infty}^{\infty} |w_k| \frac{1}{1+\left(\alpha k+\beta-\frac{z_i}{\Delta x}+j\right)^2} \\ &\leq 4C_X\Delta t \sum_{k=-\infty}^{\infty} \frac{C_w}{1+k^2} \frac{1}{1+\left(\alpha k+\beta-\frac{z_i}{\Delta x}+j\right)^2} \\ &\leq \frac{C_B\Delta t}{(1+\alpha)^2+\left(\beta-\frac{z_i}{\Delta x}+j\right)^2}, \end{aligned} \tag{6.72}$$

where we have applied again Lemma 6.8 and collected all multiplicative constant into C_B.

Conclusions. To sum up, recalling (6.68) and applying Lemma 6.9 to the estimates (6.70) and (6.72), we obtain that

$$\left\| B^{ALG}(\Delta;t_n) - B^{LG}(\Delta;t_n)\right\|_\infty \leq \sup_i \sum_j (A_{ij}+B_{ij}) \leq C\Delta t,$$

$$\left\| B^{ALG}(\Delta;t_n) - B^{LG}(\Delta;t_n)\right\|_1 \leq \sup_j \sum_i (A_{ij}+B_{ij}) \leq C\Delta t,$$

so that the bound (6.49) holds. □

Remark 6.11. *It is immediate to see that, according to this result, the class of basis functions for which the ALG scheme is stable includes all bounded, piecewise Lipschitz continuous functions with bounded support. In this respect, this result includes and generalizes the corresponding theorem in* [MPS88].

Application to Lagrange interpolation We turn back to the case of Lagrange interpolation of odd order. As we said, the solution of (6.23) is not explicitly known, but rather characterized in terms of its Fourier transform. So what is needed is to relate properties (6.50)–(6.54) to properties of the Fourier transform $\widehat{\phi}(\omega)$.

- Concerning the rate of decay of $\phi(y)$, it is known that ϕ decays like $|y|^{-k}$, provided $\widehat{\phi}^{(k)} \in L^1(\mathbb{R})$. Therefore, assumption (6.50) is satisfied if $\widehat{\phi}''' \in L^1(\mathbb{R})$.
- Checking assumption (6.51) requires singling out a periodic component in $i\omega\widehat{\phi}(\omega)$ (that is, in the transform $\mathscr{F}[\phi']$). This component generates the singular part (sum of evenly spaced Dirac distributions) in the inverse Fourier transform. Once $i\omega\widehat{\phi}_s$ is defined as the difference between $i\omega\widehat{\phi}$ and its periodic component, the decay assumption (6.52) requires its second derivative with respect to ω to be in $L^1(\mathbb{R})$.
- Assumption (6.53) is related to the decay of Fourier coefficients for periodic functions. It turns out that this assumption is satisfied, provided the periodic component of $i\omega\widehat{\phi}(\omega)$ has a locally L^1 second derivative with respect to ω.

Example: $\mathbb{P}_1$ interpolation We reconsider this example, already solved in the constant-coefficient case, by discussing the use of the two different solutions of the integral equation (6.23). First, using (6.25), we get

$$\widehat{\phi}^{[1]}(\omega) = \frac{\left|\sin\frac{\omega}{2}\right|}{\left|\frac{\omega}{2}\right|} \tag{6.73}$$

and the solution could be defined accordingly as the inverse transform of (6.73) (see the upper line of Figure 6.1). Although this solution would be fine for the constant-coefficient case, it does not satisfy the required decay assumptions. For example, the transform (6.73) is not three times differentiable, even in the sense of distributions, so it is not expected to satisfy (6.50).

On the other hand, using the other possibility, that is,

$$\widehat{\phi}^{[1]}(\omega)=\frac{\sin\frac{\omega}{2}}{\frac{\omega}{2}}, \tag{6.74}$$

and taking the inverse transform as

$$\phi^{[1]}(y)=\mathcal{F}^{-1}\left\{\frac{\sin\frac{\omega}{2}}{\frac{\omega}{2}}\right\}=\begin{cases}1 & \text{if } -1/2\le y\le 1/2,\\ 0 & \text{elsewhere,}\end{cases}$$

we obtain a solution which satisfies all the basic assumptions, as is easy to see. Note that, even with the lack of an explicit inverse transform, this could be extracted from the properties of the Fourier transform (6.74). First, (6.74) defines an analytic function, so the inverse transform $\phi^{[1]}(y)$ decays faster than any negative power of y and (6.50) is satisfied (actually, it has bounded support). Second, the transform of $\phi^{[1]'}$ reads as

$$\mathcal{F}\left\{\phi^{[1]'}\right\}=2i\sin\frac{\omega}{2}.$$

Here, we have only a periodic component and $\phi_s^{[1]'}\equiv 0$, so (6.52) is satisfied. In turn, the periodic component is also analytic, so its Fourier coefficients have a fast decay and (6.53) holds (in fact, the only nonzero coefficient is related to the "harmonic" $\sin(\omega/2)$, and $\phi^{[1]'}(y)=\delta(y+1/2)-\delta(y-1/2)$).

Lagrange interpolation of higher (odd) order Although, for $r\ge 3$, the problem of differentiability of $\widehat{\phi}^{[r]}$ outlined for $r=1$ no longer appears, defining a solution through (6.25) would not allow one to single out a periodic component in $\mathcal{F}[\phi']$, and as a result the solution in the y-domain would have a slower decay. A way to circumvent this problem is to define a solution in the ω-domain as

$$\begin{aligned}\widehat{\phi}^{[r]}(\omega)&=\frac{\sqrt{a_0+a_2\omega^2+\cdots+a_{r-1}\omega^{r-1}}}{\omega|\omega|^{\frac{r-1}{2}}}\cdot\sin\frac{\omega}{2}\left|\sin\frac{\omega}{2}\right|^{\frac{r-1}{2}}\\ &=A_r(\omega)\cdot B_r(\omega).\end{aligned} \tag{6.75}$$

In (6.75), A_r and B_r denote, respectively, the algebraic and the trigonometric terms of $\widehat{\phi}^{[r]}$. Note that in (6.75) all derivatives are continuous at $\omega=0$, whereas due to the structure of $B_r(\omega)$ it occurs that derivatives are continuous only up to a certain order (depending on r) for ω coinciding with multiples of 2π.

To check the basic assumptions, we also need to compute $\mathcal{F}\left\{\phi^{[r]'}\right\}$:

$$\mathcal{F}\left\{\phi^{[r]'}\right\}(\omega)=i\omega\widehat{\phi}^{[r]}(\omega)=i\frac{\sqrt{a_0+a_2\omega^2+\cdots+a_{r-1}\omega^{r-1}}}{|\omega|^{\frac{r-1}{2}}}\cdot\sin\frac{\omega}{2}\left|\sin\frac{\omega}{2}\right|^{\frac{r-1}{2}}, \tag{6.76}$$

which contains again an algebraic and a periodic term. The first one has a finite imaginary limit $i\sqrt{a_{r-1}}$ for $\omega\to\pm\infty$, so that we can rewrite (6.76) as the sum of a term vanishing for $\omega \to \pm\infty$, plus a periodic term giving the asymptotic behavior, that is,

$$\begin{aligned}\mathcal{F}\left\{\phi^{[r]'}\right\}(\omega) &= i\left(\frac{\sqrt{a_0 + a_2\omega^2 + \cdots + a_{r-1}\omega^{r-1}}}{|\omega|^{\frac{r-1}{2}}} - \sqrt{a_{r-1}}\right)\sin\frac{\omega}{2}\left|\sin\frac{\omega}{2}\right|^{\frac{r-1}{2}} \\ &\quad + i\sqrt{a_{r-1}}\sin\frac{\omega}{2}\left|\sin\frac{\omega}{2}\right|^{\frac{r-1}{2}} \\ &= iC_r(\omega) + iD_r(\omega).\end{aligned}$$

As it has been explained at the start of the section, in order to satisfy the basic assumptions we have to check that

(i) the transform (6.75) has its third derivative (with respect to ω) in $L^1(\mathbb{R})$;

(ii) the term $C_r(\omega)$ has its second derivative in $L^1(\mathbb{R})$; and

(iii) the term $D_r(\omega)$ has a locally L^1 second derivative.

Let us start with point (i). We have, from elementary differentiation rules,

$$\frac{d^3}{d\omega^3}\widehat{\phi}^{[r]}(\omega) = A_r'''B_r + 3A_r''B_r' + 3A_r'B_r'' + A_rB_r'''. \tag{6.77}$$

Now, the terms B_r, B_r', and B_r'' are always bounded. The term B_r''' is bounded for all $r > 3$, whereas, for $r = 3$, it has the form of a sequence of Dirac distributions of constant weight. On the other hand, we have that $A_r, \ldots, A_r''' = O(1/\omega^2)$, so that the right-hand side of (6.77) is in $L^1(\mathbb{R})$ and point (i) is satisfied for any $r \geq 3$. Concerning point (ii), using the same arguments of point (i), it suffices that the algebraic term in brackets, along with its first and second derivative, be $O(1/\omega^2)$. This is in fact the case, so the requirement of point (ii) is also satisfied. Last, point (iii) is also trivially satisfied since D_r'' is bounded on $\mathbb{R}$. All symbolic computations required for this subsection, along with the Fourier transforms, have been carried out with *Mathematica* for $r \leq 13$.

We show in Figure 6.6 the reference LG basis functions obtained by solving numerically (6.23) by FFT, in both the form (6.25) (left column) and the form (6.75) (right column), for the linear, cubic, and quintic Lagrange interpolations. Note that, since D_r has a period of 4π in the ω-domain, the discontinuities in $\phi^{[r]}$ (that is, Dirac distributions in $\phi^{[r]'}$) appear in the y-domain at multiples of $1/2$, so that (6.54) holds with $\alpha = 1/2$, $\beta = 0$ (in fact, even harmonics are missing in D_r, so $w_k = 0$ for k even). If $\frac{r-1}{2}$ is even (e.g., in the linear and quintic cases), then

$$\sin\frac{\omega}{2}\left|\sin\frac{\omega}{2}\right|^{\frac{r-1}{2}} = \left(\sin\frac{\omega}{2}\right)^{\frac{r+1}{2}},$$

so that $D_r(\omega)$ contains a finite number of harmonics, and as a result the LG reference basis function has a finite number of discontinuities.

6.1.4 ▪ Convergence

We sum up the convergence analysis in a theorem. Note that the technical assumptions (6.55)–(6.56) are already included as assumptions on the smoothness of f and the consistency rate of time discretization.

Theorem 6.12. *Let $f, g \in W^{p,\infty}(\mathbb{R})$, let u be the solution of* (6.1), *and let v_j^n be defined by* (6.3), *with an interpolation basis ψ_i satisfying the assumptions of Theorem* 6.1. *Assume,*

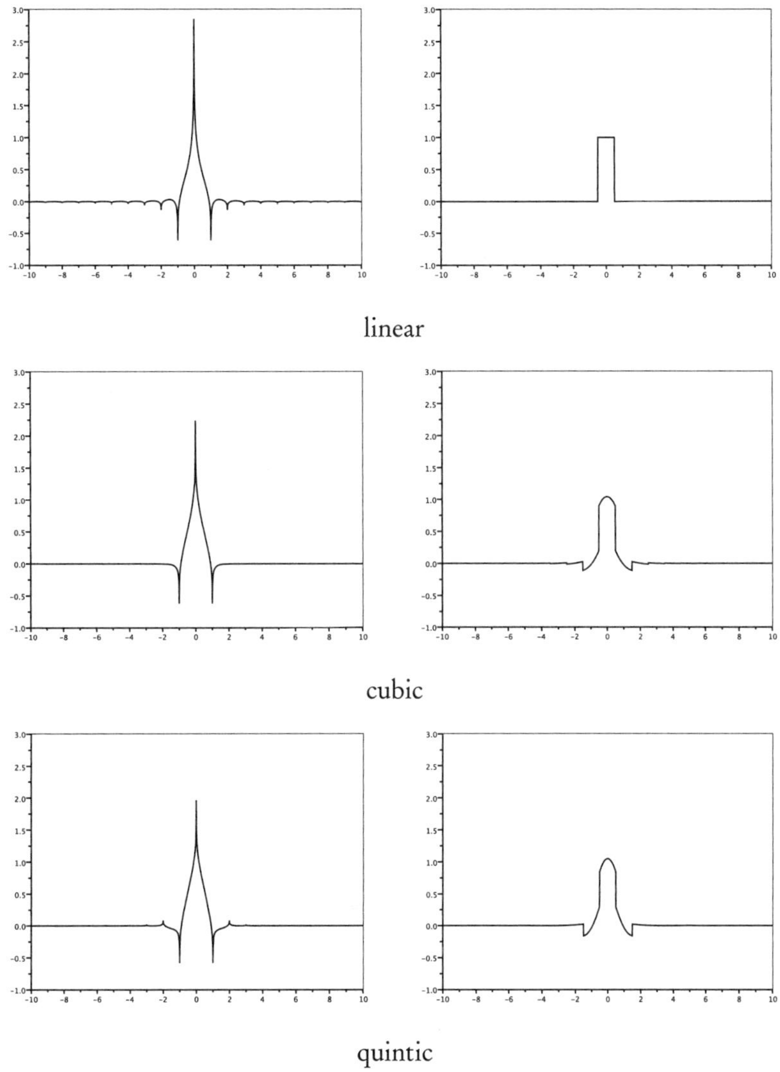

Figure 6.6. *Equivalent LG reference basis functions obtained via* (6.25) *(left) and* (6.75) *(right) for the linear (upper), cubic (middle), and quintic (lower) reconstructions* [Fe13a].

moreover, that (5.58), (5.59), (6.5) *hold and that the function* ϕ, *solution of* (6.23), *satisfies* (6.50)–(6.54). *Then, for any* $j \in \mathbb{Z}$ *and* $n \in [1, T/\Delta t]$,

$$\left| v_j^n - u(x_j, t_n) \right| \to 0 \tag{6.78}$$

as $\Delta \to 0$.

Moreover, if $u \in L^\infty([0,T], W^{s,\infty}(\mathbb{R}))$, *then*

$$\|V^n - U(t_n)\|_2 \le C\left(\Delta t^p + \min\left(\Delta x^{\min(s-1,r)} \frac{\Delta x^{\min(s,r+1)}}{\Delta t} \right)\right). \tag{6.79}$$

We also discuss here a modified equation and the dispersive behavior of the SL scheme, along with the interactions between discretization steps at the level of tuning of the scheme.

Numerical dispersion

Repeating the analysis of Chapter 5, we consider again the equation in the form (6.2) and a scheme in the form (6.12), and we assume v to be a regular extension of the numerical solution of the scheme. If we replace the error estimate for $\mathbb{P}_1$ interpolation with the complete estimate for Lagrange interpolation of a generic order r, we can express the value $I_r[V^n](x_j - c\Delta t)$ as

$$I_r[V^n](x_j - c\Delta t) = v(x_j - c\Delta t, t_n) - \frac{\prod_{x_k \in \mathcal{S}} (x_j - c\Delta t - x_k)}{(r+1)!} \frac{\partial^{r+1}}{\partial x^{r+1}} v(\xi_j),$$

where $\mathcal{S}$ is the stencil of points involved in the reconstruction, and ξ_j is an unknown point located between the extreme points of $\mathcal{S}$. Next, retracing the computations of the first-order case, we obtain from (6.12)

$$v(x_j, t_{n+1}) = v(x_j - c\Delta t, t_n) + \nu(\Delta) \left[\frac{\partial^{r+1}}{\partial x^{r+1}} v(x_j - c\Delta t) + O(\Delta x) \right],$$

which is in the form (4.86), with $k = r+1$, once we set the dispersivity coefficient ν as

$$\nu(\Delta) = -\frac{\prod_{x_k \in \mathcal{S}} (x_j - c\Delta t - x_k)}{(r+1)!\,\Delta t}$$

(note that $\operatorname{sgn}(\nu(\Delta)) = (-1)^{\frac{r-1}{2}}$ and that all terms of the product are $O(\Delta x)$). We obtain therefore a modified equation in the form (4.85), and more precisely

$$v_t + c v_x = (-1)^{\frac{r-1}{2}} c_r \frac{\Delta x^{r+1}}{\Delta t} \frac{\partial^{r+1}}{\partial x^{r+1}} v + o\left(\frac{\Delta x^{r+1}}{\Delta t} \right), \tag{6.80}$$

with c_r a bounded nonnegative function of λ'. Then, the dispersivity of the scheme can be analyzed by applying standard self-similarity arguments to the modified equation. This procedure shows that the resolution of discontinuities is of the order of $O(\Delta x / \Delta t^{1/(r+1)})$.

Interplay between discretization steps

As it has been noticed treating the CIR scheme, the convergence estimate (6.79) shows an interaction between discretization steps, which requires relating Δx and Δt in a less trivial form with respect to usual difference schemes. Once we express Δt as a function of Δx in the form $\Delta t = \Delta x^\alpha$, we briefly discuss the optimization of the convergence rate in (6.79), as well as the resolution of discontinuities, with respect to the parameter α.

Maximization of the convergence rate Using the relationship between the two steps in (6.79), we obtain

$$\|V^n - U(t_n)\|_2 \le C\left(\Delta x^{\alpha p} + \Delta x^{\min(s,r+1)-\alpha} \right)$$

(we are assuming here that the Courant number does not vanish, but the adaptation of these arguments to the complete estimate is straightforward and gives similar results).

Since the first exponent increases with α, whereas the second decreases, the convergence rate is maximized when they coincide, that is, for

$$\alpha = \frac{\min(s, r+1)}{p+1}. \tag{6.81}$$

Note that the $\Delta t/\Delta x$ relationship depends on the regularity. If the solution is smooth (more precisely, if $s > r+1$), then we obtain

$$\alpha = \frac{r+1}{p+1},$$

a choice maximizing the consistency rate. Note that if $r > p$ (e.g., coupling a second-order time integration with a cubic reconstruction), the relationship optimizing the consistency rate entails a vanishing Courant number. In all cases, α decreases as the regularity decreases, which means that less regular solutions are better approximated with larger time steps.

Resolution of singularities In analyzing the behavior of the scheme on discontinuous solutions, still two opposite effects appear: on one hand, large Courant numbers cause a lesser number of reconstructions to be performed (this causing a smaller numerical dispersion); on the other hand, they cause larger errors in computing characteristics (this causing a less precise location of singularities).

Assuming as before that $\Delta t = \Delta x^\alpha$, the analysis of the modified equation shows that the dispersion of discontinuities is of the order of $\Delta x^{1-\alpha/(r+1)}$. Hence, taking into account that characteristics are computed with an error of $\Delta x^{\alpha p}$, it turns out that the resolution of the scheme is maximized when the two orders coincide, that is, with the choice

$$\alpha = \frac{r+1}{p(r+1)+1}. \tag{6.82}$$

Note that this relationship implies $\alpha < 1/p$, so that $\Delta x = o(\Delta t^p)$, and in practice $\alpha \approx 1/p$ (for example, for a second-order time and space discretization $p = r = 2$, and we would obtain $\alpha = 3/7$). In general, this choice could lead to relatively large time discretization errors. However, it confirms the indication that high Courant numbers are more suitable for treating nonsmooth solutions.

6.1.5 ▪ Examples

In order to compare the performances of the various schemes, we consider again the simple one-dimensional problem (5.69), which is rewritten here as

$$\begin{cases} u_t(x,t) - (x - \overline{x})u_x(x,t) = 0, & (x,t) \in (0,1) \times (0,1), \\ u(x,0) = u_0(x). \end{cases} \tag{6.83}$$

As in Chapter 5, we have set $\overline{x} = 1.1$ and used three initial conditions u_0 of different regularities with bounded support. The first is the characteristic function (5.70), the second is a $W^{1,\infty}$ function in the form

$$u_0(x) = \max(\sin(2\pi x), 0), \tag{6.84}$$

and the third is in $W^{2,\infty}$:

$$u_0(x) = \max(\sin(2\pi x), 0)^2. \tag{6.85}$$

Note that, in addition to what has been remarked in Chapter 5, the initial conditions have been changed to have a nonpolynomial structure, so as to evaluate the efficiency of the reconstruction procedure.

The test has been carried out combining a second-order Runge–Kutta scheme for following characteristics with $\mathbb{P}_2$ finite element, cubic, and third-order ENO reconstructions. $\Delta t/\Delta x$ relationships in the form $\Delta t = \Delta x^\alpha$ have been considered for three different values of α: the first, $\alpha = 1/2$, is close to the value optimizing the resolution of singularities, the second is the classical linear relationship $\alpha = 1$ (more precisely, $\Delta t = 5\Delta x$ following the choice made for the monotone case), and the third is the value of α optimizing the consistency rate (in the $\mathbb{P}_2$ case, this criterion gives again $\alpha = 1$). The ENO scheme has been tested on the most efficient relationship used in the cubic case.

We list in Table 6.1 numerical errors and overall convergence rates in the 2-norm for the various schemes. Figure 6.7 compares the schemes on the advection of the characteristic function (5.70). Here, the first three figures are obtained with $\alpha = 1/2$ which (nearly) optimizes the resolution of discontinuities, whereas the fourth is obtained with $\alpha = 4/3$ which optimizes the consistency rate for the cubic reconstruction.

As a first general remark, note that the error tables confirm the general indication that less regular solutions are better approximated with large Courant numbers, i.e., with lower values of α. In cubic approximation, the relationship maximizing the consistency rate proves to be relatively inefficient, partly because it would require a smoother solution (in $W^{4,\infty}$) to achieve its best performance. This is also apparent from Figure 6.7, which shows the remarkable increase in numerical dispersion caused by the choice $\alpha = 4/3$. Note also that the difference in convergence rates should be considered against complexity. Under the relationships shown in the tables, a test with 100 nodes requires 10 time steps with $\alpha = 1/2$, 20 with $\alpha = 1$, and 464 with $\alpha = 4/3$. In terms of absolute error, the best performances are obtained by the $\mathbb{P}_2$ scheme, but we point out once more that the cubic scheme would require a smoother solution to increase its convergence rate.

6.2 ▪ SL schemes for the convex HJ equation

In order to illustrate the convergence theory of high-order SL schemes for HJ equations, we go back to our one-dimensional model problem:

$$\begin{cases} u_t(x,t) + H(u_x(x,t)) = 0, & (x,t) \in \mathbb{R} \times [0,T], \\ u(x,0) = u_0(x), & x \in \mathbb{R}. \end{cases} \tag{6.86}$$

Throughout this section, we will make the standing assumption that H is a strictly convex function, and more precisely

$$H''(p) \geq m_H. \tag{6.87}$$

We recall that this implies that if u_0 is a Lipschitz continuous function, then $u(x,t)$ is semiconcave for all $t > 0$, and if u_0 is semiconcave itself, then $u(x,t)$ is uniformly semiconcave for $t \geq 0$. Moreover, in terms of the Legendre transform, condition (6.87) implies that

$$H^{*''}(\alpha) \leq \frac{1}{m_H}. \tag{6.88}$$

6.2.1 ▪ Construction of the scheme

As it has been shown for the linear case, turning to high order the first-order SL scheme (5.95) requires only replacing the monotone $\mathbb{P}_1$ interpolation I_1 with a reconstruction of

Table 6.1. *Errors in the 2-norm for the $\mathbb{P}_2$, cubic, and third-order ENO schemes, with a second-order time discretization.*

L^∞ solution						
	$\mathbb{P}_2$		Cubic			ENO3
n_n	$\alpha=1/2$	$\alpha=1$	$\alpha=1/2$	$\alpha=1$	$\alpha=4/3$	$\alpha=1/2$
25	$1.44\cdot10^{-1}$	$1.44\cdot10^{-1}$	$1.43\cdot10^{-1}$	$1.43\cdot10^{-1}$	$2.35\cdot10^{-1}$	$2.1\cdot10^{-1}$
50	$6.82\cdot10^{-2}$	$5.96\cdot10^{-2}$	$1.01\cdot10^{-1}$	$1.04\cdot10^{-1}$	$1.29\cdot10^{-1}$	$1.06\cdot10^{-1}$
100	$5.38\cdot10^{-2}$	$5.52\cdot10^{-2}$	$6.6\cdot10^{-2}$	$7.74\cdot10^{-2}$	$9.39\cdot10^{-2}$	$8.22\cdot10^{-2}$
200	$4.09\cdot10^{-2}$	$4.09\cdot10^{-2}$	$5.36\cdot10^{-2}$	$6.06\cdot10^{-2}$	$7.65\cdot10^{-2}$	$5.94\cdot10^{-2}$
400	$2.57\cdot10^{-2}$	$3.56\cdot10^{-2}$	$3.72\cdot10^{-2}$	$4.63\cdot10^{-2}$	$5.83\cdot10^{-2}$	$4.39\cdot10^{-2}$
Rate	0.62	0.50	0.49	0.41	0.50	0.56

$W^{1,\infty}$ solution						
	$\mathbb{P}_2$		Cubic			ENO3
n_n	$\alpha=1/2$	$\alpha=1$	$\alpha=1/2$	$\alpha=1$	$\alpha=4/3$	$\alpha=1/2$
25	$2.04\cdot10^{-2}$	$2.04\cdot10^{-2}$	$8.37\cdot10^{-2}$	$8.37\cdot10^{-2}$	$1.30\cdot10^{-1}$	$8.94\cdot10^{-2}$
50	$9.92\cdot10^{-3}$	$9.62\cdot10^{-3}$	$2.03\cdot10^{-2}$	$2.44\cdot10^{-2}$	$4.99\cdot10^{-2}$	$2.58\cdot10^{-2}$
100	$3.44\cdot10^{-3}$	$3.15\cdot10^{-3}$	$7.21\cdot10^{-3}$	$9.11\cdot10^{-3}$	$1.83\cdot10^{-2}$	$1.01\cdot10^{-2}$
200	$1.82\cdot10^{-3}$	$1.98\cdot10^{-3}$	$3.26\cdot10^{-3}$	$4.39\cdot10^{-3}$	$8.18\cdot10^{-3}$	$4.47\cdot10^{-3}$
400	$7.96\cdot10^{-4}$	$7.92\cdot10^{-4}$	$1.29\cdot10^{-3}$	$1.87\cdot10^{-3}$	$3.59\cdot10^{-3}$	$1.86\cdot10^{-3}$
Rate	1.17	1.17	1.50	1.37	1.29	1.40

$W^{2,\infty}$ solution						
	$\mathbb{P}_2$		Cubic			ENO3
n_n	$\alpha=1/2$	$\alpha=1$	$\alpha=1/2$	$\alpha=1$	$\alpha=4/3$	$\alpha=1$
25	$1.80\cdot10^{-2}$	$1.80\cdot10^{-2}$	$1.06\cdot10^{-1}$	$1.06\cdot10^{-1}$	$1.40\cdot10^{-1}$	$1.06\cdot10^{-1}$
50	$5.76\cdot10^{-3}$	$3.70\cdot10^{-3}$	$1.96\cdot10^{-2}$	$2.01\cdot10^{-2}$	$6.47\cdot10^{-2}$	$2.56\cdot10^{-2}$
100	$2.72\cdot10^{-3}$	$1.19\cdot10^{-3}$	$3.84\cdot10^{-3}$	$4.32\cdot10^{-3}$	$1.65\cdot10^{-2}$	$5.99\cdot10^{-3}$
200	$1.35\cdot10^{-3}$	$2.96\cdot10^{-4}$	$1.41\cdot10^{-3}$	$9.83\cdot10^{-4}$	$3.38\cdot10^{-3}$	$1.64\cdot10^{-3}$
400	$6.63\cdot10^{-4}$	$8.42\cdot10^{-5}$	$6.67\cdot10^{-4}$	$2.49\cdot10^{-4}$	$7.65\cdot10^{-4}$	$4.73\cdot10^{-4}$
Rate	1.19	1.93	1.83	2.18	1.88	1.95

degree $r > 1$:

$$\begin{cases} v_j^{n+1} = \min_{\alpha\in\mathbb{R}}\left[\Delta t H^*(\alpha) + I_r[V^n](x_j - \alpha\Delta t)\right], \\ v_j^0 = u_0(x_j). \end{cases} \tag{6.89}$$

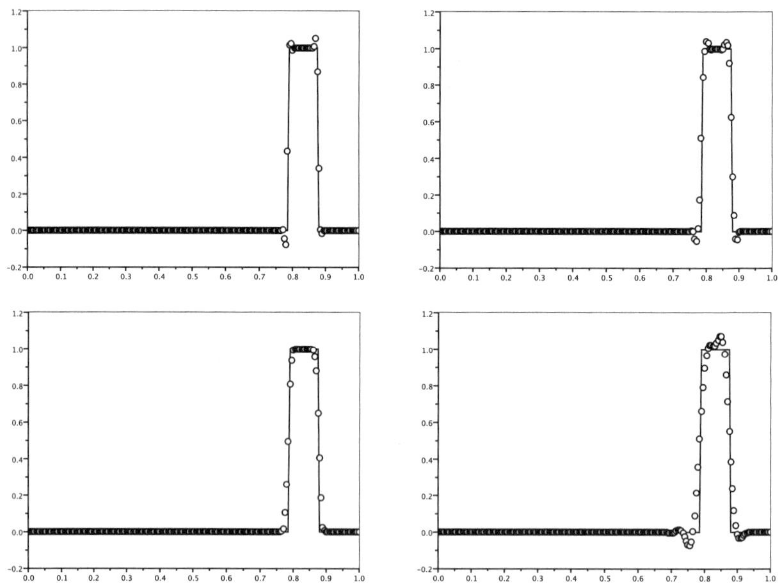

Figure 6.7. *Numerical results for the advection of a characteristic function obtained via the* $\mathbb{P}_2$ *(upper left), cubic (upper right), third-order ENO (lower left), and cubic with maximum consistency rate (lower right) schemes, with a second-order time discretization and* 200 *nodes.*

In this section, we will develop a convergence theory which allows one to treat the cases of Lagrange, finite element, ENO, and WENO reconstructions. Strictly speaking, the scheme (6.89) is not monotone, but in fact the Barles–Souganidis theorem requires a weaker condition, that is, that (6.89) should be monotone up to a term $o(\Delta t)$. As we will show, this generalized form of monotonicity can occur under increasing Courant numbers if the scheme is Lipschitz stable, so this will be the main stability issue.

6.2.2 ▪ Consistency

Consistency analysis clearly follows the same lines of the low-order scheme, once the $\mathbb{P}_1$ reconstruction is replaced by a high-order version and the definition of S^{SL} is changed accordingly. We can repeat the steps of the corresponding estimate, except for using (6.5) in place of (5.44), obtaining at last for the high-order SL scheme (6.89) the consistency estimate

$$\left\| L^{SL}(\Delta; t, U(t)) \right\| \leq C \frac{\Delta x^{r+1}}{\Delta t}. \tag{6.90}$$

Clearly, the consistency estimate can be adapted to the case of vanishing Courant numbers, following the guidelines of the linear problem. However, as we will see, the case of small time steps seems unsuitable to prove convergence and will therefore be dropped.

Remark 6.13. *As it has already been noticed about the first-order SL scheme, since characteristics are straight lines, the bound* (6.90) *does not take into account the accuracy of time discretization. This is related to the specific structure of the HJ equation* (6.86), *but if a transport term appears in the equation and is discretized with order* p, *then the consistency estimate* *coincides with the one obtained in the linear case, that is,*

$$\left\| L^{SL}(\Delta; t, U(t)) \right\| \le C \left(\Delta t^p + \frac{\Delta x^{r+1}}{\Delta t} \right).$$

6.2.3 ▪ Lipschitz stability

In order to be as general as possible, we start with some basic assumption on the reconstruction operator. First, we define the stencil of reconstruction as

$$\mathscr{S}(x) = (x - h_- \Delta x, x + h_+ \Delta x), \tag{6.91}$$

so that the reconstruction at the point x uses the nodes inside $\mathscr{S}(x)$. For example, a quadratic Lagrange reconstruction can be performed taking one node on the left and two nodes on the right of the point x (in this case, $h_- = 1$, $h_+ = 2$), or two nodes on the left and one on the right (and in this case, $h_- = 2$, $h_+ = 1$). In a second-order finite element or ENO/WENO reconstruction, both cases are possible, and thus $h_- = h_+ = 2$. A third-order Lagrange reconstruction is performed using two nodes on the left and two on the right, so that $h_- = h_+ = 2$. In the third-order finite element or ENO/WENO case, $h_- = h_+ = 3$, and so forth.

Then, we assume that, given a Lipschitz continuous function $v(x)$ and the sequence $V = \{v_j\}_{j\in\mathscr{I}} = \{v(x_j)\}_{j\in\mathscr{I}}$, the operator $I_r[V]$ satisfies

$$|I_r[V](x) - I_1[V](x)| \le C_r \max_{x_{j-1}, x_j, x_{j+1} \in \mathscr{S}(x)} |v_{j+1} - 2v_j + v_{j-1}| \qquad (C_r < 1). \tag{6.92}$$

The key point of assumption (6.92) consists in requiring that $C_r < 1$. In fact, as it is immediate to check, any polynomial reconstruction satisfies (6.92) with a constant depending on the interpolation degree r. Besides a technical reason which will become clear in the proof of Lemma 6.14, the bound on C_r amounts to assuming that I_r is not "too sensitive" with respect to large second increments which can occur on singularities of the solution.

It will also be useful to define the function

$$F_j(\alpha) := \Delta t H^*(\alpha) + I_r[V^n](x_j - \alpha \Delta t) \tag{6.93}$$

and denote again by $\overline{\alpha}_j$ the value of α achieving the minimum in (6.89) and in F_j. In order to stress the "locality" in the above definition, the notation explicitly shows the dependence of F on the node index j, although for simplicity the dependence on the time step n has been dropped.

We give now a bound on the second increment of the numerical solution. The bound is globally one-sided, but becomes two-sided at the foot of characteristics, that is, in a neighborhood $\overline{U}(x_j + \overline{\alpha}_j \Delta t)$, with $\overline{U}$ defined as

$$\overline{U}(x) = (x - h\Delta x, x + h\Delta x), \tag{6.94}$$

with a fixed $h > \max(h_+, h_-)$, so that $\mathscr{S}(x) \subset \overline{U}(x)$. More precisely, we have the following technical result.

Lemma 6.14. *Let V^n be defined by the scheme* (6.89). *If* (6.88) *holds, then, for any $j \in \mathbb{Z}$ and $n \ge 1$,*

$$v_{j+1}^n - 2v_j^n + v_{j-1}^n \le \frac{\Delta x^2}{m_H \Delta t}. \tag{6.95}$$

Moreover, assuming, in addition, that (6.92) *holds, then*

$$\max_{x_{i-1},x_i,x_{i+1}\in\overline{U}(x_j+\overline{\alpha}_j\Delta t)} |v_{i+1}^n - 2v_i^n + v_{i-1}^n| \le C\frac{\Delta x^2}{\Delta t} \tag{6.96}$$

with $\overline{U}$ *defined by* (6.94) *and for some positive constant* C *depending on* C_r, h, *and* m_H.

Proof. We start by proving (6.95). By (6.89) we have, for $n \ge 1$,

$$v_j^n = \Delta t H^*(\overline{\alpha}_j) + I_r\left[V^{n-1}\right](x_j - \overline{\alpha}_j\Delta t),$$

$$\begin{aligned} v_{j-1}^n &= \Delta t H^*(\overline{\alpha}_{j-1}) + I_r\left[V^{n-1}\right](x_{j-1} - \overline{\alpha}_{j-1}\Delta t) \\ &\le \Delta t H^*\left(\overline{\alpha}_j - \frac{\Delta x}{\Delta t}\right) + I_r\left[V^{n-1}\right](x_j - \overline{\alpha}_j\Delta t), \end{aligned}$$

$$\begin{aligned} v_{j+1}^n &= \Delta t H^*(\overline{\alpha}_{j+1}) + I_r\left[V^{n-1}\right](x_{j+1} - \overline{\alpha}_{j+1}\Delta t) \\ &\le \Delta t H^*\left(\overline{\alpha}_j + \frac{\Delta x}{\Delta t}\right) + I_r\left[V^{n-1}\right](x_j - \overline{\alpha}_j\Delta t) \end{aligned}$$

(where the inequalities come from the use of $\overline{\alpha}_j \pm \Delta x/\Delta t$ instead of the actual minimizers $\overline{\alpha}_{j\pm1}$), so that

$$\begin{aligned} v_{j+1}^n - 2v_j^n + v_{j-1}^n &\le \Delta t\left[H^*\left(\overline{\alpha}_j + \frac{\Delta x}{\Delta t}\right) - 2H^*(\overline{\alpha}_j) + H^*\left(\overline{\alpha}_j - \frac{\Delta x}{\Delta t}\right)\right] \\ &\le \Delta t\left(\frac{\Delta x}{\Delta t}\right)^2 \sup H^{*''}, \end{aligned} \tag{6.97}$$

which, using (6.88), gives (6.95).

In order to prove (6.96), we take into consideration the values of $F_j(\alpha)$ for $x_j - \alpha\Delta t$ coinciding with a grid node, that is, for $\alpha = k\Delta x/\Delta t$. Let k_j be defined as

$$k_j = \operatorname{sgn}(\overline{\alpha}_j)\left\lfloor |\overline{\alpha}_j|\frac{\Delta t}{\Delta x}\right\rfloor$$

so that

$$\begin{cases} x_j - \overline{\alpha}_j\Delta t \in [x_{j-k_j}, x_{j-k_j+1}) & \text{if } \overline{\alpha}_j \le 0, \\ x_j - \overline{\alpha}_j\Delta t \in (x_{j-k_j-1}, x_{j-k_j}] & \text{if } \overline{\alpha}_j \ge 0. \end{cases}$$

Since the two cases are perfectly symmetric, we examine in detail the second, $\overline{\alpha}_j \ge 0$. In this case, the point $x_j - \overline{\alpha}_j\Delta t$ is on the left of x_j, and more precisely

$$x_{j-k_j-1} < x_j - \overline{\alpha}_j\Delta t \le x_{j-k_j} \le x_j.$$

Moreover, the minimizer $\overline{\alpha}_j$ satisfies

$$k_j\frac{\Delta x}{\Delta t} \le \overline{\alpha}_j < (k_j+1)\frac{\Delta x}{\Delta t}.$$

We will denote by δ_j the (signed) maximal second increment of the numerical solution in the neighborhood $\overline{U}(x_j - \overline{\alpha}_j \Delta t)$ and by $j - k_j + l_j$ the index of the node at which this second increment occurs. If $\delta_j \geq 0$, then (6.96) follows from (6.95) and we have nothing else to prove. We will assume, therefore, that $\delta_j < 0$, so that

$$\begin{aligned}\delta_j &= - \max_{x_{i-1},x_i,x_{i+1} \in \overline{U}(x_j - \overline{\alpha}_j \Delta t)} |v_{i+1}^n - 2v_i^n + v_{i-1}^n| \\ &= v_{j-k_j+l_j+1}^n - 2v_{j-k_j+l_j}^n + v_{j-k_j+l_j-1}^n.\end{aligned} \tag{6.98}$$

We recall that $|l_j| < h$ and will also assume in what follows that $l_j \geq 0$, the reverse case being similar to prove.

Step 1. Upper bound of the second increment of the function F_j. We have

$$\begin{aligned}F_j&\left((k+1)\frac{\Delta x}{\Delta t}\right) - 2F_j\left(k\frac{\Delta x}{\Delta t}\right) + F_j\left((k-1)\frac{\Delta x}{\Delta t}\right) \\ &= \Delta t\left[H^*\left((k+1)\frac{\Delta x}{\Delta t}\right) - 2H^*\left(k\frac{\Delta x}{\Delta t}\right) + H^*\left((k-1)\frac{\Delta x}{\Delta t}\right)\right] \\ &\quad + v_{j-k-1}^n - 2v_{j-k}^n + v_{j-k+1}^n \\ &\leq \frac{\Delta x^2}{m_H \Delta t} + v_{j-k-1}^n - 2v_{j-k}^n + v_{j-k+1}^n \leq \frac{2\Delta x^2}{m_H \Delta t},\end{aligned} \tag{6.99}$$

where the second increment of H^* has been estimated as in (6.97), and the last inequality holds for $n \geq 1$.

Step 2. In the second step, we prove that

$$F_j(\overline{\alpha}_j) \geq \min\left[F_j\left(k_j\frac{\Delta x}{\Delta t}\right), F_j\left((k_j+1)\frac{\Delta x}{\Delta t}\right)\right] - \frac{\Delta x^2}{8m_H \Delta t} - C_r|\delta_j|. \tag{6.100}$$

In fact, we set $\overline{\alpha}_j = (1-\overline{\theta})k_j \Delta x/\Delta t + \overline{\theta}(k_j+1)\Delta x/\Delta t$, with $\overline{\theta} \in [0,1]$. Taking into account the uniform convexity of H^* implied by (6.88), we have

$$\begin{aligned}H^*(\overline{\alpha}_j) &\geq (1-\overline{\theta})H^*\left(k_j\frac{\Delta x}{\Delta t}\right) + \overline{\theta}H^*\left((k_j+1)\frac{\Delta x}{\Delta t}\right) \\ &\quad - \frac{1}{2}\overline{\theta}(1-\overline{\theta})\left(\frac{\Delta x}{\Delta t}\right)^2 \sup H^{*''} \\ &\geq (1-\overline{\theta})H^*\left(k_j\frac{\Delta x}{\Delta t}\right) + \overline{\theta}H^*\left((k_j+1)\frac{\Delta x}{\Delta t}\right) \\ &\quad - \frac{\Delta x^2}{8m_H \Delta t^2}\end{aligned} \tag{6.101}$$

in which the third term has been minimized with respect to $\overline{\theta}$.

In a similar way, we also get, using (6.92) and the inclusion of $\mathscr{S}(x)$ in $\overline{U}(x)$,

$$
\begin{aligned}
I_r[V^n](x_j-\overline{\alpha}_j\Delta t) &\geq I_1[V^n](x_j-\overline{\alpha}_j\Delta t) \\
&\qquad - C_r \max_{x_{i-1},x_i,x_{i+1}\in\mathscr{S}(x_j-\overline{\alpha}_j\Delta t)} |v_{i+1}^n-2v_i^n+v_{i-1}^n| \\
&\geq I_1[V^n](x_j-\overline{\alpha}_j\Delta t) \\
&\qquad - C_r \max_{x_{i-1},x_i,x_{i+1}\in\overline{U}(x_j-\overline{\alpha}_j\Delta t)} |v_{i+1}^n-2v_i^n+v_{i-1}^n| \\
&= (1-\overline{\theta})v_{j-k_j}^n+\overline{\theta}v_{j-k_j-1}^n - C_r|\delta_j| \\
&= (1-\overline{\theta})I_r[V^n]\left(x_j-k_j\frac{\Delta x}{\Delta t}\Delta t\right) \\
&\qquad +\overline{\theta}I_r[V^n]\left(x_j-(k_j+1)\frac{\Delta x}{\Delta t}\Delta t\right)-C_r|\delta_j|,
\end{aligned}
$$

which, combined with (6.101), gives

$$
F_j(\overline{\alpha}_j)\geq(1-\overline{\theta})F_j\left(k_j\frac{\Delta x}{\Delta t}\right)+\overline{\theta}F_j\left((k_j+1)\frac{\Delta x}{\Delta t}\right)-\frac{\Delta x^2}{8m_H\Delta t}-C_r|\delta_j|,
$$

which in turn implies (6.100).

Step 3. Upper bound of the increment of F_j between $k_j\Delta x/\Delta t$ and $(k_j+1)\Delta x/\Delta t$. Assume first that $F_j(k_j\Delta x/\Delta t)<F_j((k_j+1)\Delta x/\Delta t)$. Using (6.100) and the optimality of $\overline{\alpha}_j$, we obtain

$$
F_j\left((k_j-1)\frac{\Delta x}{\Delta t}\right)\geq F_j(\overline{\alpha}_j)\geq F_j\left(k_j\frac{\Delta x}{\Delta t}\right)-\frac{\Delta x^2}{8m_H\Delta t}-C_r|\delta_j|,
$$

that is, considering the extreme terms

$$
F_j\left(k_j\frac{\Delta x}{\Delta t}\right)-F_j\left((k_j-1)\frac{\Delta x}{\Delta t}\right)\leq\frac{\Delta x^2}{8m_H\Delta t}+C_r|\delta_j|.
$$

On the other hand, from (6.99) written with $k=k_j$, we also have

$$
F_j\left((k_j+1)\frac{\Delta x}{\Delta t}\right)-F_j\left(k_j\frac{\Delta x}{\Delta t}\right)\leq F_j\left(k_j\frac{\Delta x}{\Delta t}\right)-F_j\left((k_j-1)\frac{\Delta x}{\Delta t}\right)+\frac{2\Delta x^2}{m_H\Delta t},
$$

and therefore, combining the last two inequalities, we get the desired bound on the first increment:

$$
F_j\left((k_j+1)\frac{\Delta x}{\Delta t}\right)-F_j\left(k_j\frac{\Delta x}{\Delta t}\right)\leq\frac{17\Delta x^2}{8m_H\Delta t}+C_r|\delta_j|. \tag{6.102}
$$

Moreover, if $F_j((k_j+1)\Delta x/\Delta t)<F_j(k_j\Delta x/\Delta t)$, then

$$
F_j\left((k_j+1)\frac{\Delta x}{\Delta t}\right)-F_j\left(k_j\frac{\Delta x}{\Delta t}\right)\leq 0
$$

and (6.102) is also trivially satisfied.

Step 4. We show that, in order for $\overline{\alpha}_j$ to be a minimizer for F_j, δ_j must satisfy (6.96). To this end, we first assume that $F_j(k_j\Delta x/\Delta t) \le F_j((k_j+1)\Delta x/\Delta t)$ and bound the values $F_j(k\Delta x/\Delta t)$ from above using a function $\widetilde{F}_j(k\Delta x/\Delta t)$ constructed so as to coincide with F_j at $k_j\Delta x/\Delta t$ and to have first and second increments greater than or equal to the corresponding increments of F_j. Taking into account the bounds on the first and second increments of F_j obtained in the previous steps of the proof, we could define $\widetilde{F}_j$ so that its first increment between $k_j\Delta x/\Delta t$ and $(k_j+1)\Delta x/\Delta t$ would be given by the right-hand side of (6.102), and the second increment would be obtained in view of (6.98), (6.99) as

$$\widetilde{F}_j\left((k+1)\frac{\Delta x}{\Delta t}\right) - 2\widetilde{F}_j\left(k\frac{\Delta x}{\Delta t}\right) + \widetilde{F}_j\left((k-1)\frac{\Delta x}{\Delta t}\right) = \begin{cases} \dfrac{2\Delta x^2}{m_H \Delta t} & \text{if } \; k \ne k_j + l_j, \\[2ex] \dfrac{\Delta x^2}{m_H\Delta t} + \delta_j & \text{if } \; k = k_j + l_j. \end{cases} \tag{6.103}$$

We recall that, as it is easy to prove by induction, if a sequence f_i has a constant second increment,

$$f_{i+2} - 2f_{i+1} + f_i \equiv d,$$

then the values of the elements and of the first increments are given by

$$\begin{aligned} f_{k+l} &= f_k + l(f_{k+1} - f_k) + (1 + 2 + \cdots + (l-1))d, \\ f_{k+l+1} - f_{k+l} &= (f_{k+1} - f_k) + ld. \end{aligned}$$

Using the previous equalities, a function $\widetilde{F}_j$ suitable for our purpose could be defined more explicitly as

$$\begin{aligned}
\widetilde{F}_j\left(k_j\frac{\Delta x}{\Delta t}\right) &= F_j\left(k_j\frac{\Delta x}{\Delta t}\right), \\
\widetilde{F}_j\left((k_j+1)\frac{\Delta x}{\Delta t}\right) &= F_j\left(k_j\frac{\Delta x}{\Delta t}\right) + \frac{17\Delta x^2}{8m_H\Delta t} + C_r|\delta_j|, \\
\widetilde{F}_j\left((k_j+2)\frac{\Delta x}{\Delta t}\right) &= F_j\left(k_j\frac{\Delta x}{\Delta t}\right) + 2\left(\frac{17\Delta x^2}{8m_H\Delta t} + C_r|\delta_j|\right) + \frac{2\Delta x^2}{m_H\Delta t}, \\
&\;\;\vdots \\
\widetilde{F}_j\left((k_j+l_j)\frac{\Delta x}{\Delta t}\right) &= F_j\left(k_j\frac{\Delta x}{\Delta t}\right) + l_j\left(\frac{17\Delta x^2}{8m_H\Delta t} + C_r|\delta_j|\right) \\
&\quad + (1 + \cdots + (l_j - 1))\frac{2\Delta x^2}{m_H\Delta t}, \\
\widetilde{F}_j\left((k_j+l_j+1)\frac{\Delta x}{\Delta t}\right) &= F_j\left(k_j\frac{\Delta x}{\Delta t}\right) + l_j\left(\frac{17\Delta x^2}{8m_H\Delta t} + C_r|\delta_j|\right) \\
&\quad + (1 + \cdots + (l_j - 1))\frac{2\Delta x^2}{m_H\Delta t} \\
&\quad + \left(\frac{17\Delta x^2}{8m_H\Delta t} + C_r|\delta_j|\right) + l_j\frac{2\Delta x^2}{m_H\Delta t} + \delta_j
\end{aligned}$$

(note that the computation of $\widetilde{F}_j((k_j+l_j+1)\Delta x/\Delta t)$ is "restarted" because of (6.103)), and last, for any integer $m>0$,

$$\begin{aligned}\widetilde{F}_j\left((k_j+l_j+m)\frac{\Delta x}{\Delta t}\right) &= F_j\left(k_j\frac{\Delta x}{\Delta t}\right)\\ &\quad+(l_j+m)\left(\frac{17\Delta x^2}{8m_H\Delta t}+C_r|\delta_j|\right)\\ &\quad+(1+\cdots+(l_j+m-1))\frac{2\Delta x^2}{m_H\Delta t}+m\delta_j\\ &\le F_j\left(k_j\frac{\Delta x}{\Delta t}\right)+(h+m)\left(\frac{17\Delta x^2}{8m_H\Delta t}+C_r|\delta_j|\right)\\ &\quad+(1+\cdots+(h+m-1))\frac{2\Delta x^2}{m_H\Delta t}+m\delta_j. \end{aligned} \tag{6.104}$$

On the other hand, by (6.100) and the optimality of $\overline{\alpha}_j$, we also have, for any $m>0$,

$$\begin{aligned}\widetilde{F}_j\left((k_j+l_j+m)\frac{\Delta x}{\Delta t}\right) &\ge F_j\left((k_j+l_j+m)\frac{\Delta x}{\Delta t}\right)\\ &\ge F_j(\overline{\alpha}_j)\\ &\ge \min\left[F_j\left(k_j\frac{\Delta x}{\Delta t}\right),F_j\left((k_j+1)\frac{\Delta x}{\Delta t}\right)\right]\\ &\quad-\frac{\Delta x^2}{8m_H\Delta t}-C_r|\delta_j|. \end{aligned} \tag{6.105}$$

We explicitly note that in (6.105) the first inequality follows from the construction of $\widetilde{F}_j$ as an upper bound for F_j, the second one from the optimality of $\overline{\alpha}_j$ in F_j, and the third one from the lower bound (6.100).

Recalling that we have assumed $F_j(k_j\Delta x/\Delta t)\le F_j((k_j+1)\Delta x/\Delta t)$ and $\delta_j<0$, and using (6.104), we obtain from the two extreme terms of (6.105)

$$\begin{aligned}F_j\left(k_j\frac{\Delta x}{\Delta t}\right)+(h+m)\left(\frac{17\Delta x^2}{8m_H\Delta t}-C_r\delta_j\right)&\\ +(1+\cdots+(h+m-1))\frac{2\Delta x^2}{m_H\Delta t}+m\delta_j &\ge F_j\left(k_j\frac{\Delta x}{\Delta t}\right)\\ &\quad-\frac{\Delta x^2}{8m_H\Delta t}+C_r\delta_j;\end{aligned}$$

that is,

$$\left[\frac{17}{8}(h+m)+2(1+\cdots+(h+m-1))+1\right]\frac{\Delta x^2}{m_H\Delta t}\ge[C_r(h+m)-m+C_r]\delta_j,$$

which gives, solving for δ_j,

$$-\frac{m^2(1+o(1))}{(1-C_r)m-C_r(h+1)}\frac{\Delta x^2}{m_H\Delta t}\le\delta_j<0 \tag{6.106}$$

in which it is easy to recognize that the supremum for $m > 0$ of the left-hand side (which is in fact a maximum) is $O(\Delta x^2/\Delta t)$, provided $C_r < 1$. Using the reverse estimate (6.95), we get at last (6.96).

If otherwise $F_j((k_j + 1)\Delta x/\Delta t) < F_j(k_j \Delta x/\Delta t)$, then it is possible to redefine the function $\widetilde{F}_j$ so that $\widetilde{F}_j((k_j+1)\Delta x/\Delta t) = F_j((k_j+1)\Delta x/\Delta t)$, and the same upper bounds are used on the first and second increments. In this way, we replace (6.104) with

$$\begin{aligned}\widetilde{F}_j\left((k_j + l_j + m)\frac{\Delta x}{\Delta t}\right) &= F_j\left((k_j + 1)\frac{\Delta x}{\Delta t}\right)\\ &\quad + (l_j + m - 1)\left(\frac{17\Delta x^2}{8m_H \Delta t} + C_r|\delta_j|\right)\\ &\quad + (1 + \cdots + (l_j + m - 1))\frac{2\Delta x^2}{m_H \Delta t} + m\delta_j,\end{aligned}$$

which again implies (6.106). This construction completely parallels the previous one and is therefore left to the reader. □

Before proving Lipschitz continuity, we derive a couple of useful consequences of Lemma 6.14. First, given a Lipschitz continuous function $v(x)$ and the sequence $V = \{v_j\}_{j\in\mathcal{J}} = \{v(x_j)\}_{j\in\mathcal{J}}$, the condition

$$|I_r[V](x) - I_1[V](x)| = O(\Delta x),$$

typical in the interpolation of a function with a bounded derivative, is written by (6.92) as

$$\begin{aligned}|I_r[V](x) - I_1[V](x)| &\leq C_r \max_{x_{i-1},x_i,x_{i+1}\in\mathcal{S}(x)} |v_{i+1} - 2v_i + v_{i-1}|\\ &\leq C_r\left(\sup|v_{i+1} - v_i| + \sup|v_{i-1} - v_i|\right)\\ &\leq 2L\Delta x.\end{aligned} \tag{6.107}$$

Also, at the foot of a characteristic the stronger condition

$$|I_r[V^n](x_i + \overline{a}_j\Delta t) - I_1[V^n](x_i + \overline{a}_j\Delta t)| \leq C_r C\frac{\Delta x^2}{\Delta t} \tag{6.108}$$

holds for any $j \in \mathbb{Z}$, $n \geq 1$ and for any node $i = j \pm 1$. In fact, (6.108) follows from (6.92) and Lemma 6.14, since $\overline{U}(x_j + \overline{a}_j\Delta t)$ contains all the nodes involved in the reconstructions $I_r[V^n](x_{j\pm1} + \overline{a}_j\Delta t)$.

We can now state the result of Lipschitz stability.

Theorem 6.15. *Let V^n be defined by the scheme* (6.89). *Assume that* (6.88), (6.92) *hold, that $\Delta x = O(\Delta t^2)$, and that u_0 is Lipschitz continuous with Lipschitz constant L_0. Then, V^n satisfies, for any i and j, the discrete Lipschitz estimate*

$$\frac{|v_i^n - v_j^n|}{|x_i - x_j|} \leq L$$

for a constant L independent of Δx and Δt, and for $0 \leq n \leq T/\Delta t$, as $\Delta t \to 0$.

Proof. It clearly suffices to prove the claim for i, j such that $i = j \pm 1$. Assume that at the previous step the discrete solution satisfies

$$\frac{|v_i^{n-1} - v_j^{n-1}|}{\Delta x} \leq L_{n-1}.$$

Making the argmin explicit and using (6.108), we have

$$\begin{aligned} v_j^n &= \Delta t H^*(\overline{\alpha}_j) + I_r\left[V^{n-1}\right](x_j + \overline{\alpha}_j \Delta t) \\ &\geq \Delta t H^*(\overline{\alpha}_j) + I_1\left[V^{n-1}\right](x_j + \overline{\alpha}_j \Delta t) - C_r C \frac{\Delta x^2}{\Delta t}. \end{aligned} \tag{6.109}$$

In order to estimate the discrete incremental ratio of V^n, we give on v_i^n the bound

$$\begin{aligned} v_i^n &= \Delta t H^*(\overline{\alpha}_i) + I_r\left[V^{n-1}\right](x_i + \overline{\alpha}_i \Delta t) \\ &\leq \Delta t H^*(\overline{\alpha}_j) + I_r\left[V^{n-1}\right](x_i + \overline{\alpha}_j \Delta t) \\ &\leq \Delta t H^*(\overline{\alpha}_j) + I_1\left[V^{n-1}\right](x_i + \overline{\alpha}_j \Delta t) + C_r C \frac{\Delta x^2}{\Delta t}, \end{aligned} \tag{6.110}$$

which results from both the optimality of $\overline{\alpha}_i$ and (6.108) and holds for any $n \geq 2$. If $n = 1$, applying (6.107) instead of (6.108), we obtain

$$\begin{aligned} v_i^1 &= \Delta t H^*(\overline{\alpha}_i) + I_r\left[V^0\right](x_i + \overline{\alpha}_i \Delta t) \\ &\leq \Delta t H^*(\overline{\alpha}_j) + I_r\left[V^0\right](x_i + \overline{\alpha}_j \Delta t) \\ &\leq \Delta t H^*(\overline{\alpha}_j) + I_1\left[V^0\right](x_i + \overline{\alpha}_j \Delta t) + 2L_0 \Delta x. \end{aligned} \tag{6.111}$$

From (6.109) and (6.110) we obtain, for $n \geq 2$, the unilateral estimate

$$\begin{aligned} \frac{v_i^n - v_j^n}{\Delta x} &\leq \frac{1}{\Delta x}\left(I_1\left[V^{n-1}\right](x_i + \overline{\alpha}_j \Delta t) - I_1\left[V^{n-1}\right](x_j + \overline{\alpha}_j \Delta t)\right) + 2C_r C \frac{\Delta x^2}{\Delta t} \\ &\leq L_{n-1} + 2C_r C \frac{\Delta x}{\Delta t}, \end{aligned} \tag{6.112}$$

in which we have used the fact that the first-order reconstruction I_1 at step $n-1$ also has Lipschitz constant L_{n-1}. Interchanging the roles of $\overline{\alpha}_j$ and $\overline{\alpha}_{j+1}$, we get the reverse estimate

$$\frac{v_j^n - v_i^n}{\Delta x} \leq L_{n-1} + 2C_r C \frac{\Delta x}{\Delta t},$$

and therefore

$$\frac{|v_j^n - v_i^n|}{\Delta x} \leq L_{n-1} + 2C_r C \frac{\Delta x}{\Delta t}. \tag{6.113}$$

A similar computation yields, for $n = 1$,

$$\frac{|v_j^1 - v_i^1|}{\Delta x} \leq L_0 + 4L_0 = 5L_0, \tag{6.114}$$

so that combining (6.113) with (6.114) and iterating back, we have

$$\begin{aligned} L_n &\leq L_{n-1} + 2C_r C \frac{\Delta x}{\Delta t} \leq \cdots \\ &\leq L_1 + \frac{T - \Delta t}{\Delta t} 2C_r C \frac{\Delta x}{\Delta t} \\ &\leq 5L_0 + 2TC_r C \frac{\Delta x}{\Delta t^2}. \end{aligned} \tag{6.115}$$

Last, it is possible to get a finite limit in (6.115), if and only if $\Delta x = O(\Delta t^2)$. □

Applications to various reconstruction operators

In order to check the applicability of this theory to reconstruction operators of practical interest, we first test assumption (6.88) on the situation of an interpolating polynomial with a stencil including the reconstruction point x (this is directly applicable to the case of finite element and ENO interpolations). Then, starting with this result, we discuss the cases of symmetric Lagrange and WENO interpolations.

Finite element and ENO interpolations Here, we make no assumptions on the structure of the stencil $\mathscr{S}(x)$, except that it must include one node on each side of x. The reconstruction is assumed to be in the Newton form

$$\begin{aligned} I_r[V](x) = {} & V[x_{j_0}] + V[x_{j_0}, x_{j_1}](x - x_{j_0}) + \cdots \\ & + V[x_{j_0}, \ldots, x_{j_r}](x - x_{j_0}) \cdots (x - x_{j_{r-1}}), \end{aligned} \tag{6.116}$$

where $x_{j_0}, \ldots, x_{j_r}$ are $r+1$ adjacent nodes so that $\max(x_{j_0}, \ldots, x_{j_r}) - \min(x_{j_0}, \ldots, x_{j_r}) = r\Delta x$ and, moreover,

$$x \in (\min(x_{j_0}, \ldots, x_{j_r}), \max(x_{j_0}, \ldots, x_{j_r})) \subset \mathscr{S}(x). \tag{6.117}$$

This definition includes both finite element interpolations, for which the reconstruction stencil is fixed (but not necessarily symmetric around x), and ENO reconstructions, for which it depends on the solution itself. The divided differences are defined, as usual, by

$$\begin{aligned} V[x_{j_0}] &= v_{j_0}, \\ &\vdots \\ V[x_{j_0}, \ldots, x_{j_k}] &= \frac{V[x_{j_1}, \ldots, x_{j_k}] - V[x_{j_0}, \ldots, x_{j_{k-1}}]}{x_{j_k} - x_{j_0}} \qquad (k = 1, \ldots, r). \end{aligned}$$

Note that, although in principle the nodes $x_{j_0}, \ldots, x_{j_k}$ neither need be adjacent nor satisfy $x \in (\min x_{j_i}, \max x_{j_i})$, it is possible to reorder the nodes so that both conditions would be satisfied.

According to its definition, we can bound a generic kth order divided difference as follows:

$$|V[x_{j_0}, \ldots, x_{j_k}]| \le \frac{2 \max |V[x_{j_i}, \ldots, x_{j_{k+i-1}}]|}{k\Delta x} \tag{6.118}$$

in which the max is performed for $x_{j_i}, \ldots, x_{j_{k+i-1}} \in \mathscr{S}(x)$. To prove (6.92), we start from the second divided difference

$$|V[x_{j_0}, x_{j_1}, x_{j_2}]| = \frac{|v_{j_0} - 2v_{j_1} + v_{j_2}|}{2\Delta x^2},$$

and, hence,

$$|V[x_{j_0},x_{j_1},x_{j_2},x_{j_3}]| \le \frac{2\max|v_{j_i}-2v_{j_{i+1}}+v_{j_{i+2}}|}{3!\Delta x^3},$$

$$\vdots$$

$$|V[x_{j_0},\dots,x_{j_r}]| \le \frac{2^{r-2}\max|v_{j_i}-2v_{j_{i+1}}+v_{j_{i+2}}|}{r!\Delta x^r}.$$

Plugging such bounds into (6.116), we get an estimate in the form (6.92); that is,

$$\begin{aligned} |I_r[V](x)-I_1[V](x)| &\le |V[x_{j_0},x_{j_1},x_{j_2}](x-x_{j_0})(x-x_{j_1})+\cdots \\ &\quad + V[x_{j_0},\dots,x_{j_r}](x-x_{j_0})\cdots(x-x_{j_{r-1}})| \\ &\le \frac{|v_{j_0}-2v_{j_1}+v_{j_2}|}{2\Delta x^2}M_2\Delta x^2+\cdots \\ &\quad + \frac{2^{r-2}\max|v_{j_i}-2v_{j_{i+1}}+v_{j_{i+2}}|}{r!\Delta x^r}M_r\Delta x^r \\ &\le \max|v_{j_i}-2v_{j_{i+1}}+v_{j_{i+2}}|\sum_{k=2}^{r}\frac{M_k 2^{k-2}}{k!}, \end{aligned} \tag{6.119}$$

where

$$\begin{aligned} M_k &:= \frac{1}{\Delta x^k}\max_{x\in(x_{j_0},x_{j_{k-1}})}|(x-x_{j_0})\cdots(x-x_{j_{k-1}})| \\ &= \max_{t\in(0,k-1)}|t(t-1)\cdots(t-k+1)|. \end{aligned}$$

It remains to check that $C_r < 1$; that is,

$$\sum_{k=2}^{r}\frac{M_k 2^{k-2}}{k!} < 1. \tag{6.120}$$

Indeed, the first values M_k may be either computed by algebraic manipulations (up to M_5) or estimated by simply plotting the polynomials $t(t-1)\cdots(t-k+1)$ on the interval $[0,k-1]$. It turns out that $M_2 = 1/4$, $M_3 = 2\sqrt{3}/9 \approx 0.3849$, $M_4 = 1$, $M_5 \approx 3.6314$, $M_6 < 17$, and so on. Accordingly, the computation of the left-hand side of (6.120) for various values of r gives $C_2 = 1/8 = 0.125$, $C_3 \approx 0.2533$, $C_4 \approx 0.42$, $C_5 \approx 0.6621$, $C_6 \approx 1.04$. We can conclude that by this technique, finite element and ENO reconstructions up to the fifth order can be proved to satisfy (6.92).

Symmetric Lagrange and WENO interpolations It is possible to extend this theory to Lagrange and WENO interpolations, which work with a symmetric reconstruction stencil, that is, $h_+ = h_-$. Assume that the interpolation $I_r[V](x)$ is obtained as a sum

$$I_r[V] = W_1(x)P_1(x)+\cdots+W_q(x)P_q(x) \tag{6.121}$$

in which the P_i are interpolating polynomials constructed on smaller stencils which include the point x and, once x is fixed, the $W_i(x)$ are coefficients of a convex combination ($W_i(x)\ge 0$, $\sum_i W_i(x) = 1$). Then, we have

$$\min(P_1(x),\dots,P_q(x)) \le I_r[V](x) \le \max(P_1(x),\dots,P_q(x)).$$

If $\deg P_i \leq 5$, all the polynomials P_i satisfy (6.92), and therefore, being satisfied by all P_i, (6.92) is also satisfied by $I_r[V]$.

Note that (6.121) is the form for both symmetric Lagrange and WENO interpolation. In the first case, the functions W_i play the role of linear weights as in (3.25), and it has been proved in Theorem 3.12 that they are positive and have unity sum. In the second case, the form of I_r is given by (3.28) and, although the nonlinear weights are no longer polynomials, they still perform a convex combination due to the positivity of the linear weights. Therefore, (6.92) is again satisfied, provided $\deg P_i \leq 5$. In both cases, this means a reconstruction of order $r \leq 9$.

6.2.4 ▪ Convergence

Last, we present the main convergence result for the scheme (6.89).

Theorem 6.16. *Let u be the solution of* (6.86), *and let V^n be defined by the scheme* (6.89). *Assume that* (6.107), (6.108) *hold, that $\Delta x = O(\Delta t^2)$, and that u_0 is Lipschitz continuous. Then,*

$$\|I_r[V^n] - U(t_n)\|_\infty \to 0$$

for $0 \leq n \leq T/\Delta t$ as $\Delta t \to 0$.

Proof. First, note that the scheme is invariant for the addition of constants and consistent. Then, retracing the proof of (5.107) and using (6.107) and Theorem 6.15, we obtain that the scheme is monotone up to a term $O(\Delta x)$ (therefore $o(\Delta t)$), and the generalized monotonicity condition (4.58)–(4.59) is satisfied. Then, convergence follows from the Barles–Souganidis theorem. □

Remark 6.17. *In this theorem, the condition $\Delta x = O(\Delta t^2)$ is required only to ensure Lipschitz stability. In fact (and numerical experiments confirm this idea) this could be an overly restrictive assumption, possibly owing to some intrinsic limitations in the technique of proof. Once uniform Lipschitz continuity of numerical solutions is ensured, the generalized monotonicity condition requires instead the weaker constraint $\Delta x = o(\Delta t)$.*

6.2.5 ▪ Examples

Uniformly semiconcave solution This test (also considered, for example, in [LT00, JP00]) concerns the HJ equation

$$\begin{cases} u_t(x,t) + \frac{1}{2}(u_x(x,t)+1)^2 = 0, \\ u(x,0) = u_0(x) = -\cos(\pi x), \end{cases} \tag{6.122}$$

on the interval $[0,2]$, with periodic boundary conditions. The solution eventually develops a singularity in the derivative, and the approximate solution is computed before the singularity, at $T = 0.8/\pi^2$, with 4 time steps, and after the singularity, at $T = 1.5/\pi^2$, with 5 time steps. Plots of exact vs. approximate solutions are shown in Figure 6.8, and L^∞ errors obtained with different schemes are compared in Table 6.2.

Note that, according to the consistency estimate (6.90), the convergence rates should be $r+1 = 4$ for the WENO3 scheme and $r+1 = 6$ for the WENO5 scheme, since the time step has been kept constant. Roughly speaking, this behavior is confirmed by

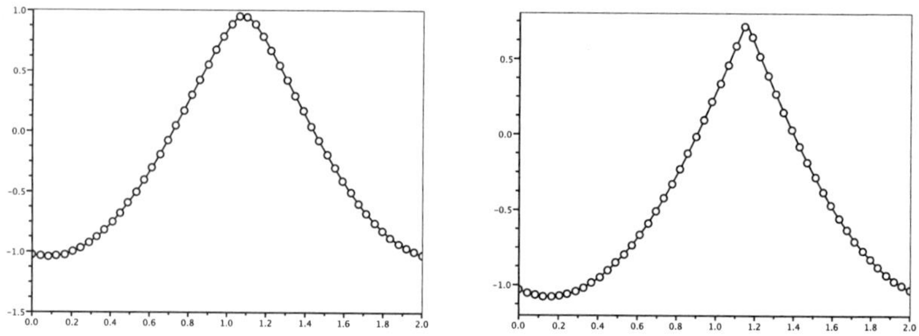

Figure 6.8. *Approximate vs. exact solution for test* (6.122)*: before singularity (left) and after singularity (right)* [CFR05].

Table 6.2. *Errors before and after the singularity for test* (6.122) *for the cubic, ENO3-5, and WENO3-5 schemes.*

Before singularity					
n_n	Cubic	ENO3	ENO5	WENO3	WENO5
25	$2.38 \cdot 10^{-3}$	$2.36 \cdot 10^{-3}$	$8.68 \cdot 10^{-4}$	$2.52 \cdot 10^{-3}$	$1.29 \cdot 10^{-3}$
50	$4.36 \cdot 10^{-5}$	$9.46 \cdot 10^{-5}$	$7.04 \cdot 10^{-6}$	$8.77 \cdot 10^{-5}$	$1.87 \cdot 10^{-5}$
100	$1.48 \cdot 10^{-5}$	$1.48 \cdot 10^{-5}$	$6.57 \cdot 10^{-7}$	$1.53 \cdot 10^{-5}$	$9.13 \cdot 10^{-7}$
200	$1.01 \cdot 10^{-6}$	$1.01 \cdot 10^{-6}$	$1.61 \cdot 10^{-8}$	$9.63 \cdot 10^{-7}$	$2.01 \cdot 10^{-8}$
Rate	3.73	3.73	5.24	3.78	5.32

After singularity					
n_n	Cubic	ENO3	ENO5	WENO3	WENO5
25	$5.67 \cdot 10^{-3}$	$6.11 \cdot 10^{-4}$	$1.41 \cdot 10^{-4}$	$2.88 \cdot 10^{-3}$	$3.05 \cdot 10^{-3}$
50	$5.25 \cdot 10^{-5}$	$3.69 \cdot 10^{-5}$	$2.52 \cdot 10^{-6}$	$5.12 \cdot 10^{-5}$	$5.83 \cdot 10^{-6}$
100	$2.91 \cdot 10^{-6}$	$5.75 \cdot 10^{-6}$	$2.51 \cdot 10^{-7}$	$2.19 \cdot 10^{-6}$	$7.25 \cdot 10^{-8}$
200	$2.69 \cdot 10^{-7}$	$4.59 \cdot 10^{-7}$	$4.95 \cdot 10^{-9}$	$2.39 \cdot 10^{-7}$	$1.89 \cdot 10^{-9}$
Rate	4.79	3.46	4.93	4.52	6.87

numerical tests. Due to semiconcavity, even after the onset of the singularity the computation of the approximate solution remains remarkably accurate, more than what could be deduced from consistency analysis for a nondifferentiable solution. The influence of semiconcavity on the accuracy of numerical schemes has already been discussed in the presentation of numerical examples of Chapter 5.

6.3 ▪ Commented references

The first attempts of a convergence analysis for high-order SL schemes have appeared in a somewhat fragmentary form. As far as we can say, the first papers investigating convergence issues for (linear) SL schemes are [BM82, M84, M87]; later (and, in fact, independently) convergence issues for the linear case were collected together by the authors in [FF98]. Even this paper, however, does not solve the point of l^2 stability in a completely rigorous way: a Fourier analysis is performed, but a Von Neumann condition is enforced only graphically, whereas an explicit (and very technical) solution of the Von Neumann analysis for Lagrange reconstructions has been given later in [BM08].

Equivalence between SL and LG schemes has proved to be a simpler tool for proving l^2 stability. This approach was first studied in [Fe10a] for constant-coefficient equations and extended to variable-coefficient equations in [Fe13a]. Basic references on the LG technique are the seminal papers [DR82] and [Pi82], while the area-weighted version is proposed and studied in [MPS88] (see [BS12] and the references in [Fe13a] for more recent developments). The simplified analysis for finite element reconstructions has been developed in [P06] (the paper [FP07] gives a synthesis of this work), whereas an extensive analysis of the interaction between discretization steps is presented in [FFM01].

High-order SL schemes for HJ equations were first considered in [FF02], and a convergence analysis based on the condition $\Delta x = O(\Delta t^2)$ is carried out in [Fe03]. Convergence is directly proved in that paper starting from Lipschitz stability, whereas we have chosen here to prove it via the Barles–Souganidis theorem. The adaptation of the theory to weighted ENO reconstructions is given in [CFR05], along with a number of numerical tests comparing the various high-order versions of the scheme. Other numerical tests, mostly in higher dimensions and concerned with applications to front propagation and optimal control, are presented in [CFF04].

It is worth mentioning here that a number of different numerical approaches and techniques (many of which are of a high order) have been applied to HJ equations. Starting from the pioneering work [OS91] on ENO schemes (and without aiming at completeness), we quote the application of Godunov/central schemes [A96, A03, AA00, BrL03, BrL04], SuperBee/UltraBee [BZ07], MUSCL [LS95], Discontinuous Galerkin [HS99], sparse grid semi-Lagrangian [BGGK13], and WENO [ZS02]. Many of the works above allow for an unstructured implementation, and some of them mix this feature with nonoscillatory (ENO/WENO) reconstructions. Besides the sparse grid adaptive scheme in [BGGK13], other adaptive techniques have been proposed and analyzed in [G97, G04, SS01, BGS06].

Chapter 7
Fluid Dynamics

In a way, Computational Fluid Dynamics (CFD) is the birthplace of SL schemes. Although this monograph is focused on HJ equations, and the application of SL schemes to Fluid Dynamics would be worth a book in itself, we thought that at least some basic ideas of CFD applications could be well within the scope of this work.

This chapter treats some simple, typical applications of SL schemes in Fluid Dynamics models, mostly of environmental origin. As we have said now and then, NWP and related models have become a favorite area of application of SL schemes since the 1950s, and despite the introduction of more and more complex models, the SL approach has kept an important role in the field.

Application of the SL philosophy in CFD problems is typically related to an advection step in a more complicated model. However, the integration of this step in the model, as well as in the scheme, should be careful enough to properly address stability and accuracy issues. We will therefore show in this chapter a few classic examples, with a short discussion of the main features leading to (and resulting from) the use of SL techniques. Also, we will present a couple of additional numerical tools for constructing SL schemes for fluids.

In this chapter only, according to the usual practice in Fluid Dynamics, the gradient of a function f will be denoted as ∇f, and given a vector field u, the divergence and curl of this field will be denoted by, respectively, $\nabla \cdot u$ and $\nabla \times u$. The Laplace operator will be denoted by Δ, and this will cause no ambiguity with the discretization parameters.

7.1 ▪ The incompressible Euler equation in $\mathbb{R}^2$

The most classical model of inviscid Fluid Dynamics dates back to Euler and is named after him. In its simplest formulation, the *Euler equation* describes the evolution of an inviscid, incompressible, homogeneous fluid and takes the form of the system

$$\begin{cases} u_t + (u \cdot \nabla)u + \nabla p = 0, \\ \nabla \cdot u = 0 \end{cases} \tag{7.1}$$

(with suitable initial and boundary conditions), where, for $d \le 3$, $u = (u_1, \dots, u_d)^t$ denotes the fluid velocity and p denotes the pressure. In physical terms, the two equations state, respectively, the conservation of momentum and the conservation of mass. Writing the operator $(u \cdot \nabla)$ more explicitly, (7.1) reads as

$$\begin{cases} \dfrac{\partial u_i}{\partial t} + \displaystyle\sum_{j=1}^{d} u_j \frac{\partial u_i}{\partial x_j} + \frac{\partial p}{\partial x_i} = 0 \quad (i = 1, \ldots, d), \\ \displaystyle\sum_{i=1}^{d} \frac{\partial u_i}{\partial x_i} = 0. \end{cases} \tag{7.2}$$

We recall that the well-posedness of (7.2) is an established matter in two space dimensions, but is currently an open problem in three dimensions.

From the viewpoint of applications, the Euler equation models the behavior of an ideal fluid in the limit of low viscosity (or, in other terms, of high Reynolds numbers) whenever changes in the density of the fluid are negligible. With a lack of this latter condition, fluids with low viscosity are better modeled by the *compressible Euler equation*, which is strictly related to the *Shallow Water Equation* (this model will be discussed in what follows).

7.1.1 ▪ Vorticity–stream function formulation

As usual in Fluid Dynamics, we define the *vorticity* as the curl of velocity:

$$\omega = \nabla \times u. \tag{7.3}$$

The vorticity–stream function formulation of the incompressible Euler equation may be used in both two and three dimensions, but is especially suited for the two-dimensional model. In fact, in the special case of two space dimensions, the vorticity has only one nonzero component (the one related to x_3) and can therefore be treated as a scalar.

The starting point of this formulation is that, once we define the operator

$$\nabla^\perp := \begin{pmatrix} -\dfrac{\partial}{\partial x_2} \\ \\ \dfrac{\partial}{\partial x_1} \end{pmatrix}$$

the divergence-free vector field u (in fact, *any* divergence-free vector field) may be written as

$$u = \nabla^\perp \psi \tag{7.4}$$

for a suitable function ψ, termed as *stream function*. Plugging (7.4) into (7.3), we get

$$\Delta \psi = \omega. \tag{7.5}$$

Taking the curl of the momentum equation and recalling that ∇p has a zero curl, we obtain

$$\omega_t + u \cdot \nabla \omega = 0, \tag{7.6}$$

which states that vorticity is advected along the streamlines of the velocity u. Since u is divergence free, (7.6) might also be put in the conservative form

$$\omega_t + \nabla \cdot (u\, \omega) = 0. \tag{7.7}$$

Equations (7.4)–(7.6) represent the so-called *vorticity–stream function* formulation of the two-dimensional incompressible Euler equation. Note that the dependence on time appears only in (7.6), whereas (7.4) and (7.5) represent instantaneous relationships among ω, ψ, and u.

7.1.2 ▪ SL schemes for the vorticity–stream function model

The use of SL schemes for a problem very similar to the vorticity–stream function model dates back to Wiin-Nielsen, who proposed [W59] what is usually recognized as the first SL scheme in NWP. The scheme was originally formulated for the barotropic vorticity equation, but we rewrite it here in a conceptually equivalent form for the Euler equation.

The general idea is to use an SL technique to implement the advection of vorticity in (7.6). Once vorticity is computed at a new time step, stream function and velocity are updated via, respectively, (7.5) and (7.4). To keep a second-order accuracy in time, a midpoint scheme is used in the advective step, as shown in Chapter 5 (see (5.64)). More explicitly, we are given an initial condition u_0, a corresponding stream function ψ_0 such that

$$u_0 = \nabla^\perp \psi_0,$$

and an initial vorticity $\omega_0 = \nabla \times u_0$. Moreover, we assume that a further startup value of the vorticity ω_{-1} is available.

Let U^n, Ω^n, Ψ^n denote the three numerical approximations. A step forward in time is carried out through four substeps. First, the feet of characteristics are computed as in (5.64), that is,

$$X^\Delta(x_j, t_{n+1}; t_{n-1}) = x_j - 2\Delta t\, I[U^n]\left(\frac{x_j + X^\Delta(x_j, t_{n+1}; t_{n-1})}{2}\right). \tag{7.8}$$

Second, an advection step for the vorticity is performed as

$$\omega_j^{n+1} = I[\Omega^{n-1}]\big(X^\Delta(x_j, t_{n+1}; t_{n-1})\big). \tag{7.9}$$

Third, the updated stream function is computed by (numerically) solving the Poisson equation

$$\Delta \Psi^{n+1} = \Omega^{n+1}, \tag{7.10}$$

and last, the velocity is updated as

$$U^{n+1} = \nabla^\perp \Psi^{n+1}. \tag{7.11}$$

With some abuse of notation, in (7.10)–(7.11) we have used the same symbols Δ and $\nabla^\perp$ to denote approximate (e.g., finite difference) versions of the corresponding continuous operators.

Note that, as it has been discussed in Chapter 5, to achieve second-order accuracy in time it suffices to use a $\mathbb{P}_1$ or $\mathbb{Q}_1$ interpolation in (7.8). Space discretization error is determined by the order of interpolation in (7.9) (the original version uses $\mathbb{Q}_1$ interpolation), as well as by the accuracy of the numerical operators Δ and $\nabla^\perp$. Note also that for the continuity equation we have used the advective form (7.6), but its standard discretization (7.9) does not ensure conservation of the total vorticity, even for a divergence-free advecting vector field u. We will come back to this point in subsection 7.3.1.

7.2 ▪ The Shallow Water Equation

A more complex and useful model is given by the *Shallow Water Equation* (SWE), which models the flow of a relatively thin layer of fluid with a free surface. In its basic form, the system of equations reads as

$$\begin{cases} u_t + (u \cdot \nabla)u + g\nabla z = 0, \\ z_t + \nabla \cdot [(h+z)u] = 0. \end{cases} \tag{7.12}$$

Here, u stands again for the velocity of the fluid, h denotes the depth of the bottom at the equilibrium, and z is the difference between the actual and the equilibrium surface, so that $h+z$ is the total (bottom to surface) depth of the fluid. Depending on the physical situation, the friction at the bottom, as well as the stress operated by external forces (typically, wind), could also be included in the model.

In the SWE, the assumption of small depth of the fluid allows one to neglect (more precisely, to average) the vertical component of the fluid velocity. This is a good approximation only under conditions of hydrostatic equilibrium and when the solution has structures with horizontal scales much larger than the depth. Typically, this is the case in shallow basins, as well as large scale atmospheric models.

7.2.1 ▪ The one-dimensional case

In the particular case of a single space dimension, the SWE (7.12) takes the form

$$\begin{cases} u_t + u u_x + g z_x = 0, \\ z_t + (h+z) u_x + u z_x = -u h_x. \end{cases}$$

We start by studying the SWE from the viewpoint of characteristics. This will give an insight on the key analytical features of this model.

Characteristic decomposition

Defining the solution vector $W = (u, z)^t$, the one-dimensional SWE may be rewritten as

$$W_t + A(W) W_x = D(W), \tag{7.13}$$

with $A(W)$, $D(W)$ defined by

$$A(W) = \begin{pmatrix} u & g \\ h+z & u \end{pmatrix}, \qquad D(W) = \begin{pmatrix} 0 \\ -u h_x \end{pmatrix}.$$

In this setting, it is known that the speeds of propagation of the solution are given by the eigenvalues of A, which are easily computed as

$$c_{1,2} = u \pm \sqrt{g(h+z)}.$$

Defining the function $\eta = 2\sqrt{g(h+z)}$ and with some further manipulation, the SWE can finally be put in the form

$$\begin{cases} \left[\dfrac{\partial}{\partial t} + c_1 \dfrac{\partial}{\partial x} \right] (u+\eta) = g s, \\ \left[\dfrac{\partial}{\partial t} + c_2 \dfrac{\partial}{\partial x} \right] (u-\eta) = g s, \end{cases} \tag{7.14}$$

where $s = h_x$ is the bottom slope. In the form (7.14), it becomes apparent that $u \pm \eta$ are the variables which are transported along characteristics at speed $u \pm \eta/2$. In the propagation speeds c_1 and c_2, the terms u and $\pm\sqrt{g(h+z)}$ represent, respectively, the speed of advection and the speed of motion of the gravitational waves, also termed *celerity*. The ratio between the two speeds is an adimensional parameter known as *Froude number* and is defined by

$$Fr = \frac{u}{\sqrt{gh}}.$$

Depending on this parameter, the flow may be classified as

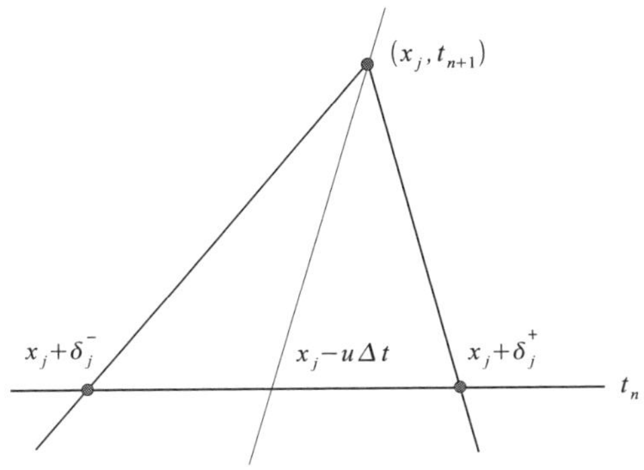

Figure 7.1. *Characteristics for the one-dimensional SL scheme.*

- *subcritical* if $Fr < 1$,
- *critical* if $Fr \approx 1$, and
- *supercritical* if $Fr > 1$.

In environmental Fluid Dynamics, subcritical flows are the main situation of interest. In fact, development of gravitational waves is a rare and relatively less relevant process, and there is little interest in having a good approximation of the fast component of the solution. However, despite being of lesser physical interest, this is the component which causes the stronger stability limits in conventional difference schemes.

We will soon show two typical SL techniques for approaching the problem in one space dimension.

SL schemes based on characteristics

A first technique to discretize (7.14) is to perform an SL upwinding along characteristics, which also amounts to a nonlinear CIR scheme. Since the speeds of propagation depend on the solution itself, they must be computed by means of a system of equations. In the case of a simple Euler scheme, we can define the approximate feet of characteristics ending at x_j as $x_j + \delta_j^\pm$, with $\delta_j^\pm$ solving the decoupled system

$$\begin{cases} \delta_j^+ = \Delta t \left(I[U^n](x_j + \delta_j^+) + \frac{1}{2} I[H^n](x_j + \delta_j^+) \right), \\ \delta_j^- = \Delta t \left(I[U^n](x_j + \delta_j^-) - \frac{1}{2} I[H^n](x_j + \delta_j^-) \right) \end{cases} \tag{7.15}$$

in which U^n and H^n denote the vectors approximating, respectively, u and η at time t_n (for simplicity, the dependence on n is not shown in the notation of the $\delta_j^\pm$). Note that (7.15) amounts to looking for two points $x_j + \delta_j^\pm$ such that the (approximate) characteristics passing through these points at time t_n arrive at x_j at t_{n+1} (see Figure 7.1). In the usual conditions of use of the SWE, we expect to have a smooth solution and hence a unique pair of characteristics.

Once we compute the $\delta_j^\pm$, the scheme takes the form

$$\begin{cases} u_j^{n+1} + \eta_j^{n+1} = I[U^n](x_j + \delta_j^+) + I[H^n](x_j + \delta_j^+) + \Delta t\, g s(x_j + \delta_j^+), \\ u_j^{n+1} - \eta_j^{n+1} = I[U^n](x_j + \delta_j^-) - I[H^n](x_j + \delta_j^-) + \Delta t\, g s(x_j + \delta_j^-). \end{cases} \tag{7.16}$$

A step of the scheme thus requires solving for each node the two nonlinear decoupled equations (7.15) plus the 2×2 linear system (7.16).

Semi-implicit SL schemes

A second technique for obtaining a stable and efficient scheme is to treat in an SL form only the slow advection term, whereas the terms related to fast modes of propagation would be treated implicitly. Note that in the form (7.13), the fast modes are generated by the off-diagonal terms, while the diagonal terms appear as an operator of a convective derivative along u, as it can be seen by rewriting (7.13) as

$$\begin{cases} \dfrac{Du}{Dt} + g\dfrac{\partial z}{\partial x} = 0, \\ \dfrac{Dz}{Dt} + (h+z)\dfrac{\partial u}{\partial x} = -su, \end{cases} \tag{7.17}$$

where

$$\frac{D}{Dt} = \frac{\partial}{\partial t} + u\frac{\partial}{\partial x}.$$

This suggests discretizing the diagonal terms in SL form (where now the advection speed is simply given by u) and the other terms implicitly. Using these ideas, the scheme could be written as

$$\begin{cases} u_j^{n+1} + g\dfrac{\Delta t}{2\Delta x}\left(z_{j+1}^{n+1} - z_{j-1}^{n+1}\right) = I[U^n](x_j + \delta_j), \\ z_j^{n+1} + \left(h(x_j) + z_j^n\right)\dfrac{\Delta t}{2\Delta x}\left(u_{j+1}^{n+1} - u_{j-1}^{n+1}\right) = I[Z^n](x_j + \delta_j) - \Delta t\, s(x_j) u_j^n, \end{cases} \tag{7.18}$$

where, for $k = j$,

$$\delta_k = \Delta t\, I[U^n](x_k + \delta_k). \tag{7.19}$$

Note that the SL upwinding (7.19) is now performed at the speed of propagation u, since it discretizes only the advective derivative term.

A step of the Semi-Implicit Semi-Lagrangian (SISL) scheme (7.18) requires first computing the upwinding δ_j for each node and the explicit terms treated in SL form. Then, using, for example, the expressions of $u_{j\pm1}^{n+1}$ from the first line of (7.18) in the second, the vector Z^{n+1} can be obtained by solving a pentadiagonal system. Finally, the vector U^{n+1} is computed from Z^{n+1}.

In discretizing the off-diagonal terms, a more accurate and efficient technique is to use a staggered grid for u and z. This is shown in Figure 7.2, in which squares correspond to nodes for u and circles correspond to nodes for z. The resulting scheme is

$$\begin{cases} u_{j+1/2}^{n+1} + g\dfrac{\Delta t}{\Delta x}\left(z_{j+1}^{n+1} - z_j^{n+1}\right) = I[U^n](x_{j+1/2} + \delta_{j+1/2}), \\ z_j^{n+1} + \left(h(x_j) + z_j^n\right)\dfrac{\Delta t}{\Delta x}\left(u_{j+1/2}^{n+1} - u_{j-1/2}^{n+1}\right) = I[Z^n](x_j + \delta_j) \\ \qquad -\Delta t\, s(x_j)\dfrac{u_{j+1/2}^n + u_{j-1/2}^n}{2} \end{cases} \tag{7.20}$$

with the δ_k defined again by (7.19) for $k = j, j \pm 1/2$. The scheme (7.20) requires the same steps of (7.18), except for solving a tridiagonal (instead of a pentadiagonal) system.

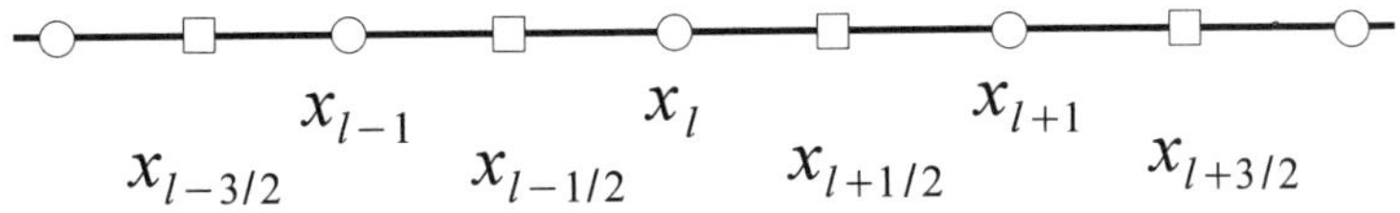

Figure 7.2. *The staggered space grid for one-dimensional SISL schemes.*

A further improvement is in a better resolution of the gravity waves, due to the use of a staggered grid (see, e.g., [Ra94]).

A similar semi-implicit discretization can also be derived starting from the conservative form

$$z_t + [(h+z)u]_x = 0$$

of the continuity equation. This yields the so-called Eulerian–Lagrangian formulation, in which only the momentum equation is discretized in an SL fashion. This approach, first proposed in [Ca90], is also employed in some atmospheric models, e.g., in [D05, TBR13]. However, the use of two different approaches for momentum and continuity equations may entail some loss of accuracy. An alternative option is the use of conservative SL schemes (see subsection 7.3.1).

7.2.2 ▪ The two-dimensional case

In order to write more explicitly the two-dimensional SWE, the velocity of the fluid will be denoted here by the vector $(u_1, u_2)^t$. In parallel with (7.17), the system may be rearranged in the form

$$\begin{cases} \dfrac{Du_1}{Dt} + g\dfrac{\partial z}{\partial x_1} = 0, \\ \dfrac{Du_2}{Dt} + g\dfrac{\partial z}{\partial x_2} = 0, \\ \dfrac{Dz}{Dt} + (h+z)\left(\dfrac{\partial u_1}{\partial x_1} + \dfrac{\partial u_2}{\partial x_2}\right) = -u_1 h_x - u_2 h_y, \end{cases} \tag{7.21}$$

where the convective derivative is written now as

$$\frac{D}{Dt} = \frac{\partial}{\partial t} + u_1\frac{\partial}{\partial x_1} + u_2\frac{\partial}{\partial x_2}.$$

When turning to the two-dimensional SWE, a crucial feature arises: characteristics ending at a point (x_j, t_{n+1}) are no longer in finite number, but belong to a cone around the speed of convection (u_1, u_2) (see Figure 7.3). Thus, propagation of the solution cannot be treated by following all characteristics, and the semi-implicit strategy remains the usual choice.

We omit a detailed construction of the scheme, which follows the same guidelines of the one-dimensional case. In the case of a staggered grid, a typical choice is the C-grid, shown in Figure 7.4. Here, the nodes are denoted by circles for z, by light squares for the x_1 component u_1 of the velocity, and by dark squares for the x_2 component u_2.

As it has been remarked for the one-dimensional case, the kind of discretization used for the continuity equation affects the mass conservation of the complete scheme.

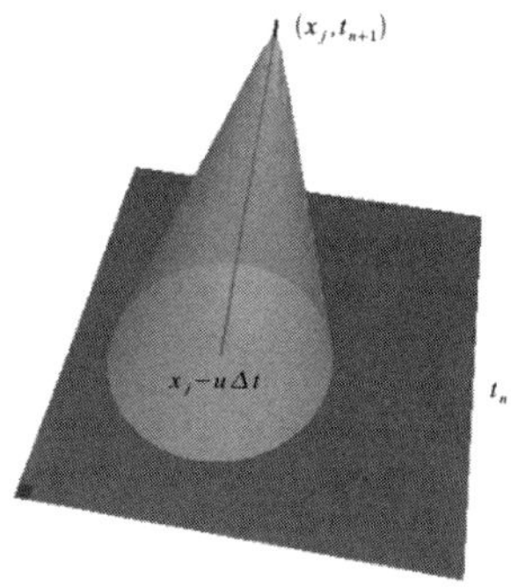

Figure 7.3. *Characteristic cone for the two-dimensional SWE.*

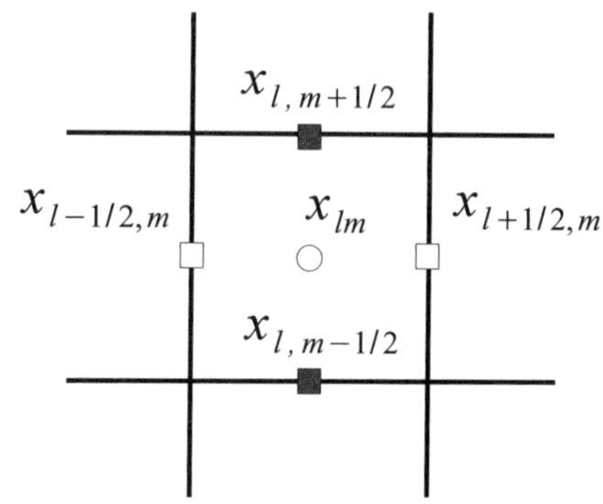

Figure 7.4. *The C-grid for two-dimensional SISL schemes.*

7.3 ▪ Some additional techniques

We discuss here a couple of techniques which can be useful in Fluid Dynamics problems. The first is related to mass conservation in SL schemes, whereas the second is related to the treatment of second-order terms. In both cases, we simply sketch the basic idea, while a more in-depth discussion will be found in the references.

7.3.1 ▪ Conservative SL schemes

When applied to Fluid Dynamics problems, a major drawback of SL schemes is that they are not strictly mass conserving. This is due to their structure, which is not originally in conservative form and cannot be recast as such. In order to overcome this problem and try to put together the advantages of both approaches, Flux-Form versions of SL schemes (also denoted as FFSL) have been proposed starting from the mid-1990s.

We briefly outline the underlying idea, using the simplified model of the constant-coefficient advection equation on $\mathbb{R}$, which will be formally put in the form of a continuity equation:

$$u_t + (cu)_x = 0. \tag{7.22}$$

First, as usual in conservative schemes, we divide the x-axis in *cells* of the form $[x_{j-1/2}, x_{j+1/2}]$ around x_j, as shown in Figure 7.5. A numerical solution V^n represents an

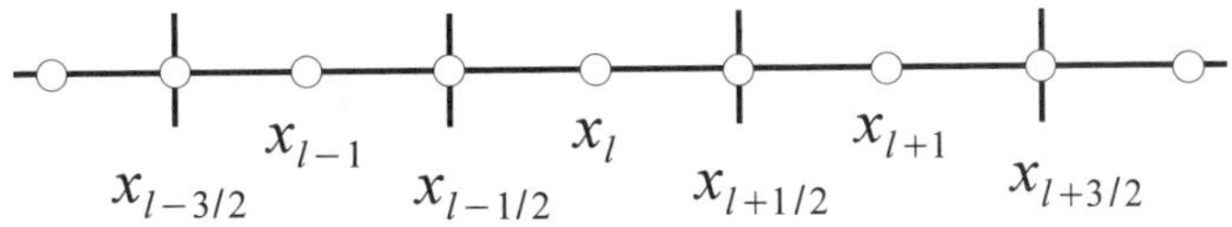

Figure 7.5. *The space grid for FFSL schemes.*

approximation of the cell averages of the solution at time t_n, so that

$$v_j^n \approx \frac{1}{\Delta x}\int_{x_{j-1/2}}^{x_{j+1/2}} u(x,t_n)dx,$$

and, in particular, the value v_j^0 is the average over the jth cell of the initial condition u_0. We also need a reconstruction operator $R[V^n](x)$ to compute the numerical solution V^n at a generic x. The reconstruction operator R is assumed to preserve the cell averages of V^n. In the simplest case, it is implemented in piecewise constant form by taking its value equal to v_j^n on the whole cell, although many higher-order examples may be borrowed from the literature on conservative schemes.

In order to update the cell averages at the time step $n+1$, we start from the mass balance of the jth cell $[x_{j-1/2}, x_{j+1/2}]$. Assuming, for example, that $c > 0$, the mass entering the cell from the interface $x_{j-1/2}$ in the time step $[t_n, t_{n+1}]$ is

$$\int_{z_{j-1/2}}^{x_{j-1/2}} v(x,t_n)dx,$$

where $z_{j\pm1/2} = x_{j\pm1/2} - c\Delta t$ denotes the foot of the characteristic starting at $(x_{j\pm1/2}, t_{n+1})$. In the same way, the mass leaving the cell may be written as

$$\int_{z_{j+1/2}}^{x_{j+1/2}} v(x,t_n)dx.$$

Applying this general idea to a numerical solution, we define a *numerical flux* by integrating over a time step the mass of numerical solution crossing the interface:

$$F_{j\pm1/2}(V^n) := \int_{z_{j\pm1/2}}^{x_{j\pm1/2}} R[V^n](x)dx. \tag{7.23}$$

With this definition, a conservative version of the SL scheme can be defined as

$$v_j^{n+1} = v_j^n + \frac{1}{\Delta x}\left[F_{j-1/2}(V^n) - F_{j+1/2}(V^n)\right]. \tag{7.24}$$

Note that the form (7.23)–(7.24) performs the correct mass balance independently of the sign of c.

In the computation of the numerical flux (7.23), any cell $[x_{k-1/2}, x_{k+1/2}]$ which is fully included in the integration domain gives a contribution of $\Delta x\, v_k$ to the integral. The only contribution which really has to be computed is the term related to the j^*th cell which contains $z_{j-1/2}$, so that, if, for example, $c > 0$,

$$F_{j-1/2}(V^n) = \Delta x \sum_{k=j^*+1}^{j-1} v_k + \int_{z_{j-1/2}}^{x_{j^*+1/2}} R[V^n](x)dx. \tag{7.25}$$

The last integral depends on the kind of reconstruction used. If, for example, a piecewise constant reconstruction R has been chosen, the numerical flux is finally computed as

$$F_{j-1/2}(V^n) = \Delta x \sum_{k=j^*+1}^{j-1} v_k + (x_{j^*+1/2} - z_{j-1/2}) v_{j^*}, \tag{7.26}$$

and the same ideas clearly apply to the numerical flux $F_{j+1/2}(V^n)$. Some simple algebra shows that in this case the FFSL scheme coincides with the CIR method, with the obvious difference that while in the CIR scheme the elements of the vector V^n are pointwise values, in the FFSL scheme they should be interpreted as cell averages. This relationship, however, becomes more complex in variable-coefficient equations and for higher-order schemes.

This setting can be generalized to the case of the variable-coefficient linear continuity equation, namely

$$u_t + (f(x,t)u)_x = 0,$$

in which the construction of the scheme follows the same form (7.24), with the flux given by (7.23) and the feet of characteristics defined as

$$z_{j\pm 1/2} = X^{\Delta}\left(x_{j\pm 1/2}, t_{n+1}; t_n\right).$$

Clearly, other choices of the reconstruction (and, ultimately, of the flux) are possible. In particular, various high-order reconstructions of use in Eulerian conservative schemes have also been proposed in the FFSL setting. In this case, if, for example, $f(x,t) > 0$, the expression (7.26) for the flux is replaced by the more general form (7.25), in which the last term is computed by actually integrating a polynomial of degree $r > 0$.

The extension of this approach to multiple dimensions must take into account the more complicated geometrical features. A brief discussion of this point will be given in the commented references.

7.3.2 ▪ Treatment of diffusive terms

Although this book is essentially addressed to inviscid models, it is worth briefly giving some hint on how to adapt SL techniques to the case of second-order equations, which arise from models with diffusive terms.

A standard idea is to separate the diffusive from the advective part, treating only the latter in SL form. In general, this implies using an implicit discretization for the second-order term and suffers somewhat from degeneracy of the diffusion operator. However, in addition to this classical technique, there exists a possibility to treat the diffusive part in an SL form, which remains explicit and stable also for degenerate operators.

To sketch the basic ideas of this latter approach, we consider here the constant-coefficient advection-diffusion equation on $\mathbb{R}$:

$$u_t + c u_x - \nu u_{xx} = 0. \tag{7.27}$$

We claim that (7.27) can be approximated via the scheme

$$v_j^{n+1} = \frac{1}{2} I[V^n]\left(x_j - c\Delta t - \sqrt{2\nu\Delta t}\right) + \frac{1}{2} I[V^n]\left(x_j - c\Delta t + \sqrt{2\nu\Delta t}\right). \tag{7.28}$$

First, note that (7.28) performs an average of two different upwind points $x_j - c\Delta t \pm \sqrt{2\nu\Delta t}$. This clearly preserves the stability properties of the inviscid version; e.g., if the interpolation $I[\cdot]$ is monotone, then (7.28) is also monotone. To show that it is consistent, we first neglect the interpolation step and write

$$\begin{aligned} u\left(x_j - c\Delta t \pm \sqrt{2\nu\Delta t}, t_n\right) &= u(x_j, t_n) + \left(-c\Delta t \pm \sqrt{2\nu\Delta t}\right) u_x(x_j, t_n) \\ &\quad + \frac{1}{2!}\left(-c\Delta t \pm \sqrt{2\nu\Delta t}\right)^2 u_{xx}(x_j, t_n) \\ &\quad + \frac{1}{3!}\left(-c\Delta t \pm \sqrt{2\nu\Delta t}\right)^3 u_{xxx}(x_j, t_n) + O(\Delta t^2) \\ &= u(x_j, t_n) - \left(c\Delta t \mp \sqrt{2\nu\Delta t}\right) u_x(x_j, t_n) \\ &\quad + \left(\nu\Delta t \mp c\Delta t\sqrt{2\nu\Delta t}\right) u_{xx}(x_j, t_n) \\ &\quad \pm \frac{\sqrt{2\nu\Delta t}^{\,3}}{6} u_{xxx}(x_j, t_n) + O(\Delta t^2). \end{aligned}$$

We have therefore

$$\begin{aligned} &\frac{1}{2}u\left(x_j - c\Delta t + \sqrt{2\nu\Delta t}, t_n\right) + \frac{1}{2}u\left(x_j - c\Delta t - \sqrt{2\nu\Delta t}, t_n\right) \\ &= u(x_j, t_n) + \Delta t\left[-c u_x(x_j, t_n) + \nu u_{xx}(x_j, t_n)\right] + O(\Delta t^2). \end{aligned}$$

Recalling that the interpolation error is in general $O(\Delta x^{r+1})$ for a smooth solution, so that

$$u\left(x_j - c\Delta t \pm \sqrt{2\nu\Delta t}, t_n\right) = I[U^n]\left(x_j - c\Delta t \pm \sqrt{2\nu\Delta t}\right) + O(\Delta x^{r+1}),$$

we can conclude that the scheme is consistent with order $O(\Delta t) + O(\Delta x^{r+1}/\Delta t)$.

This approach allows us to obtain an explicit scheme, but the price to pay is an expansion of the numerical domain of dependence, which develops at each time step a "hole" of width approximately $2\sqrt{2\nu\Delta t}$ between the two upwind points $x_j - c\Delta t \pm \sqrt{2\nu\Delta t}$. Under a refinement law $\Delta t = \Delta x^\alpha$ with $\alpha < 2$, this segment contains an increasing number of nodes, and this may cause an irregular behavior of the numerical approximations for nonsmooth solutions. It turns out (see [Fe10b]) that a condition for these "holes" to be filled up at time T is that

$$\frac{\nu^{1/3}\Delta t}{T^{2/3}\Delta x^{2/3}} \ll 1. \tag{7.29}$$

Despite introducing a sort of compatibility condition among discretization steps, this condition allows us to work at constant or slightly increasing Courant numbers and has a weak dependence on the viscosity.

The strategy can be adapted to the multidimensional case in a straightforward way. Given a d-dimensional constant-coefficient advection-diffusion equation in the form

$$u_t + c \cdot Du - \mathrm{tr}(A D^2 u) = 0, \tag{7.30}$$

where $\mathrm{tr}(B)$ stands for the trace of the matrix B, and the matrix A is supposed to be positive semidefinite, we define a matrix σ such that $\sigma\sigma^t = A$ (for example, a Cholesky factor of

A) and, for $k = 1, \ldots, 2^d$, displacements δ_k in the form

$$\delta_k = -c\Delta t + \sigma \begin{pmatrix} \pm\sqrt{2\Delta t} \\ \vdots \\ \pm\sqrt{2\Delta t} \end{pmatrix}$$

for all possible combinations of the signs. Then, the d-dimensional version of the scheme reads as

$$v_j^{n+1} = \frac{1}{2^d} \sum_{k=1}^{2^d} I[V^n]\left(x_j + \delta_k\right). \tag{7.31}$$

The idea is to discretize the second-order part by considering independent increments $\pm\sqrt{2\Delta t}$ on each variable and then generate the correct strength and direction of the diffusion by means of the matrix σ. Note that the right-hand side of (7.31) remains a convex combination of interpolations and hence is stable if the inviscid scheme is stable. The consistency analysis could be carried out as in the one-dimensional case.

In the approximation of degenerate diffusion operators, the matrix σ is of reduced rank, and in this case it can be replaced by a matrix with a reduced number of columns. For a one-dimensional diffusion, for example, σ is simply a vector spanning the subspace where diffusion operates, and only two displacements δ_1 and δ_2 (as in the one-dimensional scheme) are to be considered. This situation occurs, e.g., in treating the vertical turbulent diffusion in atmospheric models. A rectangular matrix is also used to treat the degenerate diffusion occurring in level set models with curvature-dependent speed (see Chapter 9).

As a last remark, we point out that the extension of this technique to higher orders of consistency (with respect to Δt) is somewhat cumbersome, especially for diffusion operators with variable coefficients.

7.4 ▪ Examples

In this section, we present some numerical tests for SL schemes in CFD problems, starting with two problems of passive advection and advection-diffusion and moving to the nonlinear models presented in this chapter. Much more could be shown about CFD applications—once more, this section is intended to provide only a few basic examples.

Linear advection: The Zalesak disc

One of the most classical benchmark tests for CFD schemes is passive linear advection of a characteristic functions, a usual choice being the so-called Zalesak disc. In this test, the solution is advected along a divergence-free vector field which causes a rotation around the origin:

$$f(x) = \begin{pmatrix} x_2 \\ -x_1 \end{pmatrix}.$$

The initial condition is the characteristic function of a slotted disc, which is shown in the upper left plot of Figure 7.6. Numerical solutions have been computed on $[-2,2]^2$ after five turns around the origin, that is, at $T = 10\pi$. We have used a second-order time stepping (Heun) coupled with, respectively, $\mathbb{P}_1$, cubic, and third-order ENO interpolations, with a 100×100 space grid and 200 time steps, at a maximum Courant number of about 8. The solutions are plotted at the final time along with their level curves at $u = 0.1, 0.5, 0.9$.

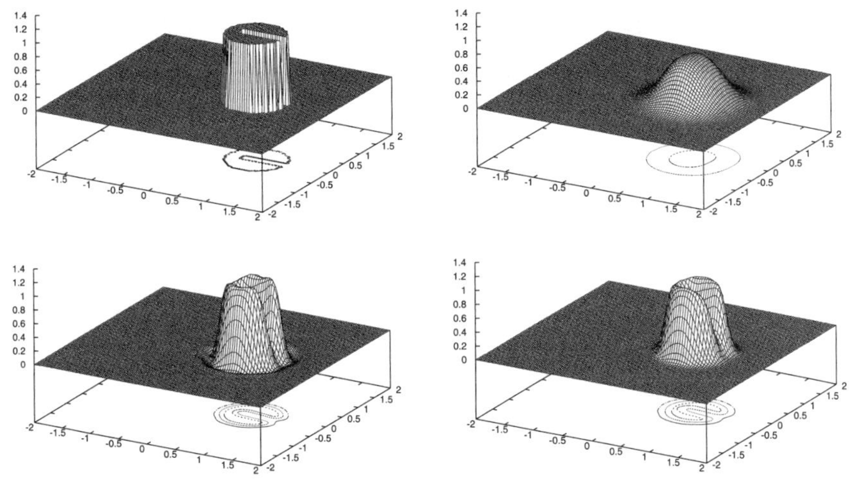

Figure 7.6. *Two-dimensional linear advection of the Zalesak disc: initial condition (upper left), $\mathbb{P}_1$ solution (upper right), cubic solution (lower left), and ENO3 solution (lower right).*

We have added a comparison with the SL-$\mathbb{P}_1$ scheme to show how the use of monotone, more diffusive schemes can make it impossible to succeed in such a discriminating test. Lagrange interpolation exhibits over- and undershoots, while ENO or other nonlinear techniques can cut out oscillations without introducing too much numerical dispersion. The increase in complexity related to their use is also to be considered, however.

As a final remark, we point out that a conservation error of about 3% appears in the $\mathbb{P}_1$ and in the cubic schemes, while this error is halved in the ENO3 scheme.

Advection-diffusion

To show the effectiveness of the technique discussed in subsection 7.3.2, we perform here a test in two space dimensions, considering (7.30) in $\mathbb{R}^2$ with constant advection speed and degenerate diffusion, and in particular

$$A = \begin{pmatrix} \frac{1}{64} & \frac{\sqrt{3}}{64} \\ \frac{\sqrt{3}}{64} & \frac{3}{64} \end{pmatrix}, \qquad f = \begin{pmatrix} -1 \\ -3 \end{pmatrix}$$

posed in $\Omega = (-2,2)^2$, with biperiodic boundary conditions. The equation is slightly advection-dominated, and diffusion is degenerate along a direction rotated by 30° with respect to the grid geometry. In this test, diffusion and advection are not aligned with one another, and neither of the two is oriented along relevant directions of the space grid. The approximate solutions are compared at $T = 4$ (that is, at the time at which a rigid advection with the same speed would bring back the solution to its original configuration).

In this example, the initial condition is the characteristic function of a unit square, rotated by 30°, so that the solution is diffused along one of the axes of the square, whereas it remains discontinuous along the other. Figure 7.7 shows the initial condition (upper left) and the numerical solutions obtained by coupling a second-order scheme with different space reconstructions, at $T = 4$, on a grid of 50×50 nodes, with $\Delta t = 0.1$. While the $\mathbb{P}_1$ scheme clearly introduces an undue (and partly isotropic) diffusion, cubic reconstruction (lower left) has a better resolution of the discontinuity (within about 3–4 mesh sizes),

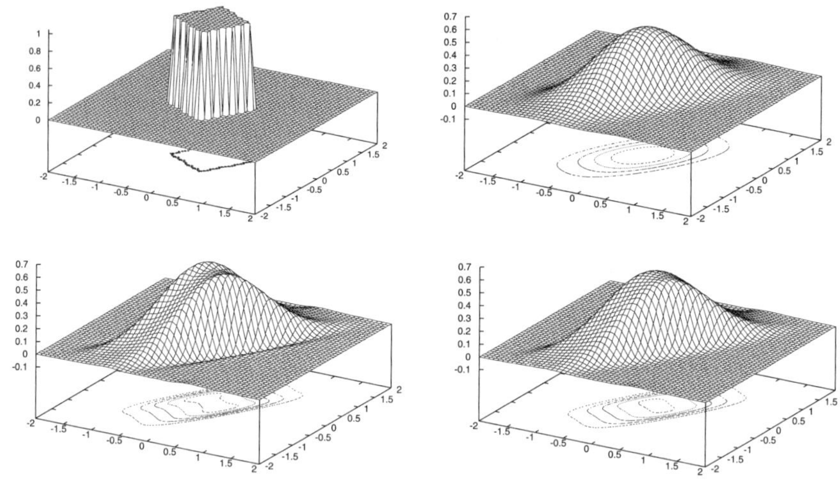

Figure 7.7. *Two-dimensional degenerate advection-diffusion equation: initial condition (upper left), $\mathbb{P}_1$ solution (upper right), cubic solution (lower left), and ENO3 solution (lower right)* [Fe10b].

although it also introduces some oscillations in the computation. Such oscillations are avoided by the third-order ENO scheme (lower right) at the price of a higher complexity.

Euler equation

We give here an example taken from [QS11]. Consider the domain $\Omega = [0, 2\pi]^2$ with periodic conditions and the initial condition on the vorticity

$$\omega_0(x) = \begin{cases} \delta \cos x_1 - \dfrac{1}{\rho} \operatorname{sech}^2 \dfrac{x_2 - \pi/2}{\rho} & \text{if } x_2 \leq \pi, \\[2ex] \delta \cos x_1 + \dfrac{1}{\rho} \operatorname{sech}^2 \dfrac{3\pi/2 - x_2}{\rho} & \text{if } x_2 < \pi \end{cases}$$

for $\delta = 0.05$ and $\rho = \pi/15$. This corresponds to the situation of a periodic shear flow with abrupt changes in the speed at two slightly oscillating interfaces placed around $x_2 = \pi/2$ and $x_2 = 3\pi/2$. The perturbation of the interface causes the onset of an instability in the form of periodic vortices, with an increasing numerical difficulty due to the development of small structures in the solution.

Figure 7.8 shows the vorticity computed at $T = 6$ and $T = 8$, with a 243×243 space grid. The advection step has been solved by, respectively, a WENO2 and a WENO3 conservative SL scheme (with an obvious improvement in the resolution for the latter), whereas a spectral solver based on the FFT has been used for the Poisson equation. Multiple dimensions have been treated in this test by directional splitting, and this has required using a small Courant number, although the development by the authors of a genuinely multidimensional implementation of the conservative SL-WENO scheme is in progress.

SWE in $\mathbb{R}$

We show here two examples, in which the one-dimensional SWE (7.13) is treated with both schemes discussed above. In the first test, the scheme (7.16) is used to simulate the propagation of a gravitational wave, whereas in the second, a stationary flow over a bump is simulated by means of a semi-implicit scheme.

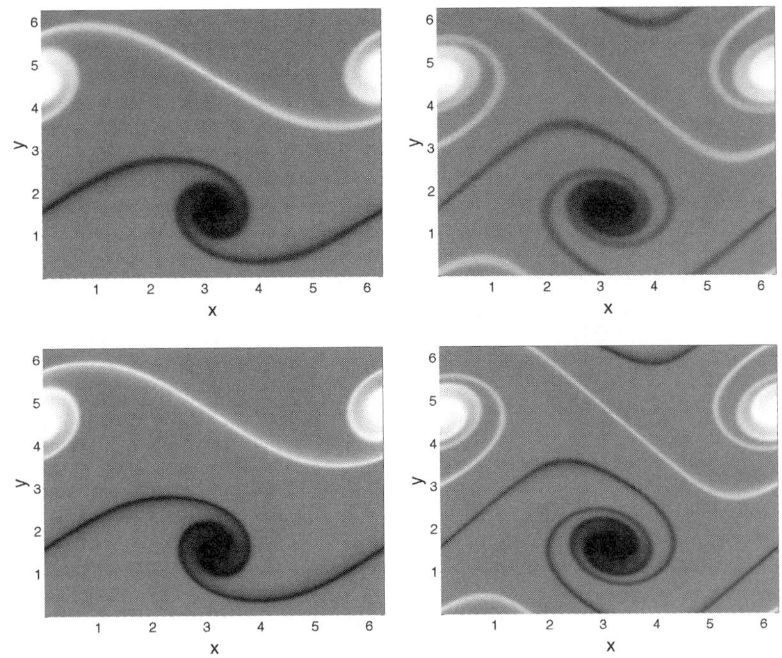

Figure 7.8. *Two-dimensional Euler equation: solutions with WENO2 (upper) and WENO3 (lower) at $T = 6$ (left) and $T = 8$ (right) (courtesy of J.-M. Qiu and C.-W. Shu)* [QS11].

Propagation of a gravitational wave In this test, the equation is posed on the interval $[-10, 10]$ with zero bottom slope, and the initial condition is

$$z_0(x) = \begin{cases} \cos \frac{\pi x}{2} & \text{if } |x| < 2, \\ 0 & \text{else,} \end{cases}$$

with an initially still fluid ($u_0 \equiv 0$). In this situation, the generation of two gravitational waves propagating right and left is expected, along with two velocity waves.

Figure 7.9 shows four snapshots of the z component of the solution each 10 time steps, as computed by the scheme (7.16), with $\Delta t = 0.2$ and 150 nodes, at a maximum Courant number of about 5.

Stationary flow over a bump In this second test (considered in [RBDG11]), a semi-implicit scheme has been used to search for the stationary solution of a flow in the presence of a bump in the bottom, via a time-marching approach. The channel bottom has the expression

$$b(x) = \max(0.2 - 0.05(x - 10)^2, 0),$$

and convection moves the fluid rightwards.

In stationary conditions, the exact solution can be computed explicitly. The exact fluid elevation is shown with a solid line in Figure 7.10, compared with two numerical solutions obtained for two different depths of water. The test is the performed in the transcritical regime (i.e., the speed of convection can be somewhere comparable to the speed of gravitational waves), and with Courant numbers ranging from 2.5 to 4.5. At the decrease of depth, a stationary shock may appear (see the right plot of Figure 7.10) and is correctly located by the scheme.

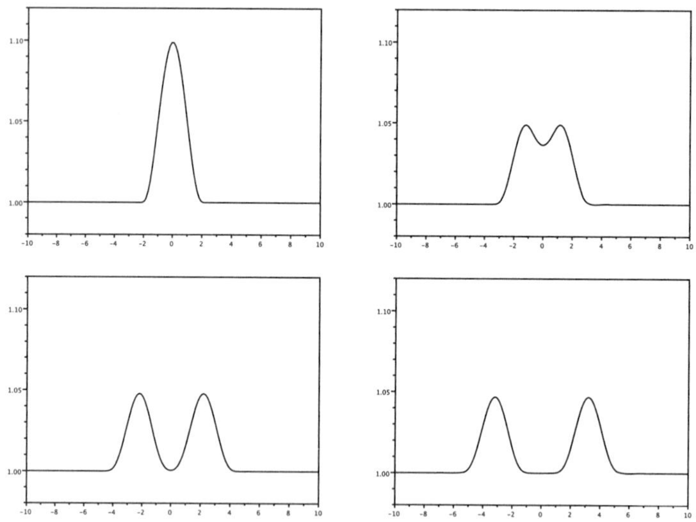

Figure 7.9. *One-dimensional SWE: propagation of a gravitational wave.*

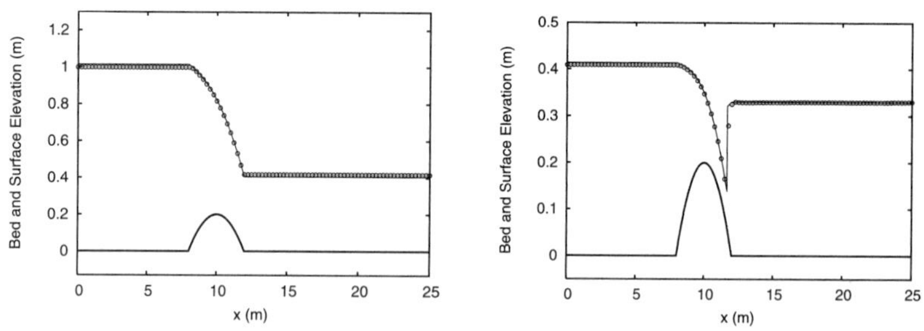

Figure 7.10. *One-dimensional SWE: stationary solution over a bump (courtesy of L. Bonaventura and G. Rosatti)* [RBDG11].

SWE in $\mathbb{R}^2$

In this last set of tests (taken from [TBR13]), the two-dimensional SWE is considered on an idealized square basin $\Omega = (0, L)^2$, with Coriolis terms. The scheme is semi-implicit and uses a p-adaptive Discontinuous Galerkin space discretization, at a Courant number ranging up to about 24.

Geostrophic adjustment A first example simulates the physical phenomenon of *geostrophic adjustment*, in which a gravitational wave starts from the center of the basin and, due to the effect of Coriolis force, evolves into a slowly varying state. This is shown in the upper graphs of Figure 7.11, which plot, respectively, the surface elevation and the velocity vector field, taken at $T = 10\,h$. The basis has a side of $L = 10^7$ m and a mean depth of 10^3 m, the initial condition for the elevation is

$$z_0(x) = 5e^{-\frac{|x-L/2|^2}{2(L/20)^2}},$$

and no-flux conditions have been enforced on the boundary.

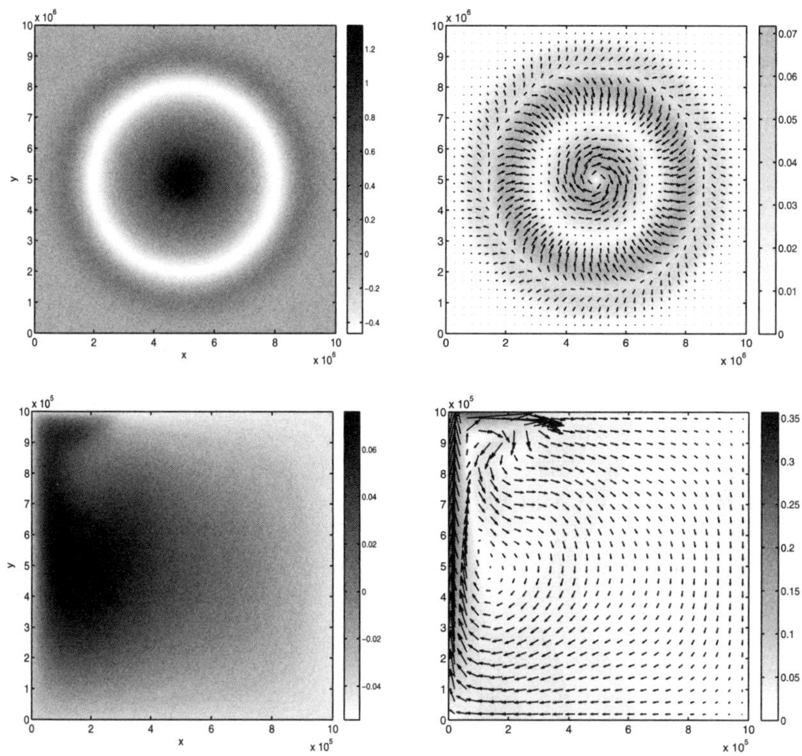

Figure 7.11. *Two-dimensional SWE: geostrophic adjustment (upper) and Stommel test (lower), free surface level (left), and velocity field (right) (courtesy of L. Bonaventura, M. Restelli, and G. Tumolo)* [TBR13].

Stommel test In the second example (the so-called *Stommel test*), a wind shear term is also included in the equation. Here, a square basin of side $L = 10^6$ m and mean depth 10^3 m is subject to wind and friction on the bottom. Wind introduces an x_1-momentum term given by

$$\phi_1^w = 10^{-7}\sin(\pi(x_2 - L/2)/L),$$

whereas the x_2 term is zero. Friction is modeled by adding a momentum term

$$\phi^f = -5 \cdot 10^{-7} u.$$

The combined effect of wind and Coriolis force causes the onset of a clockwise circulation of the fluid. The lower plots of Figure 7.11 show the stationary state computed after a time of 2 months, starting from the basin at rest.

7.5 ▪ Commented references

We give up a complete review of the theoretical literature on Fluid Dynamics models, limiting ourselves to quoting the book [MP94] among the many possible references on the Euler equation. In the field of atmospherical models, what has been presented here might be seen as simplified versions of the *system of primitive equations*. A classical reference on primitive equations, as well as other geophysical models, is [P87].

On the numerical side, after the pioneering work [W59], the idea of SL advection was further improved and applied to more general models in [K62, S63, M70], and it was recognized (first in [S63]) that it could allow using relatively large Courant numbers without stability or accuracy losses. In addition to the reduction in computational complexity, this feature was also useful for circumventing the so-called *pole problem*, a shrinking of the space grid at high latitudes caused by the use of an orthogonal grid to map the sphere.

A major breakthrough towards a balanced treatment of slow and fast modes of propagation was achieved by the semi-implicit schemes, first introduced by Robert in [Ro81], which have created a considerable amount of literature on the application of SISL techniques to environmental Fluid Dynamics. Among the various versions of SISL schemes, our presentation is based on [CC90]. For the whole NWP research line, [SC91] represents a classical (although no longer up-to-date) review paper. We also quote a book in preparation [B], mainly focused on environmental applications of SL schemes.

The lack of conservation is a well-known drawback of SL schemes, and various approaches have been suggested to correct it, ranging from a nonlocal redistribution of the mass to a genuinely conservative structure. We have presented here the latter strategy, first proposed in [LR96] (see also [RBS06] and the references therein). We also quote the recent works [CMS10, QS08, QS11] and, on the theoretical side, the one-dimensional stability analysis presented in [Fe13b]. The generalization of FFSL schemes to multiple dimensions is not straightforward, and part of the literature is based on directional splitting techniques. Among the papers quoted, for example, [LR96, QS11] present splitting-based schemes, whereas [RBS06] proposes an intrinsically multidimensional approach.

Both conservative and conventional SL schemes, as well as LG schemes and other large time-step methods, have been given a unified framework as *generalized Godunov schemes* in [M98].

The treatment of diffusion by means of the technique described in subsection 7.3.2 has a long tradition. It can be derived via a stochastic representation formula for the solution (the Feynman–Kac formula), by a stochastic discretization, and has been proposed by Kushner in the 1970s in the framework of Dynamic Programming equations (see Chapter 8). Independently, this idea has been used by Milstein and Tretyakov in a series of papers [MT00, MT01, M02] to approximate linear and semilinear parabolic problems. The extension to higher orders of consistency is studied in [Fe10b], along with a general analysis of the interplay between discretization steps, leading to condition (7.29). This technique is applied in [BM03] to Navier–Stokes equations, and in [T99] to a model of turbulent vertical diffusion for the atmosphere. Further applications are related to curvature-dependent motions and will be briefly reviewed in Chapter 9. Concerning the stochastic side of this technique, a classical (although a bit outdated) text on the numerical treatment of SDEs is [KP92].

Finally, we mention that a different strategy for the Navier–Stokes equation has also been studied in [MPR90, XK01], in which the SL approach is used to discretize the advective derivative within a high-order spectral code.

Chapter 8
Control and games

One of the most typical applications of the theory of HJ equations is in the field of optimal control problems and differential games. Via the Dynamic Programming Principle (DPP in what follows), formulated by R. Bellman in the 1950s, many optimal control problems can be characterized by means of the associated *value function*, which can be shown in turn to be the unique viscosity solution of a PDE of convex HJ type, usually called the *Bellman equation* or the *Dynamic Programming equation* (also the *HJB equation* in the rest of the chapter).

Although the direct numerical solution of the Bellman equation presents serious complexity problems related to the dimension of the state space (the so-called *curse of dimensionality*), it also allows one to recover a sharper information. Actually, in comparison with the more conventional technique given by the Pontryagin Maximum Principle (which characterizes the optimal control related to a *given* initial state as a function of *time*), the Bellman equation provides a synthesis of *feedback controls*, i.e., of controls expressed as a function of the state variable (instead of time).

In this chapter we present algorithms of SL type for the approximate solution of Dynamic Programming equations related to optimal control problems and differential games. In fact, the same approach can be applied to the analysis and approximation of zero-sum differential games where, under suitable conditions, the value function of the game solves a nonconvex HJ type-equation, usually called the *Isaacs equation*.

The SL numerical methods presented here are strongly connected to Dynamic Programing for two main reasons. The first is that both the Bellman and the Isaacs equations are derived by passing to the limit in the DPP, and the second is that the SL schemes are obtained by a discrete version of the DPP, which plays the role of the Hopf–Lax formula for this specific case. This strategy of construction gives a nice control interpretation of the schemes and also helps in their theoretical analysis. Moreover, it allows one to construct at the same time the approximations of both value function and optimal feedback, the latter being a practical solution to the original optimization problem.

While the construction and theoretical analysis of SL schemes for Dynamic Programming equations allow for an elegant and general setting, their application to real-life industrial problems still suffers from an impractical computational complexity. However, the continuous improvement of computers, as well as the development of faster and more accurate algorithms, make this field an active and promising research area.

8.1 ▪ Optimal control problems: First examples

In order to present the main ideas, we start with the classical *infinite horizon problem*. Let a controlled dynamical system be given by

$$\begin{cases} \dot{y}(s) = f(y(s), s, \alpha(s)), \\ y(t_0) = x_0, \end{cases} \tag{8.1}$$

where $x_0, y(s) \in \mathbb{R}^d$, and

$$\alpha : [t_0, T] \to A \subseteq \mathbb{R}^m,$$

with T finite or $+\infty$. In what follows, we will possibly use the less general formulation

$$\begin{cases} \dot{y}(s) = f(y(s), \alpha(s)), \\ y(t_0) = x_0. \end{cases} \tag{8.2}$$

In the applications, it is unreasonable to assume a priori any regularity of the control α. We will rather assume that the control is measurable, in which case the existence and uniqueness for the solution of (8.1) are ensured by the Carathéodory theorem.

Theorem 8.1 (Carathéodory). *Assume that*

(i) $f(\cdot,\cdot,\cdot)$ *is continuous;*

(ii) *there exists a positive constant* $L_f > 0$ *such that*

$$|f(x,t,a) - f(y,t,a))| \le L_f |x-y|$$

for all $x, y \in \mathbb{R}^d$, $t \in \mathbb{R}^+$, *and* $a \in A$; *and*

(iii) $f(x,t,\alpha(t))$ *is measurable with respect to* t.

Then,

$$y(s) = x_0 + \int_{t_0}^{s} f(y(\tau), \tau, \alpha(\tau)) d\tau \tag{8.3}$$

is the unique solution of (8.1).

Note that the solution is continuous, but only a.e. differentiable, so it must be regarded as a weak solution of (8.1). Note also that we are not interested here in the most general version of this theorem, in which a condition of *local* Lipschitz continuity of f still allows us to define a local solution y. This generalization is not of great interest in control theory, in which a controlled dynamical system is usually required to have a solution for all $s \ge t_0$.

By the theorem above, once we fix a control in the set of admissible controls

$$\alpha \in \mathcal{A} := \{a : [t_0, T] \to A, \text{measurable}\},$$

there exists a unique trajectory $y_{x_0,t_0}(s;\alpha)$ of (8.1). Changing the control policy, the trajectory will change, and we will have a family of solutions of the controlled system (8.1) depending on α. To simplify notation, when considering the autonomous dynamics (8.2) and the initial time $t_0 = 0$, we will denote this family as $y_{x_0}(s;\alpha)$ (or even with $y(s)$ if no ambiguity occurs). Moreover, it is customary in Dynamic Programming to use the notations x and t instead of x_0 and t_0 (x and t will appear as variables in the Hamilton–Jacobi–Bellman (HJB) equation), and therefore we will permanently adopt this latter notation in what follows.

The second ingredient in the definition of an optimal control problem is the cost functional to be minimized. We introduce the problem by describing a couple of classical examples.

8.1.1 ▪ The infinite horizon problem

Optimal control problems require the introduction of a *cost functional* $J : \mathscr{A} \to \mathbb{R}$ which is used to select the "optimal trajectory" for (8.2). In the case of the infinite horizon problem, we set $t_0 = 0$, $x_0 = x$, and this functional is defined as

$$J_x(\alpha) = \int_0^\infty g(y_x(s,\alpha), \alpha(s))e^{-\lambda s}\, ds \tag{8.4}$$

for a given $\lambda > 0$. The function g represents the *running cost*, and λ is the *discount factor* which allows us to compare the costs at different times by rescaling the costs at the initial time. From a technical point of view, the presence of the discount factor ensures that the integral is finite whenever g is bounded. We will assume in addition that g is globally Lipschitz continuous.

The goal of optimal control theory is to find an optimal pair (y^*, α^*) which minimizes the cost functional. If we look for optimal controls in open-loop form, i.e., as functions of t, then the main tool is given by the Pontryagin Maximum Principle, which gives the necessary conditions to be satisfied by an optimal couple (y^*, α^*).

A major drawback of an open-loop control is that, being constructed as a function of time, it cannot take into account errors in the real state of the system, due, for example, to model errors or external disturbances, which may bring the evolution far from the optimal forecasted trajectory. Another limitation of this approach is that it requires computing again the optimal control for any different initial state, so that it becomes impossible to design a static controller for the system.

For these reasons, we are interested in the so-called *feedback controls*, that is, controls expressed as functions of the state of the system. Under an optimal feedback control, if the system trajectory is modified by an external perturbation, then the system reacts by changing its control strategy according to the change in the state. One of the main motivations for using the Dynamic Programming approach is precisely that it allows us to characterize the optimal feedback, as we will see later in this chapter.

The starting point of Dynamic Programming is to introduce an auxiliary function, the *value function*, which in the case of the infinite horizon problem is defined as

$$v(x) = \inf_{\alpha \in \mathscr{A}} J_x(\alpha), \tag{8.5}$$

where, as above, x is the initial position of the system. The value function has a clear meaning: it is the optimal cost associated with the initial position x. This is a reference value which can be useful for evaluating the efficiency of a control—for example, if $J_x(\overline{\alpha})$ is close to $v(x)$, this means that $\overline{\alpha}$ is "efficient."

Bellman's DPP gives a first characterization of the value function.

Proposition 8.2 (DPP for the infinite horizon problem). *Under the assumptions of Theorem* 8.1, *for all* $x \in \mathbb{R}^d$ *and* $\tau > 0$,

$$v(x) = \inf_{\alpha \in \mathscr{A}} \left\{ \int_0^\tau g(y_x(s;\alpha), \alpha(s))e^{-\lambda s}\, ds + e^{-\lambda\tau} v(y_x(\tau;\alpha)) \right\}. \tag{8.6}$$

Proof. Denote by $\overline{v}(x)$ the right-hand side of (8.6). First, we remark that, for any $x \in \mathbb{R}^d$ and $\overline{\alpha} \in \mathscr{A}$,

$$\begin{aligned}
J_x(\overline{\alpha}) &= \int_0^\infty g(\overline{y}(s), \overline{\alpha}(s))e^{-\lambda s}\,ds \\
&= \int_0^\tau g(\overline{y}(s), \overline{\alpha}(s))e^{-\lambda s} + \int_\tau^\infty g(\overline{y}(s), \overline{\alpha}(s))e^{-\lambda s}\,ds \\
&= \int_0^\tau g(\overline{y}(s), \overline{\alpha}(s))e^{-\lambda s} + e^{-\lambda\tau}\int_0^\infty g(\overline{y}(s+\tau), \overline{\alpha}(s+\tau))e^{-\lambda s}\,ds \\
&\geq \int_0^\tau g(\overline{y}(s), \overline{\alpha}(s))e^{-\lambda s} + e^{-\lambda\tau}v(\overline{y}(\tau))
\end{aligned}$$

(here, $y_x(s, \overline{\alpha})$ is denoted for shortness as $\overline{y}(s)$). Passing to the infimum in the extreme terms of the inequality, we get

$$v(x) \geq \overline{v}(x). \tag{8.7}$$

To prove the opposite inequality, we recall that $\overline{v}$ is defined as an infimum, so that, for any $x \in \mathbb{R}^d$ and $\varepsilon > 0$, there exists a control $\overline{\alpha}_\varepsilon$ (and the corresponding evolution $\overline{y}_\varepsilon$) such that

$$\overline{v}(x) + \varepsilon \geq \int_0^\tau g(\overline{y}_\varepsilon(s), \overline{\alpha}_\varepsilon(s))e^{-\lambda s}\,ds + e^{-\lambda\tau}v(\overline{y}_\varepsilon(\tau)). \tag{8.8}$$

On the other hand, the value function v being defined also as an infimum, for any $x \in \mathbb{R}^d$ and $\varepsilon > 0$ there exists a control $\tilde{\alpha}_\varepsilon$ such that

$$v(\overline{y}_\varepsilon(\tau)) + \varepsilon \geq J_{\overline{y}_\varepsilon(\tau)}(\tilde{\alpha}_\varepsilon). \tag{8.9}$$

Plugging (8.9) into (8.8), we get

$$\begin{aligned}
\overline{v}(x) &\geq \int_0^\tau g(\overline{y}_\varepsilon(s), \overline{\alpha}_\varepsilon(s))e^{-\lambda s}\,ds + e^{-\lambda\tau}J_{\overline{y}_\varepsilon(\tau)}(\tilde{\alpha}_\varepsilon) - (1 + e^{-\lambda\tau})\varepsilon \\
&\geq J_x(\hat{\alpha}) - (1 + e^{-\lambda\tau})\varepsilon \\
&\geq v(x) - (1 + e^{-\lambda\tau})\varepsilon,
\end{aligned} \tag{8.10}$$

where $\hat{\alpha}$ is a control defined by

$$\hat{\alpha}(s) = \begin{cases} \overline{\alpha}_\varepsilon(s), & 0 \leq s < \tau, \\ \tilde{\alpha}_\varepsilon(s - \tau), & s \geq \tau. \end{cases} \tag{8.11}$$

Since ε is arbitrary, (8.10) finally yields $\overline{v}(x) \geq v(x)$. □

We note that this proof crucially relies on the fact that the control defined by (8.11) still belongs to $\mathscr{A}$, being a measurable control. The possibility of obtaining an admissible control by joining together two different measurable controls is known as *concatenation property*.

The DPP can be used to characterize the value function in terms of a nonlinear PDE. In fact, let $\alpha^* \in \mathscr{A}$ be the optimal control, and let y^* be the associated evolution (to simplify, we are assuming that the infimum is a minimum). Then,

$$v(x) = \int_0^\tau g(y^*(s), \alpha^*(s))e^{-\lambda s}\,ds + e^{-\lambda\tau}v(y^*(\tau)),$$

that is,

$$v(x)-e^{-\lambda\tau}v(y^*(\tau))=\int_0^\tau g(y^*(s),\alpha^*(s))e^{-\lambda s}ds,$$

so that adding and subtracting $e^{-\lambda\tau}v(x)$ and dividing by τ, we get

$$e^{-\lambda\tau}\frac{(v(x)-v(y^*(\tau)))}{\tau}+\frac{v(x)(1-e^{-\lambda\tau})}{\tau}=\frac{1}{\tau}\int_0^\tau g(y^*(s),\alpha^*(s))e^{-\lambda s}ds.$$

Assume now that v is regular. By passing to the limit for $\tau\to 0^+$, we have

$$\lim_{\tau\to 0^+}-\frac{v(y^*(\tau))-v(x)}{\tau}=-Dv(x)\cdot\dot{y}^*(x)=-Dv(x)\cdot f(x,\alpha^*(0)),$$

$$\lim_{\tau\to 0^+}v(x)\frac{(1-e^{-\lambda\tau})}{\tau}=\lambda v(x),$$

$$\lim_{\tau\to 0^+}\frac{1}{\tau}\int_0^\tau g(y^*(s),\alpha^*(s))e^{-\lambda s}ds=g(x,\alpha^*(0)).$$

Then, we can conclude that

$$\lambda v(x)-Dv(x)\cdot f(x,a^*)-g(x,a^*)=0,$$

where $a^*=\alpha^*(0)$. Rewriting the DPP in the equivalent form

$$v(x)+\sup_{\alpha\in\mathcal{A}}\left\{-\int_0^\tau g(y(s),\alpha(s))e^{-\lambda s}ds-e^{-\lambda\tau}v(y(\tau))\right\}=0,$$

we obtain the *HJB equation* (or *Dynamic Programming equation*) corresponding to the infinite horizon problem,

$$\lambda u(x)+\sup_{a\in A}\{-f(x,a)\cdot Du(x)-g(x,a)\}=0. \tag{8.12}$$

Note that, given x, the value of a achieving the max (assuming it exists) corresponds to the control $a^*=\alpha^*(0)$, and this makes it natural to interpret the argmax in (8.12) as the optimal feedback at x. We will get back to this point in what follows.

In short, (8.12) can be written as

$$H(x,u,Du)=0$$

with $x\in\mathbb{R}^d$, and

$$H(x,u,p)=\lambda u(x)+\sup_{a\in A}\{-f(x,a)\cdot p-g(x,a)\}. \tag{8.13}$$

Note that $H(x,u,\cdot)$ is convex (being the sup of a family of linear functions) and that $H(x,\cdot,p)$ is monotone (since $\lambda>0$), so that we are in the framework of the existence and uniqueness results presented in Chapter 2 (see Theorem 2.13).

8.1.2 ▪ Riccati equation and feedback controls for the Linear-Quadratic Regulator problem

In this problem, a linear (or linearized) controlled dynamical system must be stabilized minimizing a linear combination of, respectively, the state and the control energy. The dynamics is given by

$$\begin{cases}\dot{y}(s)=Ay(s)+B\alpha(s),\\ y(0)=x,\end{cases} \tag{8.14}$$

where $A \in \mathbb{R}^{d\times d}$ and $B \in \mathbb{R}^{d\times m}$, $x, y(s) \in \mathbb{R}^d$, and $\alpha(s) \in \mathbb{R}^m$. The cost functional is defined as

$$J_x(\alpha) = \frac{1}{2}\int_0^{+\infty} (y(s;\alpha)^t Q y(s;\alpha) + \alpha(s)^t R\alpha(s)) e^{-\lambda s} ds, \tag{8.15}$$

where $Q \in \mathbb{R}^{d\times d}$ and $R \in \mathbb{R}^{m\times m}$ are symmetric, strictly positive definite matrices, and $\lambda \ge 0$. We are not interested in giving the most general framework of the problem, but simply remark that it should be assumed that the system (8.14) is controllable, or that λ is large enough.

We show now that this special structure of the problem allows us to derive an explicit expression for the optimal control in feedback form, as $\alpha(s) = Ky(s)$, where K is an $m \times d$ matrix defined via the solution of the so-called *Riccati equation*.

To write the Bellman equation of the Linear-Quadratic Regulator (LQR), we assume that the value function has a quadratic structure:

$$v(x) = \frac{1}{2}x^t P x,$$

so that we also have $Dv(x) = Px$, for some matrix $P \in \mathbb{R}^{d\times d}$ to be determined. With this choice, (8.12) takes the form

$$\frac{\lambda}{2}x^t P x + \max_a \left\{-(Ax+Ba)^t Px - \frac{1}{2}x^t Qx - \frac{1}{2}a^t Ra\right\} = 0, \tag{8.16}$$

in which, given x, the argmax can be explicitly computed as

$$a^* = -R^{-1}B^t Px. \tag{8.17}$$

Plugging (8.17) into (8.16), we get

$$\frac{\lambda}{2}x^t Px - x^t A^t Px - \frac{1}{2}x^t Qx + \frac{1}{2}x^t P^t BR^{-1}B^t Px = 0,$$

that is,

$$\frac{1}{2}x^t \left[(\lambda I - 2A^t)P - Q + P^t BR^{-1}B^t P\right]x = 0, \tag{8.18}$$

for any $x \in \mathbb{R}^d$. Therefore, the unknown matrix P should solve the quadratic *Riccati equation*

$$P^t BR^{-1}B^t P + (\lambda I - 2A^t)P - Q = 0.$$

Once we compute the matrix P, the optimal control is obtained as a function of the current state $y(s)$ by means of (8.17), and more precisely

$$\alpha^*(s) = -R^{-1}B^t Py^*(s). \tag{8.19}$$

To check that (8.19) is indeed the optimal control, define the function

$$G(s;\alpha) = \frac{e^{-\lambda s}}{2}(y(s)^t Qy(s) + \alpha(s)^t R\alpha(s)) + e^{-\lambda s}y(s)^t P(Ay(s) + B\alpha(s)) - \frac{\lambda e^{-\lambda s}}{2}y(s)^t Py(s).$$

Now, substituting the expression (8.19) for α^*, and using the Riccati equation and some elementary algebra, we can show that

$$G(s;\alpha^*)=0 \quad \text{for any } s\in\mathbb{R}^+. \tag{8.20}$$

On the other hand, we also have

$$G(s;\alpha)=\frac{e^{-\lambda s}}{2}(y(s)^tQy(s)+\alpha(s)^tR\alpha(s))+\frac{d}{ds}\left[e^{-\lambda s}v(y(s))\right],$$

so that, using the control α^* and integrating on $[0,T]$, we get

$$\frac{1}{2}\int_0^T (y^*(s)^tQy^*(s)+\alpha^*(s)^tR\alpha^*(s))e^{-\lambda s}ds+e^{-\lambda T}v(y^*(T))-v(x).$$

Passing to the limit for $T\to\infty$ and taking into account (8.20), we obtain that, in the class of controls for which $|y(s)|^2=o(e^{\lambda s})$,

$$\frac{1}{2}\int_0^\infty (y^*(s)^tQy^*(s)+\alpha^*(s)^tR\alpha^*(s))e^{-\lambda s}ds=v(x);$$

that is, the pair (y^*,α^*) is optimal.

The results above show the relationship between the max operator appearing in the Bellman equation and the feedback form of the optimal control. They also explain the popularity of the LQR problem in control literature: the optimal feedback control can be obtained via the solution of a matrix equation and has an explicit form. Actually, in many real problems of stabilization, the first try is to linearize the dynamics and approximate the cost by quadratic terms in order to directly apply the solution of the Riccati equation to the linearized problem.

8.2 ▪ Dynamic Programming for other classical control problems

Given a controlled dynamical system of the form (8.1)–(8.2), it is possible to consider other optimal control problems with different forms of the cost functional. We briefly review other classical problems and give their corresponding Bellman equations.

8.2.1 ▪ The finite horizon problem

Assume the control system has the form (8.2), with $t_0=t$, $x_0=x$. In the *Bolza* formulation, the functional of the *finite horizon problem* has the form

$$J_{x,t}(\alpha)=\int_t^T g(y_{x,t}(s;\alpha),\alpha(s))e^{-\lambda s}ds+e^{-\lambda(T-t)}\psi(y_{x,t}(T;\alpha)) \tag{8.21}$$

in which $t\in[0,T)$ and $\lambda\geq 0$. For the finite horizon problem we can follow the same arguments already seen for the infinite horizon problem. The first step is the definition of the value function

$$v(x,t)=\inf_{\alpha\in\mathcal{A}} J_{x,t}(\alpha)$$

which in turn can be proved to satisfy the following DPP.

Proposition 8.3 (DPP for the finite horizon problem). *Under the assumptions of Theorem 8.1, for all $x \in \mathbb{R}^d$ and $t < \tau \le T$,*

$$v(x,t) = \inf_{\alpha \in \mathcal{A}} \left\{ \int_t^\tau g(y_{x,t}(s;\alpha),\alpha(s))e^{-\lambda s}\,ds + e^{-\lambda(\tau - t)} v(y_{x,t}(\tau;\alpha),\tau) \right\}. \tag{8.22}$$

Then, in a similar way, it is possible to derive the Bellman equation in the form of an evolutive HJ equation with a terminal condition:

$$\begin{cases} -u_t + \lambda u + \sup_{a \in A}\{-f(x,a)\cdot Du - g(x,a)\} = 0, & (x,t) \in \mathbb{R}^d \times (0,T), \\ u(x,T) = \psi(x), & x \in \mathbb{R}^d. \end{cases} \tag{8.23}$$

We refer the interested reader to [E98, p. 557] for a complete proof.

8.2.2 ▪ The optimal stopping problem

In this problem, besides the usual control α, it is possible to stop the dynamics at any intermediate time $\theta \in [0,+\infty]$ (respectively, $\theta \in [t,T]$ in the finite horizon case), paying a corresponding final cost given by $\psi(y(\theta))$ (respectively, $\psi(y(T \wedge \theta))$). In the infinite horizon case, the cost functional defining the *optimal stopping problem* is therefore given by

$$J_x(\theta,\alpha) = \int_0^\theta g(y_x(s;\alpha),\alpha(s))e^{-\lambda s}\,ds + e^{-\lambda\theta}\psi(y(\theta;\alpha)). \tag{8.24}$$

Then, for $x \in \mathbb{R}^d$, $\theta \in [0,+\infty]$, and $\alpha \in \mathcal{A}$, the value function is defined as

$$v(x) = \inf_{\theta,\alpha} J_x(\theta,\alpha).$$

This problem can be rephrased as a standard infinite horizon (respectively, finite horizon) problem by adding to the control set A a new control $\bar{a}$, such that $f(y,\bar{a}) = 0$ and $g(y,\bar{a}) \equiv 0$ for any $y \in \mathbb{R}^d$. A more explicit way of treating the problem is to derive a DPP which takes into account the option of stopping at a finite time and then obtain the Bellman equation, which has the form of an obstacle problem:

$$\max(H(x,u,Du), u - \psi) = 0, \tag{8.25}$$

where the Hamiltonian function H is defined as in (8.12). The set on which $u = \psi$ corresponds to the so-called *stopping set*, in which the optimal strategy is precisely to stop the system.

8.2.3 ▪ The minimum time problem

Another classical example is the *minimum time problem*. We assume a dynamics in the form (8.2), which has to be steered in the shortest possible time to a *target set* $\mathcal{T} \subset \mathbb{R}^d$. This set is given, and we will always assume the following:

The target set $\mathcal{T}$ is closed and $\mathrm{int}(\mathcal{T}) \ne \emptyset$, $\partial\mathcal{T}$ is sufficiently regular.

For every control $\alpha \in \mathcal{A}$ we can define the first time of arrival on the target $\mathcal{T}$ as

$$t_x(\alpha) = \inf_s \left\{ s \in \mathbb{R}^+ : y_x(s;\alpha) \in \mathcal{T} \right\},$$

provided the set at the right-hand side in nonempty. If the set is empty, that is, if $y_x(s;\alpha) \notin \mathcal{T}$ for any $s > 0$, it is natural to define $t_x(\alpha) = +\infty$. This leads to the following definition:

$$t_x(\alpha) = \begin{cases} +\infty & \text{if } y_x(s;\alpha) \notin \mathcal{T} \ \forall s > 0, \\ \inf_s \left\{ s \in \mathbb{R}^+ : y_x(s;\alpha) \in \mathcal{T} \right\} & \text{else.} \end{cases} \tag{8.26}$$

Since we want to minimize the time of arrival on the target, the value function for this problem is given by

$$T(x) = \inf_{\alpha \in \mathcal{A}} t_x(\alpha).$$

A basic example, discussed in Chapter 1, is the case in which

$$\dot{y}(s) = \alpha(s)$$

with $\alpha(s) \in B(0,1)$. In this case, the minimum time $T(x)$ equals the distance $d(x, \mathcal{T})$ and can be a nonsmooth function whenever $\mathcal{T}$ is nonconvex.

Dynamic Programming for the minimum time problem Consider a system driven by (8.2). Since the minimum time function (i.e., the value function for this problem) may not be finite everywhere, we need to define its domain.

Definition 8.4 (Reachable set). *The reachable set is defined as the set of initial states from which it is possible to reach the target, that is,*

$$\mathcal{R} := \left\{ x \in \mathbb{R}^d : T(x) < +\infty \right\}.$$

The reachable set clearly depends on the target, on the dynamics, and on the set of admissible controls. However, this dependence is complex even in simple situations, so that the reachable set *cannot* be regarded as a datum in our problem. This means that we have to determine the pair $(T, \mathcal{R})$ as the solution of a free boundary problem.

In order to derive the HJB equation for the minimum time problem, we state the related DPP.

Proposition 8.5 (DPP for the minimum time problem). *For all $x \in \mathcal{R}$ and $\tau \in (0, T(x))$ (so that $x \notin \mathcal{T}$),*

$$T(x) = \inf_{\alpha \in \mathcal{A}} \{\tau + T(y_x(\tau;\alpha))\}. \tag{8.27}$$

The proof of this DPP basically relies on the same arguments used for the infinite horizon case, and uses the possibility of generating new admissible controls by concatenation and/or translation. To derive the HJB equation from the DPP, we rewrite (8.27) for $t \in (0, T(x))$ as

$$T(x) - \inf_{\alpha \in \mathcal{A}} T(y_x(\tau;\alpha)) = \tau$$

and divide by τ, obtaining

$$\sup_{\alpha \in \mathcal{A}} \left\{ \frac{T(x) - T(y_x(\tau;\alpha))}{\tau} \right\} = 1.$$

In order to pass to the limit as $\tau \to 0^+$ we assume, for the moment, that T is differentiable at x and that $\lim_{\tau \to 0^+}$ commute with $\sup_\alpha$. Provided $\dot{y}_x(0;a)$ exists, we get

$$\sup_{\alpha \in \mathcal{A}} \{-DT(x) \cdot \dot{y}_x(0,\alpha)\} = 1,$$

so that, if $\lim_{\tau\to 0^+} \alpha(\tau)=a$, we obtain

$$\sup_{a\in A}\{-DT(x)\cdot f(x,a)\}=1, \tag{8.28}$$

which is the HJB equation associated with the minimum time problem. This derivation of the HJB equation basically shows that if T is regular, then it is a classical solution of (8.28), i.e., proves the following proposition.

Proposition 8.6. *If $T\in C^1$ in a neighborhood of $x\in\mathscr{R}\setminus\mathscr{T}$, then T satisfies* (8.28) *at x.*

However, (8.28) needs to be complemented with a boundary condition, the "natural" condition being

$$T(x)=0 \qquad (x\in\partial\mathscr{T}).$$

Therefore, once we define the Hamiltonian

$$H(x,p):=\sup_{a\in A}\{-p\cdot f(x,a)\}-1,$$

we can rewrite (8.28) as a boundary value problem in the compact form

$$\begin{cases} H(x,DT(x))=0, & x\in\mathscr{R}\setminus\mathscr{T},\\ T(x)=0, & x\in\partial\mathscr{T}.\end{cases} \tag{8.29}$$

Note again that $H(x,\cdot)$ is convex since it is the sup of linear operators.

We have already seen that T may not be differentiable even in the simplest cases. In addition, the continuity of T around the target may also fail to hold, unless the following property of Small-Time Local Controllability (STLC) around $\mathscr{T}$ is assumed.

Definition 8.7 (STLC). *Assume $\partial\mathscr{T}$ is smooth. Then, the property of STLC is satisfied if, for any $x\in\partial\mathscr{T}$, there exists a control vector $\widehat{a}\in A$ such that*

$$f(x,\widehat{a})\cdot\nu(x)<0,$$

where $\nu(x)$ is the exterior normal to $\partial\mathscr{T}$ at x.

STLC ensures that $\mathscr{R}$ is an open subset of $\mathbb{R}^d$ and that T is continuous up to $\partial\mathscr{T}$. Moreover, for all $z\in\partial\mathscr{R}$,

$$\lim_{x\to z} T(x)=+\infty.$$

With this last assumption, it is possible to prove that $T(\cdot)$ is a viscosity solution of the boundary value problem (8.29).

Theorem 8.8. *If $\mathscr{R}\setminus\mathscr{T}$ is open and T is continuous, then T is a viscosity solution of* (8.29).

Proof. For τ small enough, by the DPP we have that, for any $\overline{\alpha}\in\mathscr{A}$,

$$T(x)-T(\overline{y}(\tau))\le\tau \tag{8.30}$$

and that, given $\varepsilon>0$, there exists an $\alpha_\varepsilon\in\mathscr{A}$ such that

$$T(x)-T(y_\varepsilon(\tau))\ge t(1-\varepsilon), \tag{8.31}$$

where, as usual, $\overline{y}(\tau)=y_x(\tau;\overline{\alpha})$ and $y_\varepsilon(\tau)=y_x(\tau;\alpha_\varepsilon)$.

Consider now a test function $\phi \in C^1(\mathbb{R}^d \setminus \mathcal{T})$ such that $T-\phi$ has a local maximum at x. Then, for τ small enough, we have

$$T(x)-\phi(x) \geq T(\overline{y}(\tau))-\phi(\overline{y}(\tau)),$$

and by (8.30)

$$\phi(x)-\phi(\overline{y}(\tau)) \leq T(x)-T(\overline{y}(\tau)) \leq \tau,$$

so that, dividing by τ and passing to the limit for $\tau \to 0^+$ (as it has been done in the derivation of (8.28)), we get

$$H(x, D\phi(x)) \leq 0.$$

Similarly, if $\phi \in C^1$ and $T-\phi$ has a local minimum at x, for τ small enough we have

$$T(x)-\phi(x) \leq T(y_\varepsilon(\tau))-\phi(y_\varepsilon(\tau)),$$

and by (8.31)

$$\phi(x)-\phi(y_\varepsilon(\tau)) \geq T(x)-T(y_\varepsilon(\tau)) \geq \tau(1-\varepsilon).$$

Since ε is arbitrary, dividing by τ and passing to the limit, we obtain

$$H(x, D\phi(x)) \geq 0,$$

which fulfills the definition of viscosity solution.

Using the STLC property, it is also possible to prove that the boundary condition is satisfied on $\partial\mathcal{T}$. This part of the proof will be skipped. □

A different approach to the solution of (8.28) is to perform a suitable rescaling of T, obtained via the auxiliary function

$$v(x) := \begin{cases} 1/\mu & \text{if } T(x) = +\infty, \\ 1/\mu - e^{-\mu T(x)} & \text{else} \end{cases} \tag{8.32}$$

for some positive μ. It is easy to check that v is itself a value function, that is,

$$v(x) = \inf_{\alpha \in \mathcal{A}} J_x(\alpha),$$

where J_x is given by

$$J_x(\alpha) = \int_0^{t_x(\alpha)} e^{-\mu s} ds$$

and $t_x(\alpha)$ is defined by (8.26). This control problem is nothing but an infinite horizon problem (with constant running cost $g(x,a) \equiv 1$) in which the state of the system is stopped as soon as it reaches the target $\mathcal{T}$, with zero stopping cost. The Dynamic Programming equation is in the form of a boundary value problem:

$$\begin{cases} \mu v(x) + \sup_{a \in A} \{-Dv \cdot f(x,a) - 1\} = 0, & x \in \mathbb{R}^d \setminus \mathcal{T}, \\ v(x) = 0, & x \in \partial\mathcal{T}. \end{cases} \tag{8.33}$$

The change of variables (8.32) is called the *Kružkov transformation* and gives several advantages. First, $v(x)$ takes values in $[0,1]$ (whereas T is generally unbounded), and this helps in both the analysis and in the numerical approximation. Second, the formulation

(8.33) avoids any reference to the reachable set. Once one obtains v, the minimum time T and reachable set $\mathcal{R}$ can be recovered by the relationships

$$T(x) = -\frac{1}{\mu}\ln\left(\frac{1}{\mu} - v(x)\right), \qquad \mathcal{R} = \left\{x \in \mathbb{R}^d : v(x) < \frac{1}{\mu}\right\}.$$

In addition, the formulation as an infinite horizon problem allows the time-marching numerical scheme to be a contraction mapping. We will come back to this point in the section on numerical schemes.

8.3 ▪ The link between Bellman equation and Pontryagin Maximum Principle

There exists an important link between the DPP and the Pontryagin Maximum Principle (PMP in what follows), which gives a set of necessary conditions to characterize the optimal open-loop solution $(y^*(\cdot), \alpha^*(\cdot))$. Using this relationship, we derive here the PMP for the finite horizon problem (8.1), (8.21) under additional simplifying assumptions on the data. In particular, we will assume that $\psi \in C^2(\mathbb{R}^d)$, that Df is continuous, and, above all, that $v \in C^2(\mathbb{R}^d)$ so that the HJ equation (8.23) is verified in a classical sense. We also assume, for simplicity, that $g \equiv 0$ and $\lambda = 0$ so that, by the DPP, the function

$$h(s) := v(y(s;\alpha), s)$$

is nondecreasing for every α and is constant only for the optimal trajectory. Then, an optimality condition for a pair (y^*, α^*) is that, for any $s \in [0, T]$,

$$h'(s) = f(y^*(s), \alpha^*(s)) \cdot Dv(y^*(s), s) + v_s(y^*(s), s) = 0. \tag{8.34}$$

On the other hand, since at any point $(z, s) \in \mathbb{R}^d \times [0, T]$ the HJB equation gives

$$-v_t(z, s) - f(z, a) \cdot Dv(z, s) \leq 0 \qquad \text{for every } a \in A, \tag{8.35}$$

setting $z = y^*(s)$, by (8.34)–(8.35) we obtain an equivalent condition known as the Minimum Principle:

$$f(y^*(s), \alpha^*(s)) \cdot Dv(y^*(s), s) = \min_{a \in A} f(y^*(s), a) \cdot Dv(y^*(s), s).$$

Now, in order to prove the PMP, we recall that, by (8.34), the left-hand side of (8.35) vanishes at $(z, s) = (y^*(s), s)$ for $a = \alpha^*(s)$ if the control α^* is optimal. This implies that the trajectory $y^*(s)$ maximizes the map $z \to -v_t(z, s) - Dv(z, s) \cdot f(z, \alpha^*(s))$ and that, due to our regularity assumptions, we can differentiate this map with respect to z obtaining

$$-Dv_t - J_f Dv - D_v^2 f = 0, \tag{8.36}$$

in which J_f is the $d \times d$ Jacobian matrix for f and D_v^2 is the $d \times d$ Hessian matrix for v. Let us now define the co-state or adjoint vector $p(t)$ as the vector

$$p(s) := Dv(y(s)), s).$$

Differentiating with respect to s we obtain, by (8.36) and the regularity of v,

$$\dot{p} = D_v^2 f - Dv_t = -J_f p. \tag{8.37}$$

This means that the co-state associated with $(y(s),s)$ and α^* can be redefined, without making reference to the value function, as the unique solution on $[0,T]$ of the system of linear differential equations

$$\dot{p}(s) = -J_f(y(s),\alpha(s))p(s), \tag{8.38}$$

with the terminal condition

$$p(T) = D\psi(y(T)). \tag{8.39}$$

The necessary optimality conditions on the couple $(y^*(t),\alpha^*(t))$ are then written as

$$-p^*(t)\cdot f(y^*(t),\alpha^*(t)) = \max_{a\in A}\{-p(t)\cdot f(y(t),a)\}. \tag{8.40}$$

We can summarize this simplified analysis in the following theorem.

Theorem 8.9. *Assume $v \in C^2(\mathbb{R}^d)$, A is compact, $\psi \in C^2(\mathbb{R}^d)$, and f is continuously differentiable with respect to x. Let $y(\cdot) = y_{x_0}(\cdot;\alpha)$ solve* (8.2), *and let the co-state $p(\cdot)$ be the corresponding solution of* (8.38)–(8.39). *Then, a necessary condition for a solution $(y^*(t),p^*(t),\alpha^*(t))$ to be optimal is that, for almost any $t\in(0,T]$,*

$$-p^*(t)\cdot f(y^*(t),\alpha^*(t)) = \max_{a\in A}\{-p(t)\cdot f(y(t),a)\}.$$

In a remarkably technical way, the PMP can also be derived in a more general setting which, in particular, does not require the regularity assumptions on the data and on the value function. We refer the interested reader to [BCD97] for a more general proof of this result.

8.4 ▪ SL schemes for optimal control problems

The construction of SL schemes for Dynamic Programming equations follows the same guidelines shown in Chapters 5–6 for the simpler model problems. In fact, (5.73) or (5.132) may be seen as special cases of HJB equations for control problems in which

$$\begin{cases} f(x,a) = a, \\ g(x,a) = H^*(a), \\ \psi(x) = u_0(x) \end{cases} \tag{8.41}$$

(with the minor difference that (5.73) is a forward, instead of a backward, equation). When considering more general Dynamic Programming equations, it should be clear that optimal trajectories play the role of characteristics, but they are no longer expected to be straight lines. Therefore, in a sense, time discretization must combine strategies used in both advection and convex HJ equations.

For all these reasons, as well as for the sake of obtaining explicit error estimates, we are led to consider time discretization as an intermediate (semi-)discretization step. This step also has the meaning of approximating the original control problem with a problem in discrete time, and is in fact the situation in which the DPP was first formulated and applied.

Next, the space discretization is performed as in Chapter 5, except for replacing the Hopf–Lax formula with the discrete version of the DPP.

This section will review the construction and convergence analysis of SL schemes for HJB equations, with a special emphasis on the control problems examined in sections

8.1–8.2. For simplicity, we will consider schemes on the whole of $\mathbb{R}^d$ and low-order (monotone) implementations. The treatment of boundary conditions, along with the construction of high-order versions, can be performed with the techniques discussed in Chapters 5–6.

8.4.1 ▪ The infinite horizon problem

We sketch the basic ideas in the case of the infinite horizon problem. The plan is to construct a time-marching scheme, starting with time discretization.

Time discretization In order to give a discrete time approximation of the control problem, we fix a time step Δt and set $t_m = m\Delta t$ $(m \in \mathbb{N})$. The simplest way to discretize (8.2) is by using the explicit Euler scheme, which corresponds to the following discrete dynamical system:

$$\begin{cases} y_{m+1} = y_m + \Delta t f(y_m, a_m), \\ y_0 = x. \end{cases} \tag{8.42}$$

Here, the sequence of vectors $a_m \in A$ has the role of a (discrete) control. We will denote by $\alpha^{\Delta t} = \{a_m\}$ the sequence as a whole, and possibly identify this sequence (with a slight abuse of notation) with the continuous, piecewise constant control defined by

$$\alpha^{\Delta t}(s) = a_m, \qquad s \in [t_m, t_m + 1). \tag{8.43}$$

Whenever we will need to emphasize the dependence of the trajectory on x and $\alpha^{\Delta t}$ we will use the notation $y_m(x;\alpha^{\Delta t})$.

Remark 8.10. *It is also possible to implement higher-order time discretizations by using, e.g., Runge–Kutta-type schemes in* (8.42). *We refer the reader to* [FF94] *for a more extensive study of this technique, whereas an example of application will be given in the numerical tests section.*

Given the time discretization (8.42) for the controlled system, the corresponding discrete version of the cost functional J_x may be obtained by a rectangle quadrature:

$$J_x^{\Delta t}\left(\alpha^{\Delta t}\right) := \Delta t \sum_{m=0}^{+\infty} g(y_m, a_m) e^{-\lambda t_m}, \tag{8.44}$$

in which the dependence on $\alpha^{\Delta t}$ also appears via the trajectory points y_m.

The discrete dynamics (8.42) and cost functional (8.44) define a discrete time control problem whose value function is

$$v^{\Delta t}(x) := \inf_{\alpha^{\Delta t}} J_x^{\Delta t}\left(\alpha^{\Delta t}\right). \tag{8.45}$$

Adapting the arguments of the continuous case, it is possible to prove a Discrete DPP (DDPP) for the optimal control problem (8.42)–(8.44).

Proposition 8.11 (DDPP for the infinite horizon problem). *Fix $\Delta t > 0$. Then, for all $x \in \mathbb{R}^d$ and any positive integer $\overline{n}$,*

$$v^{\Delta t}(x) = \inf_{\alpha^{\Delta t}} \left\{ \Delta t \sum_{m=0}^{\overline{n}-1} g(y_m, a_m) e^{-\lambda t_m} + e^{-\lambda t_{\overline{n}}} v^{\Delta t}(y_{\overline{n}}) \right\}. \tag{8.46}$$

The discrete time version of the SL scheme can be obtained by choosing $\overline{n} = 1$ in (8.46). This gives an approximation in the fixed-point form

$$v^{\Delta t}(x) = \min_{a \in A} \left\{ \Delta t\, g(x,a) + e^{-\lambda \Delta t} v^{\Delta t}(x + \Delta t f(x,a)) \right\} \tag{8.47}$$

in which $e^{-\lambda \Delta t}$ is sometimes replaced by its first-order Taylor approximation $1 - \lambda \Delta t$ (this has no influence on the consistency rate, but introduces a constraint on Δt to get a contraction in the right-hand side of (8.47)).

Space discretization The space discretization of the semidiscrete approximation (8.47) is performed in a completely standard way by replacing the value $v^{\Delta t}(x + \Delta f(x,a))$ by a $\mathbb{P}_1$ or $\mathbb{Q}_1$ interpolation. We end up with the fully discrete scheme

$$v_j = \min_{a \in A} \left\{ \Delta t\, g(x_j, a) + e^{-\lambda \Delta t} I_1[V](x_j + \Delta t f(x_j, a)) \right\}. \tag{8.48}$$

Denoting the right-hand side of (8.48) as $S(\Delta, V)$, the scheme may be written as

$$V = S(\Delta, V)$$

or also, in iterative form, as

$$V^{(k+1)} = S(\Delta, V^{(k)}) \tag{8.49}$$

where k is the iteration index. An iterative algorithm in the form (8.49), complemented with a suitable stopping criterion, is also termed a *value iteration* in the control literature.

An analysis of the scheme along the guidelines of Chapter 5 shows the following:

- $S(\Delta, \cdot)$ is monotone.
- $S(\Delta, \cdot)$ is a contraction in l^∞, and therefore $V^{(k)}$ converges towards a unique solution $V \in l^\infty$ for any initial guess $V^{(0)}$. The contraction coefficient is easily shown to be $L_S = e^{-\lambda \Delta t}$, and hence, as usual in time-marching schemes, convergence may become very slow as $\Delta t \to 0$. Suitable acceleration techniques (exploiting monotonicity of the operator S) have been constructed to handle this problem (see subsection 8.4.7 for more details).
- The scheme is l^∞ stable if the running cost g is bounded. In fact, since the interpolation I_1 is nonexpansive, from (8.48) we have, assuming that $V^{(0)} = 0$,

$$\begin{aligned}
\left\| V^{(k+1)} \right\|_\infty &\le \Delta t \|g\|_\infty + e^{-\lambda \Delta t} \left\| V^{(k)} \right\|_\infty \\
&\le \cdots \le \Delta t \|g\|_\infty \left(1 + e^{-\lambda \Delta t} + e^{-2\lambda \Delta t} + \cdots \right) \\
&\le \frac{\Delta t \|g\|_\infty}{1 - e^{-\lambda \Delta t}} = \frac{\|g\|_\infty}{\lambda}(1 + O(\Delta t)).
\end{aligned}$$

Convergence analysis First, we point out that a first convergence analysis might be performed by means of the Barles–Souganidis theorem, the scheme being monotone. When passing from the Hopf–Lax formula to the DPP, however, consistency analysis (if carried out with the arguments of Chapter 5) requires more caution.

Rather than duplicating the analysis performed in Chapter 5, we will apply here the same arguments to estimate the rate of convergence of the approximate solution to the value function, i.e., to the solution of the Bellman equation (8.12). At this level, it can be convenient to separate time from space discretization.

Concerning time discretization, we first assume that v is Lipschitz continuous and in addition, that the following hold:

(i) $\|f\|$ is uniformly bounded, and A is convex. Moreover, $f(x,a)$ is linear with respect to a; i.e., it has the structure

$$f(x,a) = f_1(x) + F_2(x)a$$

with $f_1 : \mathbb{R}^d \to \mathbb{R}^d$ and $F_2(x)$ a $d \times m$ matrix.

(ii) $g(\cdot,a)$ is Lipschitz continuous, and $g(x,\cdot)$ is convex.

(iii) There exists an optimal control $\alpha^* \in \mathscr{A}$ (this could be derived from the two assumptions above, but we state it explicitly).

First, making explicit the minimum in (8.6) and (8.47), we can rewrite v and $v^{\Delta t}$ as

$$v(x) = \int_0^{\Delta t} g(y_x(s;\alpha^*),\alpha^*(s))e^{-\lambda s}ds + e^{-\lambda\Delta t}v(y_x(\Delta t;\alpha^*)),$$
$$v^{\Delta t}(x) = \Delta t\, g(x,a^*) + e^{-\lambda\Delta t}v^{\Delta t}(x + \Delta t f(x,a^*)).$$

Then, we can bound $v^{\Delta t}(x) - v(x)$ from above by using in the continuous problem the constant control $\alpha(s) \equiv a^*$ (which is suboptimal), obtaining

$$v(x) - v^{\Delta t}(x) \le \int_0^{\Delta t} g(y_x(s;a^*),a^*)e^{-\lambda s}ds + e^{-\lambda\Delta t}v(y_x(\Delta t;a^*))$$
$$- \Delta t\, g(x,a^*) - e^{-\lambda\Delta t}v^{\Delta t}(x + \Delta t f(x,a^*)).$$

Now, by elementary approximation arguments,

$$\int_0^{\Delta t} g(y_x(s;a^*),a^*)e^{-\lambda s}ds - \Delta t\, g(x,a^*) = O\left(\Delta t^2\right),$$

whereas

$$\begin{aligned} v(y_x(\Delta t;a^*)) - v^{\Delta t}(x + \Delta t f(x,a^*)) &\le v(y_x(\Delta t;a^*)) - v(x + \Delta t f(x,a^*)) \\ &\quad + v(x + \Delta t f(x,a^*)) - v^{\Delta t}(x + \Delta t f(x,a^*)) \\ &\le O\left(\Delta t^2\right) + \left\|v - v^{\Delta t}\right\|_\infty, \end{aligned}$$

where the first term follows from the Lipschitz continuity of v and elementary results for the Euler scheme. We can therefore deduce that

$$v(x) - v^{\Delta t}(x) \le O\left(\Delta t^2\right) + e^{-\lambda\Delta t}\left\|v - v^{\Delta t}\right\|_\infty. \tag{8.50}$$

Note that this unilateral estimation does not require assumptions (i) and (ii) above.

To perform the reverse estimate, we should replace the discrete optimal control a^* by a suboptimal $\overline{a}$, so that

$$\begin{aligned} v^{\Delta t}(x) - v(x) \le\; &\Delta t\, g(x,\overline{a}) + e^{-\lambda\Delta t}v^{\Delta t}(x + \Delta t f(x,\overline{a})) \\ &- \int_0^{\Delta t} g(y_x(s;\alpha^*),\alpha^*(s))e^{-\lambda s}ds - e^{-\lambda\Delta t}v(y_x(\Delta t;\alpha^*)). \end{aligned} \tag{8.51}$$

A suitable choice for $\overline{a}$ is to define it as the integral mean of α^* over $[0,\Delta t]$:

$$\overline{a}=\frac{1}{\Delta t}\int_0^{\Delta t}\alpha^*(s)ds.$$

In fact, with this choice we can write $y^*(\Delta t)=y_x(\Delta t;\alpha^*)$ as

$$\begin{aligned}
y^*(\Delta t)&=x+\int_0^{\Delta t}f_1(y^*(s))ds+\int_0^{\Delta t}F_2(y^*(s))\alpha^*(s)ds\\
&=x+\int_0^{\Delta t}[f_1(x)+O(\Delta t)]ds+\int_0^{\Delta t}[F_2(x)+O(\Delta t)]\alpha^*(s)ds\\
&=x+\Delta t f_1(x)+F_2(x)\int_0^{\Delta t}\alpha^*(s)ds+O\left(\Delta t^2\right)\\
&=x+\Delta t[f_1(x)+F_2(x)\overline{a}]+O\left(\Delta t^2\right)\\
&=x+\Delta t f(x,\overline{a})+O\left(\Delta t^2\right).
\end{aligned}\tag{8.52}$$

On the other hand,

$$\begin{aligned}
\int_0^{\Delta t}g(y^*(s),\alpha^*(s))e^{-\lambda s}ds&=\int_0^{\Delta t}[g(x,\alpha^*(s))+O(\Delta t)](1+O(\Delta t))ds\\
&=\int_0^{\Delta t}g(x,\alpha^*(s))ds+O\left(\Delta t^2\right)\\
&\geq \Delta t\, g(x,\overline{a})+O\left(\Delta t^2\right),
\end{aligned}\tag{8.53}$$

where the last inequality follows from Jensen's inequality, due to the convexity of $g(x,\cdot)$. Using (8.52)–(8.53) in (8.51) and retracing the proof of (8.50), we get

$$v^{\Delta t}(x)-v(x)\leq O\left(\Delta t^2\right)+e^{-\lambda\Delta t}\left\|v-v^{\Delta t}\right\|_\infty,\tag{8.54}$$

which gives, taking into account (8.50) itself,

$$\left|v^{\Delta t}(x)-v(x)\right|\leq O\left(\Delta t^2\right)+e^{-\lambda\Delta t}\left\|v-v^{\Delta t}\right\|_\infty.\tag{8.55}$$

Due to the assumptions on f and g, the $O(\Delta t^2)$ term is uniform with respect to x, so we can pass to the ∞-norm in (8.55), obtaining that, for some positive constant C,

$$\left(1-e^{-\lambda\Delta t}\right)\left\|v^{\Delta t}-v\right\|_\infty\leq C\Delta t^2,$$

and therefore

$$\left\|v^{\Delta t}-v\right\|_\infty\leq C\Delta t.$$

This estimate has been proved (with different techniques) in [CDI84], under the assumption that v is semiconcave. If v is only Lipschitz continuous and assumptions (i)–(iii) above are dropped, then

$$\left\|v^{\Delta t}-v\right\|_\infty\leq C\Delta t^{1/2}.$$

Note that a sufficient condition for the Lipschitz continuity of v is to have Lipschitz continuous data (f and g) and $\lambda>L_f$.

For the fully discrete scheme (8.48), we estimate the space discretization error comparing with the same technique above v_j and $v^{\Delta t}(x_j)$. Using the control a^*, optimal for $v^{\Delta t}$, and denoting $z_j^* = x_j + \Delta t f(x_j, a^*)$, we obtain the one-sided estimate:

$$\begin{aligned} v_j - v^{\Delta t}(x_j) &\le \Delta t\, g(x_j, a^*) + e^{-\lambda \Delta t} I_1[V](z_j^*)) \\ &\quad - \Delta t\, g(x_j, a^*) + e^{-\lambda \Delta t} v^{\Delta t}(z_j^*)) \\ &\le e^{-\lambda \Delta t} \left| I_1[V](z_j^*) - I_1\left[V^{\Delta t}\right](z_j^*) \right| \\ &\quad + e^{-\lambda \Delta t} \left| I_1\left[V^{\Delta t}\right](z_j^*) - v^{\Delta t}(z_j^*) \right| \\ &\le e^{-\lambda \Delta t} \left\| V - V^{\Delta t} \right\|_\infty + O(\Delta x), \end{aligned}$$

where $V^{\Delta t}$ is the vector of samples of $v_{\Delta t}$ and, by the Lipschitz continuity of $v_{\Delta t}$, we have bounded its interpolation error with $O(\Delta x)$. Moreover, we have used the monotonicity of the reconstruction I_1 to bound the error at z_j^* with the error on the nodes.

Giving the opposite bound and passing to the ∞-norm, we obtain

$$\left\| V^{\Delta t} - V \right\|_\infty \le C \frac{\Delta x}{\Delta t}.$$

Combining time and space discretization error estimates, we have the complete estimate:

$$|v_j - v(x_j)| \le C_1 \Delta t^\gamma + C_2 \frac{\Delta x}{\Delta t},$$

where $\gamma = 1/2$ or $\gamma = 1$, depending on the assumptions. In the former case, the best coupling is obtained for $\Delta x = \Delta t^{3/2}$, and in the latter for $\Delta x = \Delta t^2$.

8.4.2 ▪ The finite horizon problem

Let us briefly review the results (proved in [FG98]) for the finite horizon problem. Time discretization uses the time step $\Delta t = T/N$, the time horizon of the problem is $[t_n, t_N] = [n\Delta t, T]$, and the discrete dynamical system is defined by

$$\begin{cases} y_{m+1} = y_m + \Delta t f(y_m, a_m), \\ y_n = x, \end{cases} \tag{8.56}$$

whereas the cost functional is discretized by

$$J_{x,t_n}^{\Delta t}\left(\alpha^{\Delta t}\right) := \Delta t \sum_{m=n}^{N-1} g(y_m, a_m) e^{-\lambda t_m} + e^{-\lambda(t_N - t_n)} \psi(y_N), \tag{8.57}$$

in which again the dependence on the discrete control appears also through the points y_m. The value function for the discrete time problem is naturally defined as

$$v^{\Delta t, n}(x) := \inf_{\alpha^{\Delta t}} J_{x,t_n}^{\Delta t}\left(\alpha^{\Delta t}\right) \tag{8.58}$$

and satisfies the corresponding DDPP.

Proposition 8.12 (Discrete DPP for the finite horizon problem). *Set* $\Delta t = T/N$. *Then, for all* $x \in \mathbb{R}^d$, $\overline{n} \in \{n+1, \ldots, N\}$,

$$v^{\Delta t, n}(x) = \inf_{\alpha^{\Delta t}} \left\{ \Delta t \sum_{m=n}^{\overline{n}-1} g(y_m, a_m) e^{-\lambda t_m} + e^{-\lambda(t_{\overline{n}} - t_n)} v^{\Delta t, \overline{n}}(y_{\overline{n}}) \right\}. \tag{8.59}$$

The discrete time version of the SL scheme is obtained setting $\overline{n} = n+1$ in (8.59) and enforcing the final condition at time T, that is,

$$\begin{cases} v^{\Delta t,n}(x) = \min_{a\in A}\left\{\Delta t\, g(x,a) + e^{-\lambda\Delta t} v^{\Delta t,n+1}(x+\Delta t f(x,a))\right\}, \\ v^{\Delta t,N}(x) = \psi(x), \end{cases} \tag{8.60}$$

so that, as for the continuous problem, the solution is obtained backward in time from $t_N = T$ to $t_0 = 0$. Last, a $\mathbb{P}_1$ or $\mathbb{Q}_1$ interpolation is applied to reconstruct the value $v^{\Delta t,n+1}$ at the right-hand side of (8.60), giving the fully discrete scheme

$$\begin{cases} v_j^n = \min\limits_{a\in A}\left\{\Delta t\, g(x_j,a) + e^{-\lambda\Delta t} I_1\left[V^{n+1}\right](x_j+\Delta t f(x_j,a))\right\}, \\ v_j^N = \psi(x_j). \end{cases} \tag{8.61}$$

Convergence analysis We outline the main results of convergence for the finite horizon problem. First, the scheme is monotone and consistent, and hence it is convergent by the Barles–Souganidis theorem. Concerning error bounds, if the solution v of the Bellman equation (8.22) is Lipschitz continuous, then the discrete time approximation satisfies the estimate

$$\|v^{\Delta t} - v\|_\infty \le C\Delta t^{1/2},$$

where C is a positive constant which in general depends on T. Note that Lipschitz continuity of the data f and g imply again Lipschitz continuity of the value function. For the fully discrete scheme (8.61) we have (see [FG98])

$$|v_j^n - v(x_j,t_n)| \le C_1\Delta t^{1/2} + C_2\frac{\Delta x}{\Delta t^{1/2}},$$

and the best coupling is obtained by choosing $\Delta x = \Delta t$. A different analysis could be performed by applying the same assumptions and arguments of the infinite horizon problem. In this case, we obtain the estimate

$$|v_j^n - v(x_j,t_n)| \le C_1\Delta t + C_2\frac{\Delta x}{\Delta t},$$

which is optimized under the relationship $\Delta x = \Delta t^2$.

8.4.3 ▪ The optimal stopping problem

In the (infinite horizon) optimal stopping problem, it is possible to use the arguments given in the continuous formulation to derive the time discrete approximation

$$v^{\Delta t}(x) = \min\left(\min_{a\in A}\left\{\Delta t\, g(x,a) + e^{-\lambda\Delta t} v^{\Delta t}(x+\Delta t f(x,a))\right\}, \psi(x)\right). \tag{8.62}$$

In (8.62) it is possible to recognize that, in addition to the continuous control, the controller has the further choice of stopping the system, in which case a price ψ is paid. The optimal strategy clearly corresponds to the lower payoff, this being selected by the outer min operator.

In the fully discrete version, the value of $v^{\Delta t}$ is replaced as usual with a $\mathbb{P}_1$ or $\mathbb{Q}_1$ interpolation, and therefore the scheme is written as

$$v_j = \min\left(\min_{a\in A}\left\{\Delta t\, g(x_j,a) + e^{-\lambda\Delta t} I_1[V](x_j+\Delta t f(x_j,a))\right\}, \psi(x_j)\right). \tag{8.63}$$

Denoting again by $S(\Delta, V)$ the right-hand side of (8.48), and by Ψ the vector of samples of the function ψ, the scheme can be put in the compact form

$$V = \min(S(\Delta, V), \Psi), \tag{8.64}$$

which allows for the iterative version

$$V^{(k+1)} = \min\left(S(\Delta, V^{(k)}), \Psi\right).$$

An equivalent form of the scheme (8.64) is also

$$\max(V - S(\Delta, V), V - \Psi) = 0,$$

which, dividing the first argument by Δt, may be rewritten as

$$\max\left(\frac{V - S(\Delta, V)}{\Delta t}, V - \Psi\right) = 0.$$

In this form, it is immediate to recognize that the scheme is consistent with (8.25) in the sense of Remark 4.25.

To check that (8.64) is monotone, assume that $U \geq W$ component by component. Then, the scheme is monotone if

$$\min(S(\Delta, U), \Psi) - \min(S(\Delta, W), \Psi) \geq 0. \tag{8.65}$$

Now, in the case

$$\min(S(\Delta, U), \Psi) = \min(S(\Delta, W), \Psi) = \Psi,$$

as well as in the case

$$\begin{aligned} \min(S(\Delta, U), \Psi) &= S(\Delta, U), \\ \min(S(\Delta, W), \Psi) &= S(\Delta, W), \end{aligned}$$

condition (8.65) is clearly satisfied, S being monotone. On the other hand, if

$$\begin{aligned} \min(S(\Delta, U), \Psi) &= S(\Delta, U), \\ \min(S(\Delta, W), \Psi) &= \Psi, \end{aligned}$$

we can write

$$S(\Delta, W) \geq \min(S(\Delta, W), \Psi) = \Psi,$$

and hence

$$\begin{aligned} \min(S(\Delta, U), \Psi) - \min(S(\Delta, W), \Psi) &= S(\Delta, U) - \Psi \\ &\geq S(\Delta, U) - S(\Delta, W) \geq 0, \end{aligned}$$

so that (8.65) is satisfied again. Since the same argument proves the symmetric case, we can conclude that the scheme is also monotone and therefore convergent by the Barles–Souganidis theorem.

The finite horizon optimal stopping can be treated in a similar way.

8.4.4 • The minimum time problem

In the previous section, we have shown how the Kružkov change of variables (8.32) allows us to work on the (rescaled) value function as the unique viscosity solution of (8.33). This will be the standing line of work for the numerical approximation of the minimum time problem.

Time discretization In order to introduce the time discrete minimum time problem for the dynamics (8.42), we start by defining the time discrete analogue of the reachable set,

$$\mathscr{R}^{\Delta t} := \left\{ x \in \mathbb{R}^d : \exists \alpha^{\Delta t} \text{ and } m \in \mathbb{N} \text{ such that } y_m\left(x;\alpha^{\Delta t}\right) \in \mathscr{T} \right\},$$

and of the number of steps of first arrival,

$$N\left(x,\alpha^{\Delta t}\right) = \begin{cases} +\infty, & x \notin \mathscr{R}^{\Delta t}, \\ \min\{m \in \mathbb{N} : y_m\left(x;\alpha^{\Delta t}\right) \in \mathscr{T}\}, & x \in \mathscr{R}^{\Delta t}. \end{cases} \tag{8.66}$$

The role of the minimum time is played now by the minimal number of steps necessary to reach the target starting at x,

$$N(x) := \min_{\alpha^{\Delta t}} N\left(x,\alpha^{\Delta t}\right), \tag{8.67}$$

so that the discrete approximation of the minimum time function $T(x)$ is now $\Delta t N(x)$. By an adaptation of the usual arguments, it is possible to prove the following.

Proposition 8.13 (DDPP for the minimum time problem). *Let $\Delta t > 0$ be fixed. For all $x \in \mathscr{R}^{\Delta t}$ and $0 \le \overline{n} < N(x)$,*

$$N(x) = \inf_{\alpha^{\Delta t}} \left\{ \overline{n} + N\left(y_x\left(\overline{n};\alpha^{\Delta t}\right)\right)\right\}. \tag{8.68}$$

However, rather than performing a time discretization on the basis of (8.68), we further apply the Kružkov change of variables (for simplicity, with $\mu = 1$):

$$v^{\Delta t}(x) = 1 - e^{-\Delta t N(x)}.$$

Note that $0 \le v^{\Delta t} \le 1$, as in the continuous case, and that $v^{\Delta t}$ has constant value on the set of initial points x which can be driven to $\mathscr{T}$ by the discrete dynamical system in the same number of steps.

Writing (8.68) for $\overline{n} = 1$ and changing variables, we characterize $v^{\Delta t}$ as the solution of

$$\begin{cases} v^{\Delta t}(x) = \min_{a \in A} \left\{ e^{-\Delta t} v^{\Delta t}(x + \Delta t f(x,a))\right\} + 1 - e^{-\Delta t}, & x \in \mathbb{R}^d \setminus \mathscr{T}, \\ v^{\Delta t}(x) = 0, & x \in \mathscr{T}, \end{cases} \tag{8.69}$$

where the boundary condition on $\mathscr{T}$ stems from the definition of $N(x,\alpha^{\Delta t})$.

Space discretization The space discretization of (8.69) is performed as above by a $\mathbb{P}_1$ or $\mathbb{Q}_1$ interpolation. The iterative fully discrete scheme reads as

$$\begin{cases} v_j^{(k+1)} = e^{-\Delta t} \min_{a \in A} \left\{ I_1\left[V^{(k)}\right](x_j + \Delta t f(x_j,a))\right\} + 1 - e^{-\Delta t}, & x_j \in \mathbb{R}^d \setminus \mathscr{T}, \\ v_j^{(k+1)} = 0, & x_j \in \mathscr{T}, \end{cases} \tag{8.70}$$

where k is the iteration index. Except for the second condition, which works as a boundary condition, the scheme has the same structure of (8.49), so that denoting the right-hand side of (8.70) as $S(\Delta, V^{(k)})$, the following can easily be shown:

- $S(\Delta,\cdot)$ is monotone.
- Numerical solutions obtained via (8.70) are uniformly bounded.

- $S(\Delta, \cdot)$ is a contraction in l^∞ (with contraction coefficient $L_S = e^{-\Delta t}$), and therefore $V^{(k)}$ converges towards a unique solution $V \in l^\infty$ for any initial guess $V^{(0)}$. If $V^{(0)}$ is chosen as a numerical supersolution, then the sequence is monotonically decreasing, and if in particular

$$v_j^{(0)} = \begin{cases} 0, & x_j \in \mathcal{T}, \\ 1, & x_j \notin \mathcal{T}, \end{cases}$$

then at the step k the scheme updates only the nodes which are driven to the target in precisely k steps. This lends itself to a fast implementation (this point will be further discussed in Chapter 9).

Clearly, the presence of a boundary condition requires some extra effort in treating possibly complex geometries (e.g., by means of unstructured meshes as discussed in Chapter 3), as well as in implementing a variable step technique (as shown in Chapter 5). Moreover, when restricting ourselves to a bounded computational domain, a further, suitable boundary condition must be imposed on the external boundary. We skip the problem here and refer the reader to [F97] for a discussion of this point.

Convergence analysis The general features of the scheme, as outlined above, ensure convergence of the numerical solution to the value function v via the Barles–Souganidis theorem. By a deeper analysis, explicit error estimates can also be obtained. The basic ideas and technique for this analysis will be shown in the section on Pursuit-Evasion games (which can be considered as a generalization of the minimum time problem). We just point out here that two additional ingredients come into play:

- Lipschitz continuity of the value function in the neighborhood of $\partial\mathcal{T}$, which results from STLC, and
- bounded variation for optimal controls on the computational domain.

The final error estimate for the fully discrete scheme reads then as

$$|v_j - v(x_j)| \le C\Delta t^{1/2}\left(1 + \left(\frac{\Delta x}{\Delta t}\right)^2\right). \tag{8.71}$$

8.4.5 ▪ Problems with state constraints

Many relevant applications in control theory require the state of the system to satisfy additional constraints—typically, to remain within some prescribed set of the state space $\mathbb{R}^d$. We try now to present the general ideas for treating this case, skipping the most delicate technical points. The interested reader will find suitable references for state constrained problems in the reference section at the end of this chapter.

Consider our standard dynamics (8.2) coupled with the infinite horizon cost functional (8.4). In their most typical form, the state constraints require that, for all $s > 0$,

$$y_x(s; \alpha) \in \overline{\Omega}, \tag{8.72}$$

where we assume that

$$\Omega \text{ is a bounded, open subset of } \mathbb{R}^d. \tag{8.73}$$

While, in the case of free control problems, the cost J_x is minimized over the set $\mathcal{A}$ of measurable controls taking values in A, condition (8.72) restricts admissible controls to a new set which will depend on both the constraint Ω and the initial position x of the system:

$$\mathcal{A}_x = \{\alpha \in \mathcal{A} : y_x(s;\alpha) \in \overline{\Omega}\ \forall s > 0\}. \tag{8.74}$$

This set will be assumed to be nonempty for all $x \in \overline{\Omega}$.

Moreover, in general the "state constrained" value function

$$v(x) = \inf_{\alpha \in \mathcal{A}_x} \int_0^\infty g(y_x(s;\alpha))e^{-\lambda s}\,ds$$

can happen to be discontinuous, due to the complex relationship between the data of the problem (dynamics and constraints) and the set $\mathcal{A}_x$. Let us denote by A_x, for $x \in \partial\Omega$, the subset of controls such that the corresponding vector field f points inside the constraint, i.e.,

$$A_x = \begin{cases} A & \text{if } x \in \Omega, \\ \{a \in A : f(x,a) \cdot \nu(x) < 0\} & \text{if } x \in \partial\Omega. \end{cases}$$

Clearly, this set coincides with the usual control set A at interior points, where the constraint is not active. Note that the map $x \to A_x$ is not regular in general, as the following simple example shows.

Consider the dynamics in $\mathbb{R}^2$:

$$\dot{y}(s) = \alpha(s) \in A = B(0,1),$$

and $\Omega = (-1,1)^2$. The system can move in every direction at internal points, whereas on the right-hand side of the square the admissible controls can be chosen only in the half ball $B(0,1) \cap \{a_1 < 0\}$, on the upper side of the square in the half ball $B(0,1) \cap \{a_2 < 0\}$, and so on. At the vertices of Ω, admissible controls must be chosen in a quadrant of $B(0,1)$.

It has been shown by Soner [So86a, So86b] that the value function is continuous (and therefore uniformly continuous) on $\overline{\Omega}$ if for some positive constant β the following boundary condition on the vector field f is satisfied:

$$\text{For all } x \in \partial\Omega \text{ there exists } a \in A \text{ such that } f(x,a) \cdot \nu(x) \le -\beta < 0, \tag{8.75}$$

where $\nu(x)$ is the outward normal to Ω at the point x. Condition (8.75) is satisfied in the example above. However, if the state constraint Ω is cut along the x_2-axis to obtain the new constraint $\widetilde{\Omega} = \Omega \setminus \{(0,x_2) : -1 \le x_2 \le 0\}$ and the dynamics is defined by $\dot{y} = (a_1, 0)$ with $a_1 > 0$, it is easy to see that starting at an initial point on the left of the cut the dynamics will remain trapped on the left, and even for a linear running cost such as $g(x_1,x_2) = 1 - x_1$ we will have a discontinuity of the value function inside $\widetilde{\Omega}$. Note that condition (8.75) is not satisfied in this latter example, although the set $\mathcal{A}_x$ is always nonempty.

Under the condition (8.75), by the DPP, Soner has shown that the value function v is the unique "state constrained" viscosity solution of

$$H(x,u(x),Du(x)) = 0, \qquad x \in \overline{\Omega}, \tag{8.76}$$

which in turn means that v satisfies (in the viscosity sense) the inequalities

$$H(x,u(x),Du(x)) \le 0, \qquad x \in \Omega, \tag{8.77}$$

$$H(x,u(x),Du(x)) \ge 0, \qquad x \in \overline{\Omega}, \tag{8.78}$$

with H defined by (8.13). In other terms, the value function v is a solution inside Ω, but only a supersolution on the boundary $\partial\Omega$.

Theorem 8.14. *Let the assumptions of Theorem* 8.1 *be satisfied, assume that g is globally bounded and Lipschitz continuous, and let $v \in C(\overline{\Omega})$. Then, v is the unique "state constrained" viscosity solution of* (8.76) *on $\overline{\Omega}$.*

Rather than giving the proof of this theorem, we show the underlying heuristic argument. In our previous examples we have seen that at all internal points we have $A_x = A$, since the dynamics can move without restrictions. This implies that v should satisfy the same equation as in the unconstrained case. On the other hand, if $x \in \partial\Omega$, then the constraint reduces the set of admissible directions, so that $A_x \subset A$ and we have

$$\begin{aligned} H(x,v(x),Dv(x)) &= \lambda v(x) + \max_{a\in A}\{-f(x,a)\cdot Dv(x) - g(x,a)\} \\ &\geq \lambda v(x) + \max_{a\in A_x}\{-f(x,a)\cdot Dv(x) - g(x,a)\} = 0. \end{aligned} \tag{8.79}$$

The value function should therefore be a supersolution for (8.76).

Remark 8.15 (Necessary and sufficient conditions). *Condition* (8.75) *is known to be only a sufficient condition for the existence of trajectories which remain within $\overline{\Omega}$. However, necessary and sufficient conditions for the existence of solutions in $\overline{\Omega}$ have been extensively studied in* viability theory .

In order to simplify the presentation, let Ω be an open convex subset of $\mathbb{R}^d$. A trajectory is called viable *when*

$$y(s) \in \overline{\Omega} \qquad \forall s \geq 0.$$

Let $F : \overline{\Omega} \to \mathbb{R}^d$ be a multivalued map, assumed to be lower semicontinuous and with compact convex images as usual in the theory of multivalued maps. Define the tangent cone to $\overline{\Omega}$ at the point x as

$$T_K(x) := \overline{\left(\bigcup_{\mu>0} \mu(K-x)\right)}.$$

In fact it has been shown that the condition

$$F(x) \cap T_K(x) \neq \emptyset \qquad \forall x \in \overline{\Omega}$$

is necessary and sufficient to have viable trajectories for the multivalued Cauchy problem

$$\begin{cases} \dot{y}(s) \in F(y(s)), & s \geq 0, \\ y(0) = x. \end{cases}$$

This result has also been extended to more general sets and more general tangent cones.

Time and space discretization In order to build a discretization of (8.76) we use the standard discretization in time (8.42), (8.44) for, respectively, the dynamics and the cost functional. For $x \in \overline{\Omega}$, the corresponding value function for the discrete time problem is

$$v^{\Delta t}(x) = \inf_{\alpha^{\Delta t}} J_x^{\Delta t}(\alpha^{\Delta t}),$$

where $\alpha^{\Delta t} \in \mathscr{A}_x^{\Delta t}$, with the set of discrete control sequences $\{a_m\}$ such that

$$a_m \in A_x^{\Delta t} = \{a \in A : x + \Delta t f(x,a) \in \overline{\Omega}\}.$$

By standard arguments, we can obtain also for the state constrained problem a DPP in the form

$$v^{\Delta t}(x) = \inf_{\alpha^{\Delta t}} \left\{ \Delta t \sum_{m=0}^{\overline{n}-1} g(y_m, a_m) e^{-\lambda t_m} + e^{-\lambda t_{\overline{n}}} v^{\Delta t}(y_{\overline{n}}) \right\}$$

for all $x \in \mathbb{R}^d$ and any positive integer $\overline{n}$. Setting as usual $\overline{n} = 1$, this gives the time-discrete scheme

$$v^{\Delta t}(x) = \inf_{a \in A_x^{\Delta t}} \left\{ \Delta t\, g(x,a) + e^{-\lambda \Delta t} v^{\Delta t}(x + \Delta t f(x,a)) \right\}$$

and finally, replacing the computation of $v^{\Delta t}(x + \Delta t f(x,a))$ with an interpolation, the fully discrete scheme

$$v_j = \min_{a \in A_{x_j}^{\Delta t}} \left\{ \Delta t\, g(x_j,a) + e^{-\lambda \Delta t} I_1[V](x_j + \Delta t f(x_j,a)) \right\}$$

written at the grid node x_j. An iterative version can also be derived, as in the unconstrained case.

Convergence analysis It can be proved that, for a Lipschitz continuous solution v and bounded variation optimal controls, the following estimate holds true:

$$\left\| v - v^{\Delta t} \right\|_\infty \le C_1 \Delta t.$$

Finally, also taking into account space discretization, we obtain the complete estimate

$$|v_j - v(x_j)| \le C_1 \Delta t + C_2 \frac{\Delta x}{\Delta t}.$$

8.4.6 • Construction of the approximate feedback

We have focused so far on the approximation of the value function v. However, from the control point of view, the interest is rather on the construction of the (approximate) optimal control given the information on the value function. We show therefore in this section an approximate strategy which leads to constructing a quasi-optimal control in feedback form, using the information from both the numerical value function V and a uniform sampling of the state. The procedure does not depend on the specific scheme used to compute V, and is outlined for the case of the infinite horizon problem.

Once we set a time grid $t_0, t_1, \ldots, t_k, \ldots$ with $t_k = k\delta$ for some additional "small" time discretization parameter δ (to be suitably chosen in what follows depending on Δ), the general idea is to sample the state at the time t_k and define on the interval $[t_k, t_{k+1})$ a constant control, as sketched in Figure 8.1.

This approximate feedback is defined on the basis of the approximate value function and of the point $x = y(t_k)$, with a discrete Dynamic Programming Procedure. More precisely, we define

$$a^\delta = \operatorname*{argmin}_a \left\{ \delta g(x,u) + e^{-\lambda\delta} I[V](x + \delta f(x,a)) \right\} \tag{8.80}$$

and apply the control $\alpha^\delta(t) \equiv a^\delta$ for $t \in [t_k, t_{k+1})$.

Our aim is to show that this strategy is quasi-optimal. To this end, we need more strict assumptions, as it has been done in subsection 8.4.1. In particular, we assume that both the exact and the approximate value functions are Lipschitz continuous and that the following hold:

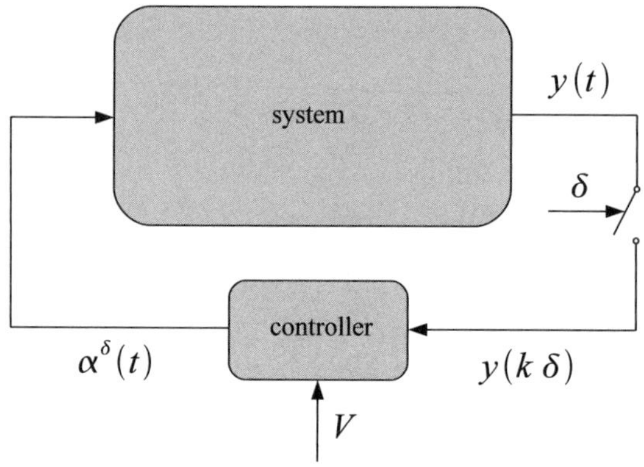

Figure 8.1. *Structure of the feedback controller.*

(i) $||f||$ is uniformly bounded, and A is convex. Moreover, $f(x,a)$ has the structure

$$f(x,a) = f_1(x) + F_2(x)a$$

with $f_1 : \mathbb{R}^d \to \mathbb{R}^d$ and $F_2(x) \in \mathbb{R}^{d\times m}$.

(ii) $g(\cdot,a)$ is Lipschitz continuous, and $g(x,\cdot)$ is convex.

(iii) There exists an optimal control $\alpha^* \in \mathscr{A}$.

We introduce a more compact notation. Let α^* denote the optimal control for the exact system restricted to the time interval $[0,\delta)$, and let α^δ denote the (constant) control obtained by the approximate feedback outlined above on the same time interval. We denote by $\gamma(\delta,\alpha)$ the cost accumulated on the interval $[0,\delta)$ starting from the state x with a control α, that is,

$$\gamma(\delta,\alpha) = \int_0^\delta g(y_x(t),\alpha(t))e^{-\lambda t}\,dt.$$

Once written on the interval $[0,\delta)$, the DPP reads as

$$\begin{aligned} v(x) &= \min_\alpha \left\{ \gamma(\delta,\alpha) + e^{-\lambda\delta} v(y_x(\delta,\alpha)) \right\} \\ &= \gamma(\delta,\alpha^*) + e^{-\lambda\delta} v(y_x(\delta,\alpha^*)), \end{aligned}$$

while, on the basis of the approximate feedback defined above, we also define

$$\begin{aligned} w(x) &:= \min_a \left\{ \delta\, g(x,a) + e^{-\lambda\delta} I[V](x+\delta f(x,a)) \right\} \\ &= \delta\, g(x,a^\delta) + e^{-\lambda\delta} I[V](x+\delta f(x,a^\delta)). \end{aligned}$$

In what follows, we will need a unilateral estimate on $w - v$. Replacing in w the actual minimizer a^δ with a different control $\overline{a}^*$ (to be properly defined), we have

$$\begin{aligned} w(x) - v(x) &\le \delta\, g(x,\overline{a}^*) - \gamma(\delta,\alpha^*) \\ &\quad + e^{-\lambda\delta} \left[I[V](x+\delta f(x,\overline{a}^*)) - v(y_x(\delta,\alpha^*)) \right] \\ &= \delta\, g(x,\overline{a}^*) - \gamma(\delta,\alpha^*) \\ &\quad + e^{-\lambda\delta} [(I[V] - v)(x+\delta f(x,\overline{a}^*)) \\ &\quad + v(x+\delta f(x,\overline{a}^*)) - v(y_x(\delta,\alpha^*))], \end{aligned} \tag{8.81}$$

so that, defining $\overline{a}^*$ as the integral mean of α^* and working as in subsection 8.4.1, we obtain

$$w(x) - v(x) \le \|I[V] - v\|_\infty + O(\delta^2).$$

In order to evaluate the increase in the cost introduced by the suboptimal strategy α^δ, consider an evolution in which this strategy is applied only in the first time step, whereas at successive times the exact optimal control α^* is used. The (nonnegative) optimality gap of this control strategy is given by

$$\Delta_{opt} := \gamma(\delta, \alpha^\delta) + e^{-\lambda\delta} v(y_x(\delta, \alpha^\delta)) - v(x) \ge 0, \tag{8.82}$$

and if we prove that, for $\Delta, \delta \to 0$, it satisfies the condition

$$\Delta_{opt} = o(\delta), \tag{8.83}$$

then we obtain that the global cost associated with the control strategy α^δ converges to the optimal cost, i.e., that α^δ is an asymptotically optimal control.

We start by giving a general upper bound on (8.82). Using the function w defined above, we can write

$$\begin{aligned}\Delta_{opt} &= \gamma(\delta, \alpha^\delta) + e^{-\lambda\delta} v(y_x(\delta, \alpha^\delta)) + (w(x) - v(x)) - w(x) \\ &\le \gamma(\delta, \alpha^\delta) - \delta g(x, a^\delta) + \|I[V] - v\|_\infty + O(\delta^2) \\ &\quad + e^{-\lambda\delta}\left[v(y_x(\delta, \alpha^\delta)) - I[V](x + \delta f(x, a^\delta))\right] \\ &\le \|I[V] - v\|_\infty + O(\delta^2) + e^{-\lambda\delta}\left[v(y_x(\delta, \alpha^\delta)) - I[V](x + \delta f(x, a^\delta))\right],\end{aligned}$$

where the term $\gamma(\delta, \alpha^\delta) - \delta g(x, a^\delta)$ has been collected into the $O(\delta^2)$ by (8.53). On the other hand, using (8.52), we can estimate the last term as

$$\begin{aligned}&v(y_x(\delta, \alpha^\delta)) - I[V](x + \delta f(x, a^\delta)) \\ &\qquad \le (v - I[V])(y_x(\delta, \alpha^\delta)) + I[V](y_x(\delta, \alpha^\delta) - I[V](x + \delta f(x, a^\delta)) \\ &\qquad \le \|v - I[V]\|_\infty + O(\delta^2),\end{aligned}$$

so that we finally obtain

$$\Delta_{opt} \le 2\|v - I[V]\|_\infty + O(\delta^2).$$

To sum up, if δ is chosen so that $\|v - I[V]\|_\infty = o(\delta)$, then the approximate optimal control defined by (8.80) is a minimizing sequence as $\Delta, \delta \to 0$.

8.4.7 ▪ Policy iteration and other acceleration techniques

We have already noticed that convergence of the value iteration (8.49) can be very slow, since the contraction constant $e^{-\lambda\Delta t}$ is close to the unity for $\Delta t \to 0$. This means that a higher accuracy would also require more iterations. Some acceleration techniques have been proposed to overcome this difficulty, and we will briefly describe two effective acceleration methods based on different ideas and resulting in a monotone convergence.

Policy iteration The first acceleration technique is the *approximation in policy space* (or *policy iteration*) and is based on a linearization of the Bellman equation. First, an initial guess for the control (policy), rather than for the value function, is chosen. Once the control has been fixed, the Bellman equation becomes linear (no search for the minimum in the control space is performed) and is solved as an advection equation. Then, a new policy is computed and a new iteration starts. Let us sketch the procedure for the scheme related to the infinite horizon problem.

1. Set an initial guess a^0 for the control as a function of the state. Set $k=0$.
2. For $k \geq 0$, until convergence, compute for any node x_i the solution v_i^k of
$$v_i^k = \Delta t\, g\left(x_i, a_i^k\right) + e^{-\lambda \Delta t} I\left[V^k\right]\left(x_i + \Delta t f\left(x_i, a_i^k\right)\right). \tag{8.84}$$
3. Compute a new policy on the solution V^k as
$$a_i^{k+1} = \operatorname*{argmin}_a \left\{\Delta t\, g(x_i, a) + e^{-\lambda \Delta t} I\left[V^k\right](x_i + \Delta t f(x_i, a))\right\}.$$
4. Increment k, and go to 2.

Note that the solution of (8.84) can be obtained either by a linear system or as the limit
$$V^k = \lim_{m \to +\infty} V^{k,m} \tag{8.85}$$
of the linear time-marching scheme
$$v_i^{k,m+1} = \Delta t\, g\left(x_i, a_i^k\right) + e^{-\lambda \Delta t} I\left[V^{k,m}\right]\left(x_i + \Delta t f\left(x_i, a_i^k\right)\right). \tag{8.86}$$
Although this scheme is still iterative, the lack of a minimization phase makes it faster than the original value iteration.

The sequence $\{V^k\}$ turns out to be monotone decreasing at every node of the grid. In fact, by construction,
$$\begin{aligned}
v_i^k &= \Delta t\, g\left(x_i, a_i^k\right) + e^{-\lambda \Delta t} I\left[V^k\right]\left(x_i + \Delta t f\left(x_i, a_i^k\right)\right) \\
&\geq \min_a \left\{\Delta t\, g(x_i, a) + e^{-\lambda \Delta t} I\left[V^k\right](x_i + \Delta t f(x_i, a))\right\} \\
&= \Delta t\, g\left(x_i, a_i^{k+1}\right) + e^{-\lambda \Delta t} I\left[V^k\right]\left(x_i + \Delta t f\left(x_i, a_i^{k+1}\right)\right) \\
&= v_i^{k+1}.
\end{aligned}$$
At a theoretical level, policy iteration can be shown to be equivalent to a Newton method, and therefore, under appropriate assumptions, it converges with quadratic speed. On the other hand, convergence is local and this may represent a drawback with respect to value iterations.

Iteration in the set of subsolutions A different monotone acceleration technique can be constructed on the idea of following the direction provided by the value iteration up to the boundary of the set of (numerical) subsolutions. This idea is justified by the property of the value function of being the maximal subsolution of the Dynamic Programming equation.

Consider the fixed-point iteration associated with the infinite horizon problem, written in short as

$$V^{(k+1)} = S(\Delta, V^{(k)}).$$

Since the operator S is a contraction mapping, we have convergence starting from any initial guess $V^{(0)}$. Moreover, if the initial guess is in the set of subsolutions Σ for the discrete problem, i.e., if

$$V^{(0)} \in \Sigma := \{W : W \le S(\Delta, W)\},$$

then convergence is monotone. In fact, by the monotonicity of S,

$$V^{(k)} \le S(\Delta, V^{(k)}) = V^{(k+1)}.$$

The set of subsolutions Σ can be proved to be a closed convex set with a maximal element, which coincides with the fixed point V. Then, we can use the operator S to obtain a search direction (along which the components of the numerical solution increase), until we reach the boundary of Σ. The accelerated algorithm is therefore the following:

1. Set an initial guess $W^0 \in \Sigma$. Set $k = 0$.
2. For $k \ge 0$, until convergence, compute

$$W^{k+1/2} = S(\Delta, W^k)$$

and the direction

$$d^k = W^{k+1/2} - W^k \ge 0.$$

3. Compute the (unique) intersection W^{k+1} between $W^k + \mu d^k$ and $\partial\Sigma$.
4. Increment k, and go to 2.

In this case, the accelerated sequence W^k converges monotonically and for any initial guess.

8.4.8 ▪ An extension to stochastic control

Dynamic Programming techniques can also be applied to the analysis and approximation of stochastic control problems. Even if this chapter is focused on deterministic control problems, we sketch the main ideas to extend the approach to the control of diffusion processes (one of the typical problems in stochastic control) and give some references in the final section of this chapter.

Let $(\Omega, \mathscr{F}, \mathbb{P}, \{\mathscr{F}_t\}_{t\ge 0})$ be a complete filtered probability space, and suppose that, given a progressively measurable process $\alpha(t)$, there exists a unique (progressively measurable) process $X(t)$ satisfying the controlled SDE:

$$\begin{cases} dX(t) = f(X(t), \alpha(t))dt + \sigma(X(t), \alpha(t))dW(t), \\ X(0) = x \in \mathbb{R}^d, \end{cases} \tag{8.87}$$

in which $W(t)$ is a d-dimensional Wiener process and σ is a square (possibly, singular) matrix depending on X and α. The process $X(t)$ represents the state of a system evolving in $\mathbb{R}^d$, and the process $\alpha(t)$ is the control applied at time t, with values in the compact set A. Typically, the cost functional is defined by a conditional expectation; e.g., the cost related to the infinite horizon problem is defined as

$$J_x(\alpha) := \mathbb{E}\left\{ \int_0^{+\infty} g(X(t),\alpha(t))e^{-\lambda t}dt \;\middle|\; X(0) = x \right\}, \tag{8.88}$$

where g and λ play the same role as in the deterministic case. The value function of the problem is defined as

$$v(x) = \inf_{\alpha \in \mathscr{A}} J_x(\alpha),$$

and it can be proved that it is the unique viscosity solution of the HJB equation

$$\lambda v(x) = \inf_{a\in A}\{L(a)[v](x) + g(x,a)\}, \tag{8.89}$$

in which $L(a)$ stands for the second-order differential operator

$$L(a) = \frac{1}{2}\sum_{i,j=1}^{d} b_{ij}(x,a)\frac{\partial^2}{\partial x_i \partial x_j} + \sum_{i=1}^{d} f_i(x,a)\frac{\partial}{\partial x_i} \tag{8.90}$$

with $B = (b_{ij}) = \sigma\sigma^t$. We give up a formal statement of this result, which would require defining a viscosity solution for second-order equations. Some references will be given at the end of the chapter.

Various approximation strategies for (8.89)–(8.90) have been studied. In particular, we quote the pioneering work of H. Kushner, who has proposed algorithms based on the approximation of the diffusion process (8.87) by a proper Markov chain on the computational grid. We rather follow here the SL approach outlined in Chapter 7 and [CF95], by constructing first the time-discrete approximation of (8.87). In the simplest case, this has the form of the *stochastic Euler method*

$$\begin{cases} X_{m+1} = X_m + \Delta t f(X_m, a_m) + \sigma(X_m, a_m)\xi^m, \\ X_0 = x, \end{cases}$$

in which ξ^m denotes the increment of the standard Brownian motion $W(t)$ on the interval $[t_m, t_{m+1}]$. The corresponding discrete time cost functional is defined as

$$J_x^{\Delta t}(\alpha^{\Delta t}) := \mathbb{E}\left\{ \Delta t \sum_{m=0}^{+\infty} g(X_m, a_m)e^{-\lambda t_m} \;\middle|\; X_0 = x \right\} \tag{8.91}$$

and the discrete value function as

$$v^{\Delta t}(x) = \inf_{\alpha^{\Delta t}} J_x^{\Delta t}(\alpha^{\Delta t}).$$

It can be proved (see [KP92]) that, whenever the focus is on expected values (as in (8.88)), the Gaussian random variable ξ_m can be replaced by a two-point distributed variable with probability distribution

$$\mathbb{P}\left(\xi_i^m = \sqrt{\Delta t}\right) = \mathbb{P}\left(\xi_i^m = -\sqrt{\Delta t}\right) = \frac{1}{2}.$$

Therefore, in a form similar to what has been shown in Chapter 7, we can define for $k = 1, \ldots, 2^d$ a set of displacements

$$\delta_k(x,a) = \Delta t f(x,a) + \sigma(x,a)\begin{pmatrix} \pm\sqrt{\Delta t} \\ \vdots \\ \pm\sqrt{\Delta t} \end{pmatrix},$$

and then, discretizing the stochastic DPP for the cost functional (8.88), we obtain the scheme in the usual fixed-point form

$$v_j = \min_{a\in A}\left\{\Delta t\, g(x,a) + \frac{e^{-\lambda\Delta t}}{2^d}\sum_{k=1}^{2^d} I[V](x_j+\delta_k(x_j,a))\right\}. \tag{8.92}$$

It is easy to check that, in the special case of the constant-coefficient advection-diffusion equation, (8.92) reduces to a stationary version of (7.30).

Since, even in the controlled case, the diffusion process (8.87) leaves in general every bounded domain in finite time, additional requirements such as reflection or absorption (Dirichlet) conditions should be enforced on the boundary. We leave these more technical aspects to the specific literature (see the commented references).

8.4.9 ▪ Examples

While one-dimensional, academic examples are intended to complement the theoretical presentation of Chapters 4 and 5, this chapter rather collects examples and tests related either to applications or to more challenging benchmarks. Among them, we mention the treatment of singular solutions in more than one space dimension, and Dynamic Programming in dimensions three and four.

Quadratic Hamiltonian, Lipschitz continuous initial data This test refers to the HJ equation

$$\begin{cases} u_t(x,t)+\frac{1}{2}|Du(x,t)|^2=0, & (x,t)\in[-2,2]^2\times[0,1], \\ v(x,0)=v_0(x)=\max(0,1-|x|^2), & x\in[-2,2]^2, \end{cases} \tag{8.93}$$

and is (along with the next example) a two-dimensional version of the test presented in Chapter 5, which may be recast as a control problem via the definitions (8.41). In this test, the initial condition has a singularity in the gradient (with an empty superdifferential) along the circle $|x|=1$. The evolution soon generates a semiconcave solution, but for $t>1/2$ a singularity in the gradient (with an empty subdifferential) appears at the origin.

The exact solution of (8.93) can be explicitly computed and for $t>1/2$ reads as

$$v(t,x)=\begin{cases}\frac{(|x|-1)^2}{2t} & \text{if } |x|\le 1,\\ 0 & \text{if } |x|\ge 1.\end{cases}$$

The solution, computed with $\Delta t = 0.1$ and a cubic reconstruction, is shown in Figure 8.2 at even time steps between $t=0$ and $t=1$. The graphs show in particular that the singularity in the gradient is resolved isotropically and without undesired numerical dispersion. Table 8.1 shows L^∞ and L^1 errors at $T=1$ for different numbers of nodes n_n on the edge of the computational domain. The solution is computed with good accuracy even with a low number of nodes, although the rate of convergence is limited by the lack of uniform semiconcavity, as it has already been remarked in the one-dimensional examples of Chapter 5.

Quadratic Hamiltonian, semiconcave initial data We consider again the equation with a quadratic Hamiltonian, but reversing the sign of the initial condition

$$\begin{cases} u_t(x,t)+\frac{1}{2}|Du(x,t)|^2=0, & (x,t)\in[-2,2]^2\times[0,1], \\ u(x,0)=u_0(x)=\min(0,|x|^2-1), & x\in[-2,2]^2. \end{cases} \tag{8.94}$$

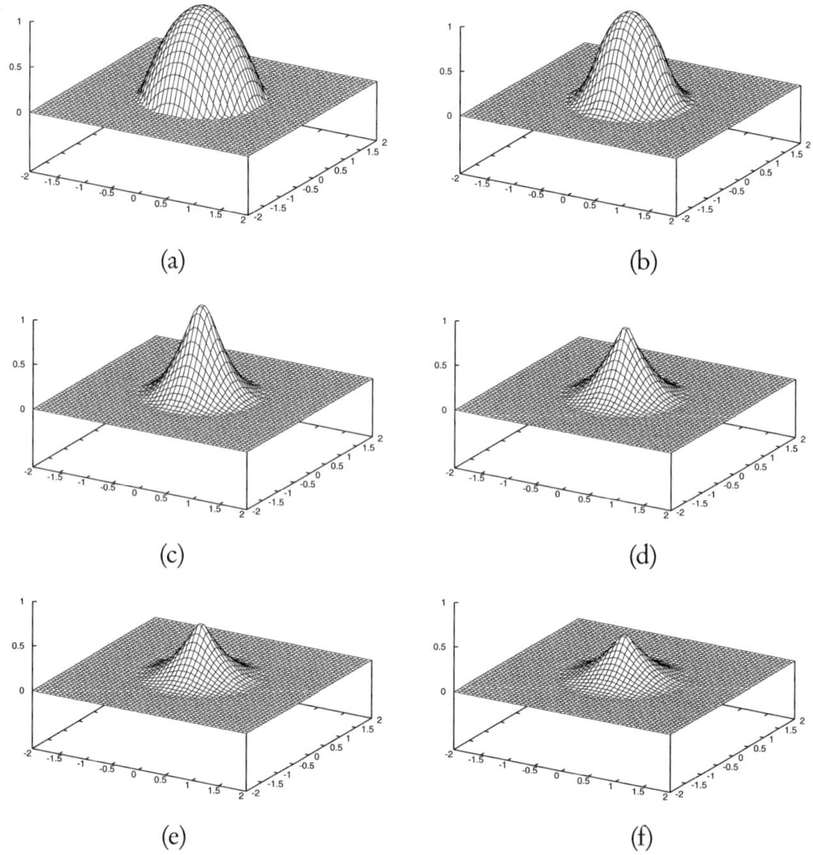

Figure 8.2. *Numerical solutions for the quadratic Hamiltonian, with Lipschitz continuous initial data* [FF02].

Table 8.1. *Numerical errors for* (8.93).

$W^{1,\infty}$ initial condition		
n_n	L^∞ error	L^1 error
25	$5.75 \cdot 10^{-2}$	$6.57 \cdot 10^{-2}$
50	$1.85 \cdot 10^{-2}$	$2.76 \cdot 10^{-2}$
100	$6.17 \cdot 10^{-3}$	$8.32 \cdot 10^{-3}$
Rate	1.61	1.49

The exact solution of this test problem is

$$u(t,x) = \begin{cases} \frac{|x|^2}{2t+1} - 1 & \text{if } |x| \le \sqrt{2t+1}, \\ 0 & \text{if } |x| \ge \sqrt{2t+1}. \end{cases}$$

In this test, starting from a semiconcave initial condition, the solution remains semiconcave for all times. Figure 8.3 shows the numerical solution computed for $t \in [0,1]$ with $\Delta t = 0.1$ and a cubic reconstruction, and Table 8.2 shows L^∞ and L^1 errors at $T = 1$. In this case, due to the uniform semiconcavity of the solution, the improvement in L^∞

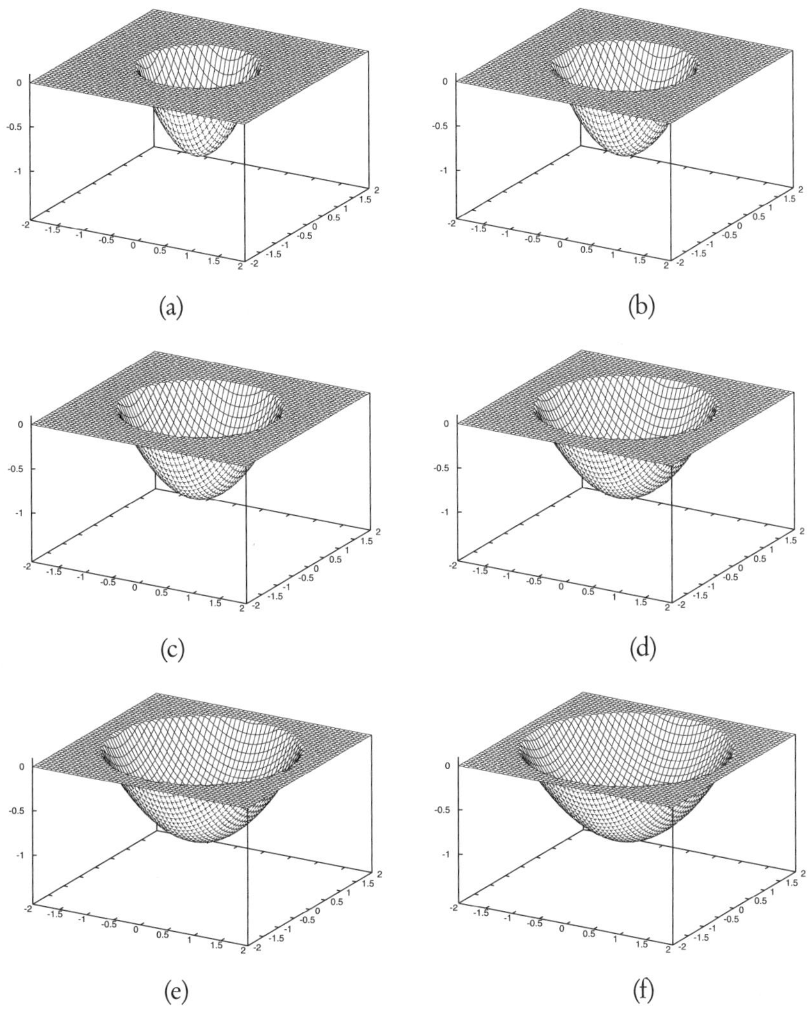

Figure 8.3. *Numerical solutions for the quadratic Hamiltonian, with semiconcave initial data* [FF02].

Table 8.2. *Numerical errors for* (8.94).

Semiconcave initial condition		
n_n	L^∞ error	L^1 error
25	$1.57 \cdot 10^{-2}$	$2.88 \cdot 10^{-2}$
50	$4.03 \cdot 10^{-3}$	$8.42 \cdot 10^{-4}$
100	$6.46 \cdot 10^{-4}$	$3.29 \cdot 10^{-5}$
Rate	2.30	4.89

error is of one order of magnitude, and in L^1 error (which is clearly measured in a weaker norm) of more than two orders of magnitude.

The moon landing problem This is a classical control problem which is taken from [FR75]. A spacecraft is attempting to land smoothly on a surface using the minimum amount of fuel. The gravity acceleration g is assumed to be constant near the surface. The motion is described by the equations

$$\begin{cases} \dot{h} = v, \\ \dot{v} = -g + m^{-1}\alpha, \\ \dot{m} = -k\alpha, \end{cases}$$

where h and v denote, respectively, the height and vertical velocity of the spacecraft, m is the mass of the spacecraft (including the variable mass of the fuel), and the control $\alpha \in [0, \widehat{\alpha}]$ denotes the thrust of the spacecraft's engine. The initial time is $t_0 = 0$, and the final time T is the first time the spacecraft touches the moon. The initial and final conditions for this problem read as

$$h(0) = h_0, \qquad v(0) = v_0, \qquad m(0) = M + F,$$
$$h(T) = 0, \qquad v(T) = 0,$$

where h_0 and v_0 are the initial height and velocity of the spacecraft, and M and F are the mass of the spacecraft without fuel and the initial mass of the fuel. Setting $(x_1, x_2, x_3) = (h, v, m)$, the problem of minimizing the fuel is equivalent to minimizing $-m(T) = -x_3(T)$ over the class of measurable controls which drive the system to the point $x_1(T) = x_2(T) = 0$ and satisfy $\alpha(t) \in A = [0, \widehat{\alpha}]$.

Treating the final condition by penalization to remove the target constraint, the control problem is in the Mayer form

$$J_{x,t}(\alpha) = \psi(y_x(T; \alpha)),$$

where

$$\psi(x) = -x_3 + \frac{x_1^2 + x_2^2}{\epsilon} \qquad (\epsilon \ll 1),$$

and its value function solves the Bellman equation

$$\begin{cases} -v_t(x,t) + \sup_{\alpha \in \mathscr{A}} \{-f(x,\alpha) \cdot Dv(x,t)\} = 0, \\ v(x,T) = \psi(x). \end{cases}$$

The test is performed using a $\mathbb{P}_2$ reconstruction in the interpolation routine and (following [FF94]) the Heun scheme to approximate the dynamics. In this version of (8.61), the point $x_j + \Delta t f(x_j, a)$ is replaced by $x_j + \Delta t \Phi(\Delta t; x_j, a_1, a_2)$, where

$$\Phi(\Delta t; x_j, a_1, a_2) = \frac{1}{2}\left[f(x_j, a_1) + f(x_j + \Delta t f(x_j, a_1), a_2)\right],$$

and the minimization is performed with respect to both a_1 and a_2. For each point of the space-time grid the approximate optimal feedback is defined by taking the $\arg\min_{a_1,a_2}$ in (8.61) and extending the two discrete controls to all of the computational domain by interpolation.

The test uses the initial condition $(x_1, x_2, x_3) = (0.3, 0, 18)$, final time $T = 1$, space steps of $(0.03, 0.15, 0.26)$, and a time step of $\Delta t = 0.03$. Figure 8.4 shows the optimal trajectory in the phase space (velocity, position), along with the discrete optimal controls a_1 and a_2 during the time interval $[0, 1]$. The spacecraft actually lands at $t = 0.99$, and the endpoint $(x_1, x_2) = (0, 0)$ is reached with some numerical error, due to the penalization strategy. All figures show that the optimal trajectory is basically obtained falling down by gravity up to a switching time t^* and then braking until the landing with maximum thrust. This is in perfect agreement with theory.

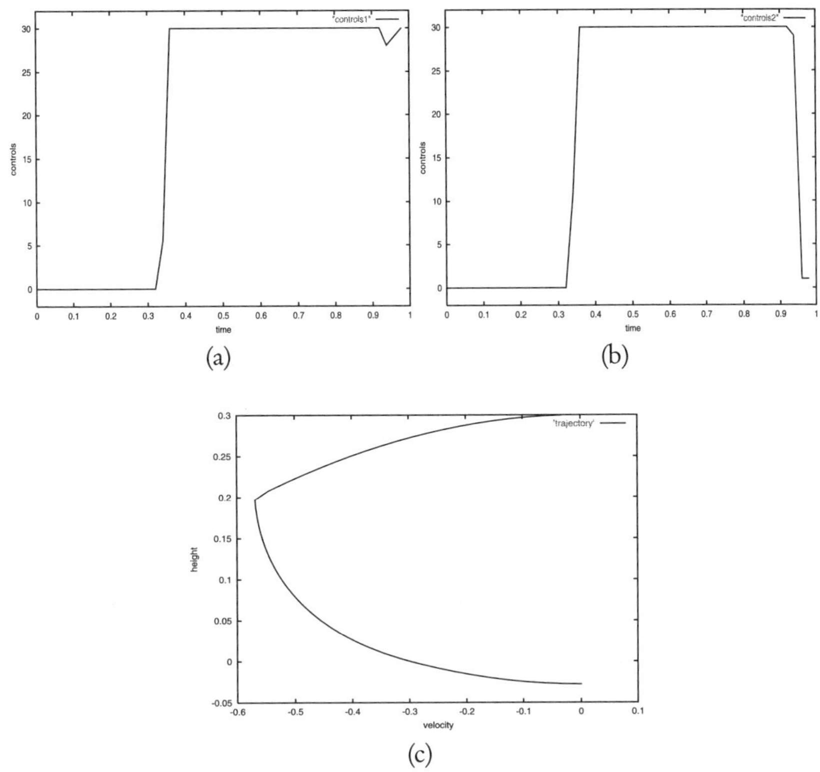

Figure 8.4. *Moon landing, with* (a) *optimal control* a_1, (b) *optimal control* a_2, (c) *optimal trajectory* [CFF04].

A Ramsey model for a multisector economy Consider an economy which has three sectors S_i, $i = 1,\dots,3$ (the model can be easily extended to many sectors), and denote by $Y_i(s)$, $K_i(s)$, and $L_i(s)$ the output rate, the capital, and the labor of sector S_i at time s. Capital and labor are connected to the production through a production function $Y_i = F_i(K_i, L_i)$. A classical example is the Cobb–Douglas function

$$F(K,L) = e^{\lambda s} K^{\alpha} L^{1-\alpha} \qquad (0 \le \alpha \le 1),$$

where λ is a coefficient connected to the technical innovation (see [A67]). For every sector, we can divide by L_i and define $y_i = \frac{Y_i}{L_i}$, $k_i = \frac{K_i}{L_i}$, $i = 1,\dots,3$, obtaining the pro-capite production functions

$$y_i = e^{\lambda_i s} k_i^{\alpha_i} \qquad (0 \le \alpha_i \le 1).$$

A typical choice is for the parameters λ_i is $\lambda_i = m(1-\alpha_i)$ (see [A67]), where m is a constant rate of technical innovation.

The rate of consumption $c(s)$ is a fraction u (the control) of a function $g(k(s))$ of the capital; i.e., for every sector the consumption rate is given by

$$c_i(s) = u(s) g_i(k_i(s)) \qquad (u_{\min} \le u(s) \le 1).$$

Then, the rate of change of the capital in every sector (assuming that the sectors are separated) is

$$\dot{k}_i = f(k_i) - u\, g(k_i) \qquad (u_{\min} \le u \le 1).$$

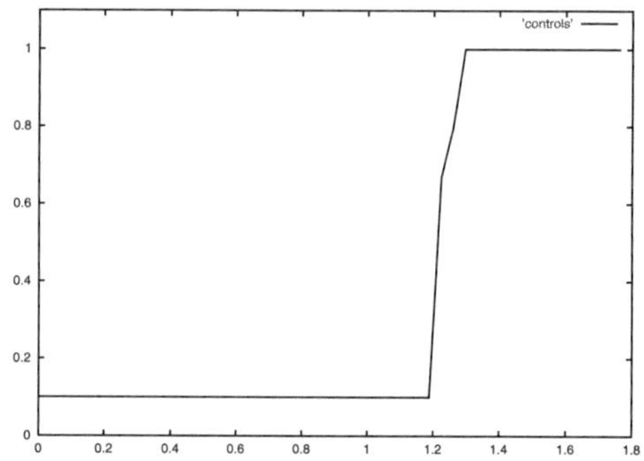

Figure 8.5. *Optimal control for the Ramsey model* [CFF04].

Let now U be a utility function which represents the utility of consuming at rate $c = c_1 + c_2 + c_3 > 0$. This function U is typically monotone increasing and bounded on the positive axis. The goal is to maximize the functional

$$\int_0^T U(c(s))ds$$

with respect to the choice of the control u under the constraints $k_i(0) = k_i^0$, $k_i(T) = k_i^1$.

Figures 8.5 and 8.6 show, respectively, the optimal control u and the evolution in time of the variables k_i and c_i $(i = 1,2,3)$. The various parameters have been set as $\alpha_1 = 0.2$, $\alpha_2 = 0.5$, and $\alpha_3 = 0.8$, $u_{\min} = 0.1$, $m = 1$, whereas g is defined as

$$g(k) = k - k_{\min},$$

and the utility function U as

$$U(c) = \frac{Mc^2}{1+c^2},$$

with $k_{\min} = k^0/3$ and $M = 10$. The initial condition for the three components of the capital is $k^0 = (6,4,2)$, and the final objective for $T = 1.8$ is $k^1 = (8,6,4)$. As in the previous test, we treat the final constraint by penalization, using the final condition

$$\psi(k) = -\frac{|k-k^1|^2}{\epsilon}.$$

For the computation, we have used $\Delta t = 0.05$, $\epsilon = 10^{-4}$ and 30 nodes in space for every dimension.

Figures 8.5 and 8.6 depict a situation in which the capital increases, with consumption at its minimum, up to a time $\bar{t} \approx 1.2$. After $\bar{t}$, the control switches from 0.1 to 1, the consumption goes to its maximum, and the first two components of the capital start decreasing. The final position of the trajectory is $k(T) \approx (7.91, 7.41, 5.68)$, and significantly differs from the target, due to both penalization and lack of controllability (a one-dimensional control has been used for a system in $\mathbb{R}^3$).

Control of the HIV-1 dynamics In the paper [PN99] Perelson and Nelson have described the dynamics of the HIV-1 virus infection (the virus that causes AIDS). In this test, we add a control term in their model in order to describe the drug therapy effect.

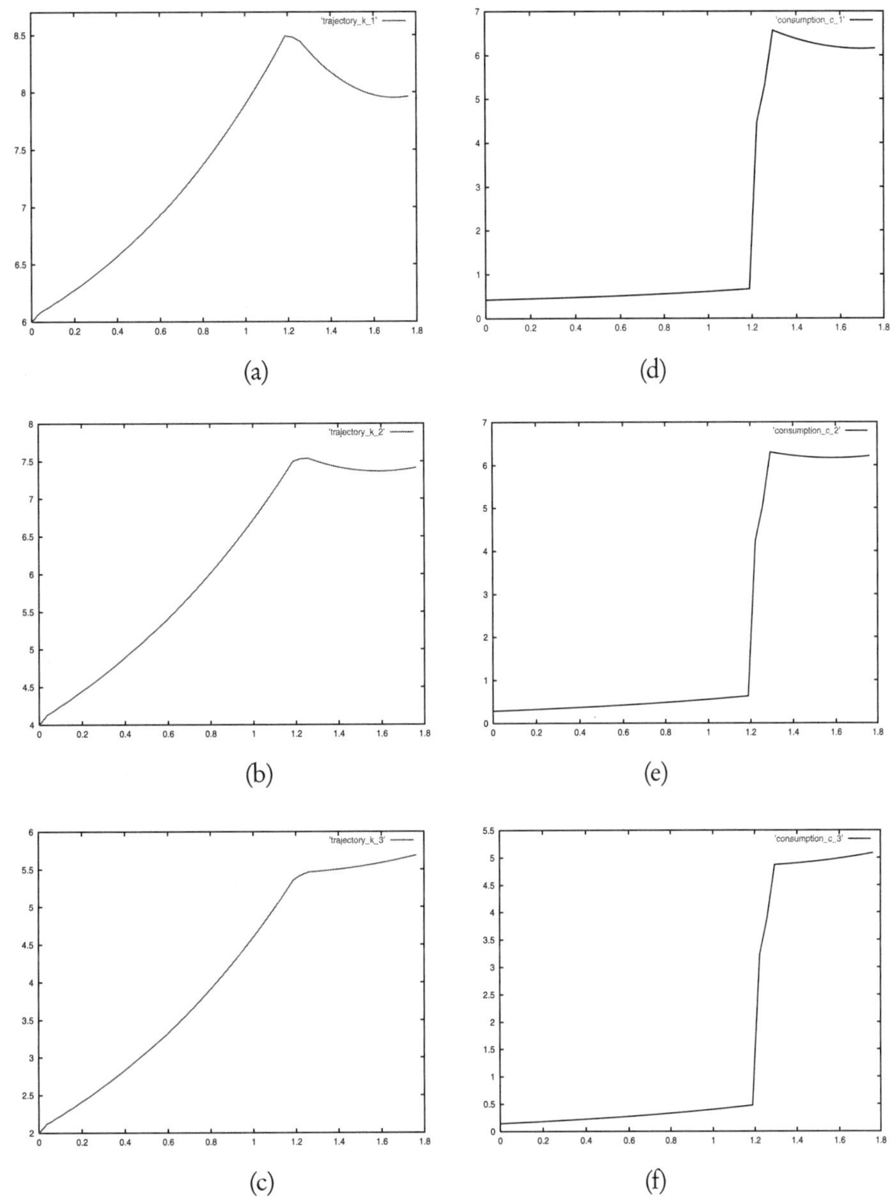

Figure 8.6. *Ramsey model, with optimal solutions;* (a)–(c) k_i, (d)–(f) c_i [CFF04].

The model considers three populations: the uninfected cells C, the productively infected cells I, and the virus V. All the populations are described by their density per volume unit. The dynamics is described by the system

$$\begin{cases} \dot{C} = s + pC\left(1 - \dfrac{C}{C_{\max}}\right) - d_C C - (1-u)kVC, \\ \dot{I} = (1-u)kVC - d_I I, \\ \dot{V} = Nd_I I - d_V V, \end{cases}$$

where s represents the rate at which new C cells are created from sources within the body,

p is the maximum proliferation rate for the C cells, $C_{\max}$ is the maximum cell population density, d_C , d_I, and d_V are the death rates for, respectively, the C, I, and V populations, Nd_I is the average rate of virion production, and u is the control driven by the drug therapy (usually, the constraint $d_C C_{\max} > s$ is added). The term kVC tells one that the rate of creation of new infected cells is proportional to the product VC (at least one virion should enter a cell to infect it).

A typical drug therapy is a treatment based on RT (reverse transcriptase) or protease inhibitors. The control u measures the effectiveness of the inhibitor: for $u = 1$ the inhibition is 100% effective, whereas for $u = 0$ there is no inhibition. Typically, $0 \le u \le u_{\max} < 1$. The functional we want to minimize has the form

$$\int_0^T \left(M\left(I^2+V^2\right)+u\right)dt$$

(for a positive constant M) and takes into account the dimension of the populations I and V and the fact that a drug excess can have negative global effects on the patient.

In the test, parameters have been set as $s=0.2$, $p=0.05$, $k=0.08$, $d_C=0.01$, $d_I=0.3$, $d_V=0.009$, $N=0.005$, and $M=8$ (a biologically realistic choice of some of the parameters can be found in [PKB93]). Figure 8.7 shows the optimal control u and the time evolution of the state variables C, I, V. The virus population decreases monotonically under the effect of the drug therapy, but since the use of large amounts of drugs is penalized, the optimal therapy suggested by the model is to use the drug at its maximum level for the first period and then stop the drug injection. After this first time interval, the evolution of infected cells and viruses naturally decreases by the uncontrolled dynamics.

The minimum time problem In the first test, the minimum time problem is related to the *computation of geodesics on a nonsmooth surface.* In this test, we set a point target

$$\mathcal{T}=(0,-0.6,0.4)$$

on the pyramidal surface

$$z(x)=\begin{cases}1-(|x_1|+|x_2|) & \text{if } |x_1|+|x_2|<1,\\ 0 & \text{elsewhere.}\end{cases}$$

The system moves on the surface with intrinsic velocity in $B(0,1)$, so that the minimum time function (restricted to the surface) is in fact the intrinsic distance, and optimal trajectories are geodesics of the surface. In this test, we compute the geodesic reaching the target from the starting point $(0,0.5,0.5)$.

It can be shown that the three-dimensional problem can be reduced to a two-dimensional problem by modifying the velocity field according to the function z. In fact, in case of a unity intrinsic velocity on the surface, it can be shown (see [Se99, Se96b]) that the velocity of the corresponding two-dimensional problem becomes

$$c(x,a)=\frac{1}{\sqrt{1+(Dz\cdot a)^2}}.$$

Figure 8.8 shows the level sets of T and its surface with the optimal trajectory on it.

In the second test, the velocity field has the form

$$f(x,a)=c(x)a,$$

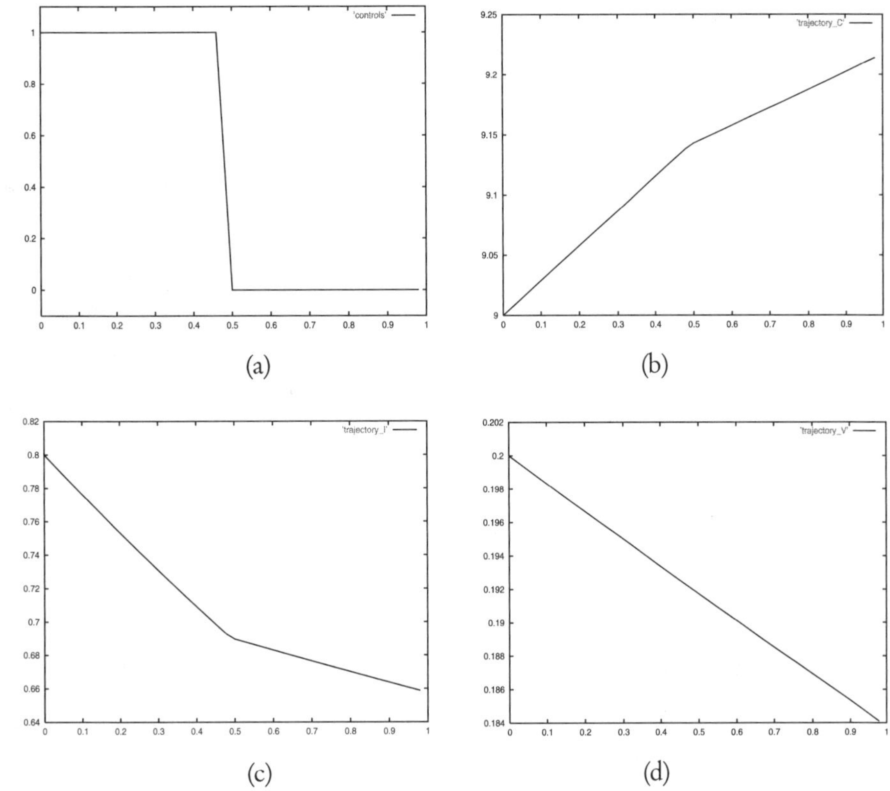

Figure 8.7. *HIV-1 dynamics with* (a) *optimal control,* (b) *uninfected cells,* (c) *infected cells,* (d) *virus* [CFF04].

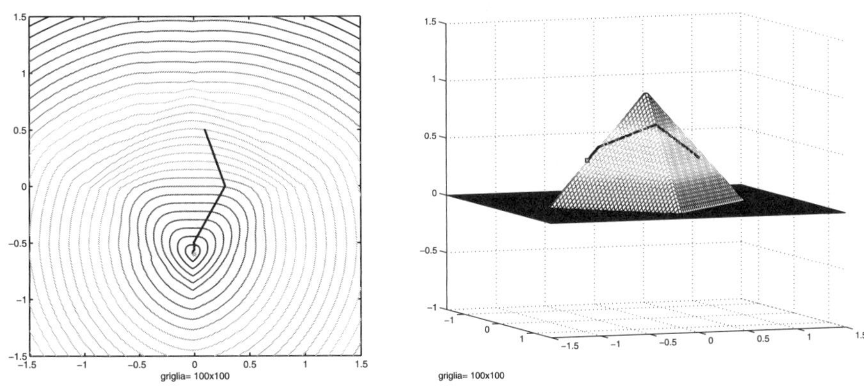

Figure 8.8. *Geodesics on a manifold: level sets of T (left) and an optimal trajectory on the surface z (right).*

with $a \in B(0,1)$, a point target $\mathcal{T} = (-1,-1)$, and $c(x)$ vanishing on two different regions (the obstacles):

$$c(x,y) = \begin{cases} 0, & (x,y) \in ([0,0.5]\times[-2,1.5]) \cup ([1,1.5]\times[-1.5,2]), \\ 1 & \text{elsewhere.} \end{cases}$$

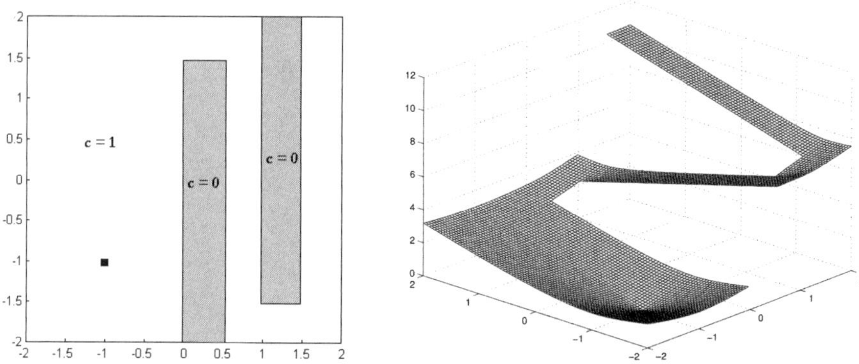

Figure 8.9. *Minimum time with obstacles: domain of the equation (left) and minimum time function T (right).*

Figure 8.9 shows the computational domain (left) and the value function $T(x)$ (right). We remark that by the use of the Kružkov transform to compute v, we do not need to modify the numerical scheme to deal with state constraints.

8.5 ▪ Dynamic Programming for differential games

The Dynamic Programming approach can also be applied to the analysis and approximation of differential games. In the general setting, we consider the nonlinear system

$$\begin{cases} \dot{y}(t) = f(y(t), a(t), b(t)), \\ y(0) = x, \end{cases} \tag{8.95}$$

where $y(t) \in \mathbb{R}^d$ is the state,

- $a \in \mathscr{A} = \{a : [0, +\infty) \to A, \text{measurable}\}$ is the control of player a,
- $b \in \mathscr{B} = \{b : [0, +\infty) \to B, \text{measurable}\}$ is the control of player b,

and $A, B \subset \mathbb{R}^m$ are given compact sets. We assume that f is uniformly Lipschitz continuous, that is,

$$|f(x,a,b) - f(y,a,b)| \leq L_f |x - y|, \tag{8.96}$$

for any $x, y \in \mathbb{R}^d$, $a \in A$, and $b \in B$. By Carathéodory's theorem, there exists a unique trajectory of (8.95) which will be denoted by $y_x(t; a, b)$. The *payoff* can be defined as in control problems. For example, in the infinite horizon case we would have

$$J_x(a,b) := \int_0^{+\infty} g(y_x(s; a, b), a(s), b(s)) e^{-\lambda s} ds.$$

In the typical situation described by *zero-sum differential games*, a gain for one player corresponds to a loss for the other; i.e., player a wants to minimize the payoff, whereas player b wants to maximize it.

Extending to the case of games, the usual definition of the value function is not immediate. For example, it is known that, in general,

$$\inf_a \sup_b J_x(a,b) \neq \sup_b \inf_a J_x(a,b).$$

Actually, a definition like

$$\inf_a \sup_b J_x(a,b)$$

would allow the player a to choose a control function with the information on the whole future strategy of player b, and this represents an obvious advantage. A less biased information pattern can be modeled by means of the notion of *nonanticipating strategies* (see [EK72] and the references therein),

$$\Delta = \{\alpha : \mathscr{B} \to \mathscr{A} : b(t) = \widetilde{b}(t)\ \forall t \le t' \Rightarrow \alpha[b](t) = \alpha[\widetilde{b}](t)\ \forall t \le t'\},$$

$$\Gamma = \{\beta : \mathscr{A} \to \mathscr{B} : a(t) = \widetilde{a}(t)\ \forall t \le t' \Rightarrow \beta[a](t) = \beta[\widetilde{a}](t)\ \forall t \le t'\}.$$

The above definition is fair with respect to both players. In fact, if player a chooses a control in Δ, it will not be influenced by future choices of player b (and vice versa).

We will present the main ideas of the Dynamic Programming approach for games in the simplified case of *Pursuit-Evasion games* (PE games in what follows), which in some sense generalize the minimum time problem.

Example 8.16 (PE Games). This is a particular class of differential games in which each player moves with its own dynamics. Player a tries to reach player b in minimum time, while player b tries to escape. More formally, we assume that the dimension d of the state space is even, and write the dynamics as

$$\begin{cases} \dot{y}_1 = f_1(y_1, a), \\ \dot{y}_2 = f_2(y_2, b), \end{cases} \tag{8.97}$$

where $y_i \in \mathbb{R}^{d/2}$. Moreover, we define a closed target set as the set where the positions of the two players coincide, that is,

$$\mathscr{T} = \{(y_1, y_2) : y_1 = y_2\},$$

or, more generally, where the two players are close enough, that is,

$$\mathscr{T} = \{|y_1 - y_2| \le \epsilon\}.$$

Then, the payoff is termed *capture time* and is defined by

$$t_x(a,b) = \min\{t : y_x(t; a, b) \in \mathscr{T}\}.$$

Clearly, $t_x(a,b)$ will be finite only under additional assumptions on the target and on the dynamics. The two players are opponents since player a (the *pursuer*) wants to minimize the payoff, whereas player b (the *evader*) wants to maximize it. ■

8.5.1 ▪ Dynamic Programming equations for PE games

Taking into account the arguments already discussed for the minimum time problem, it can be convenient to introduce again the Kružkov transform (8.32), by defining

$$v(x) := \begin{cases} 1/\mu & \text{if } t_x(a,b) = +\infty, \\ 1/\mu - e^{-\mu t_x(a,b)} & \text{else,} \end{cases} \tag{8.98}$$

which corresponds to the payoff

$$J_x(a,b)=\int_0^{t_x(a,b)} e^{-\mu t}\,dt.$$

In what follows, we will consider μ as a free parameter to be suitably chosen. For the moment, we set for simplicity $\mu=1$ and define the *lower value* of the game in terms of capture time as

$$T(x)=\inf_{\alpha\in\Delta}\sup_{b\in\mathscr{B}} t_x(\alpha[b],b)$$

or in terms of rescaled times as

$$v(x)=\inf_{\alpha\in\Delta}\sup_{b\in\mathscr{B}} J_x(\alpha[b],b).$$

Similarly, the *upper value* of the game is defined as

$$\widetilde{T}(x)=\sup_{\beta\in\Gamma}\inf_{a\in\mathscr{A}} t_x(a,\beta[a])$$

or

$$\widetilde{v}(x)=\sup_{\beta\in\Gamma}\inf_{a\in\mathscr{A}} J_x(a,\beta[a]).$$

Lower and upper values differ in general, but if they coincide, i.e., if $T=\widetilde{T}$ or $v=\widetilde{v}$, we say that *the game has a value*. Both the lower and the upper values satisfy a DPP; e.g., in the case of the lower value we have the following.

Lemma 8.17 (DPP for PE games). *For all $x\in\mathbb{R}^d\setminus\mathscr{T}$ and $0\le t<T(x)$,*

$$T(x)=\inf_{\alpha\in\Delta}\sup_{b\in\mathscr{B}}\{t+T(y_x(t;\alpha[b],b))\}$$

and

$$v(x)=\inf_{\alpha\in\Delta}\sup_{b\in\mathscr{B}}\left\{\int_0^t e^{-s}\,ds+e^{-t}v(y_x(t;\alpha[b],b))\right\}.$$

The proof is similar to the 1-player case, but more technical due to the use of nonanticipating strategies. Accordingly, we introduce the two Hamiltonians for games, termed as *Isaacs lower Hamiltonian*

$$H(x,p)=\min_{b\in B}\max_{a\in A}\{-p\cdot f(x,a,b)\}-1$$

and *Isaacs upper Hamiltonian*

$$\widetilde{H}(x,p)=\max_{a\in A}\min_{b\in B}\{-p\cdot f(x,a,b)\}-1.$$

The following existence theorem is proved in [ES84], along with an analogous result for the upper value.

Theorem 8.18.

1. *If $\mathscr{R}\setminus\mathscr{T}$ is open and $T(\cdot)$ is continuous, then T is a viscosity solution in $\mathscr{R}\setminus\mathscr{T}$ of the equation*

$$H(x,DT)=0.$$

2. *If $v(\cdot)$ is continuous, then v is a viscosity solution in $\mathbb{R}^d \setminus \mathcal{T}$ of the equation*

$$v + H(x, Dv) = 0.$$

As it has been shown in Chapter 2, uniqueness of viscosity solutions requires a comparison lemma. In the case of the Isaacs equation, a central role is played by a "structural condition" for the Hamiltonian function: it is assumed that for all x, y, p, and q,

$$|H(x,p) - H(y,q)| \le K(1+|x|)|p-q| + L|q||x-y| \tag{8.99}$$

for two positive constants K and L. More precisely, condition (8.99) implies a comparison principle whenever the solutions are in $BUC(\mathbb{R}^d \setminus \mathcal{T})$.

Theorem 8.19. *If the lower value function $v \in BUC(\mathbb{R}^d \setminus \mathcal{T})$, then v is the complete solution (maximal subsolution and minimal supersolution) of*

$$\begin{cases} u + H(x, Du) = 0, & x \in \mathbb{R}^d \setminus \mathcal{T}, \\ u = 0, & x \in \partial \mathcal{T}. \end{cases} \tag{8.100}$$

Thus, v is the unique viscosity solution.

Similar results hold for the upper value functions, replacing H with $\widetilde{H}$. Last, the following *Isaacs condition* ensures that the game has a value.

Theorem 8.20. *If $v, \widetilde{v} \in BUC(\mathbb{R}^d \setminus \mathcal{T})$, then*

$$v \le \widetilde{v},$$
$$T \le \widetilde{T}.$$

Moreover, if for all x and p

$$H(x,p) = \widetilde{H}(x,p),$$

then $v = \widetilde{v}$ and $T = \widetilde{T}$ (i.e., the game has a value).

It can be proved that the Hamiltonian functions for both the minimum time problem and the PE games satisfy condition (8.99), while usual conditions implying continuity of the value function are STLC for the minimum time problem and similar "capturability" conditions for PE games (see condition (8.107) in what follows). However, these conditions are less relevant in the case of games due to the existence of important cases in which the value functions might present discontinuities, the so-called "barriers" (an example will be presented in the numerical tests, while we refer the reader to the monograph [165] for a more extensive discussion).

Before turning to the numerical approximation, we conclude this overview by giving some information about the characterization of discontinuous value functions for games. As we have seen, to obtain uniqueness one should prove that a comparison principle holds; i.e., for every subsolution w and supersolution W we have

$$w \le W.$$

Although the usual definition of discontinuous viscosity solution (see Chapter 2) is sufficient to get uniqueness in the convex case, it will *not* provide uniqueness for nonconvex Hamiltonians (e.g., the min-max Hamiltonians under consideration).

Two new definitions have been proposed. Denote by S the set of subsolutions of our equation and by Z the set of supersolutions satisfying the Dirichlet boundary condition on $\partial\Omega$. Then we have the following.

Definition 8.21 (minmax solutions). *u is a* minmax solution *if there exist two sequences $w_n \in S$ and $W_n \in Z$ such that*

$$w_n = W_n = 0 \quad \text{on } \partial\Omega, \tag{8.101}$$

$$w_n \text{ is continuous on } \partial\Omega, \tag{8.102}$$

$$\lim_n w_n(x) = u(x) = \lim_n W_n(x), \qquad x \in \overline{\Omega}. \tag{8.103}$$

Definition 8.22 (e-solutions). *u is an* e-solution *(envelope solution) if there exist two nonempty subsets*

$$S(u) \subset S, \qquad Z(u) \subset Z$$

such that for all $x \in \overline{\Omega}$

$$u(x) = \sup_{w \in S(u)} w(x) = \inf_{W \in Z(u)} W(x).$$

By the comparison lemma there exists a unique e-solution.
In fact, if u and v are two e-solutions, then

$$u(x) = \sup_{w \in S(u)} w(x) \le \inf_{W \in Z(v)} W(x) = v(x)$$

and

$$v(x) = \sup_{w \in S(v)} w(x) \le \inf_{W \in Z(u)} W(x) = u(x).$$

It could also be proved that, under our basic assumptions, u is a minmax solution if and only if it is an e-solution.

8.5.2 ▪ SL schemes for PE games

The general ideas behind the SL discretization of Hamilton–Jacobi–Isaacs equations remain the same as for Bellman equations, and, in particular, the case of PE games appears as a generalization of the minimum time problem. A general convergence analysis may be carried out by the Barles–Souganidis theorem, although, due to the relevance of discontinuous solutions in games, the discontinuous case has been the object of more in-depth studies.

As in the case of minimum time problems, the presence of a target set makes it necessary to treat boundary conditions. We will sketch the construction and analysis of the scheme for an exterior problem in $\mathbb{R}^d$ in which the boundary conditions is imposed only on $\partial\mathcal{T}$. Assigning boundary conditions on the external boundary of the computational domain is a more delicate issue: for example, if the condition $v(x) = 1$ is enforced on the external boundary, this means that boundary points are "safe" for the evader—an advantage for this player. Some of the references in the last section will provide a deeper discussion of this point.

Time and space discretization Using the Kružkov change of variables and working as usual with Euler time discretization and $\mathbb{P}_1$ or $\mathbb{Q}_1$ interpolation, we obtain the fully discrete scheme

$$\begin{cases} v_j = \tau \max_{b\in B} \min_{a\in A} \{I_1[V](x_j + \Delta t f(x_j, a, b))\} + \dfrac{1-\tau}{\mu}, & x_j \in \mathbb{R}^d \setminus \mathscr{T}, \\ v_j = 0, & x_j \in \mathscr{T} \end{cases} \tag{8.104}$$

(with $\tau = e^{-\mu\Delta t}$), which generalizes (8.70) and can be put in iterative form in the same way. Denoting the right-hand side of (8.104) by $S(\Delta, V)$, it can be easily proved that

- $S(\Delta, \cdot)$ is monotone;
- numerical solutions obtained via (8.104) are uniformly bounded, and more precisely $\|V\|_\infty \leq 1/\mu$; and
- $S(\Delta, \cdot)$ is a contraction in l^∞ (with contraction coefficient $L_S = \tau$).

While this general framework suffices to prove convergence in the continuous case by the Barles–Souganidis theorem, we also mention that more technical convergence results can be obtained in the presence of discontinuities (barriers) for the value function.

An a priori error estimate for PE games We end this section on differential games by obtaining a convergence estimate for the case of PE games (see [Sor98]). An additional value for this error bound is that it also applies to the case of the minimum time problem, which may be seen as a special case of PE games (a situation in which the control set B is a singleton).

In the proof we will assume that the controlled vector field f is bounded and Lipschitz continuous, i.e., that for all x, z, a, and b it satisfies (8.96) and

$$|f(x,a,b)| \leq M_f. \tag{8.105}$$

The target $\mathscr{T}$ is allowed to be unbounded, but is assumed to have a piecewise smooth boundary. Moreover, we assume that the Dirichlet problem for Isaacs equation has a unique continuous, bounded, nonnegative viscosity solution v, e.g., by enforcing a suitable controllability assumption of the PE game to $\mathscr{T}$.

In order to obtain a simpler notation, we denote the upwind points in (8.104) as

$$z_i(a,b) := x_i + \Delta t f(x_i, a, b)$$

and write the interpolation of numerical solution in the form

$$I[V](z_i(a,b)) = \sum_j \lambda_{ij}(a,b) v_j,$$

where, according to the general form of the interpolation I,

$$\lambda_{ij}(a,b) = \psi_j(z_i(a,b)).$$

We recall that, in the linear or bilinear case under consideration, for a given index i the numbers λ_{ij} are coefficients of a convex combination; i.e., they are positive and with unity sum. With these notations, the scheme (8.104) can be rewritten as

$$v_i = S_i(V) = \begin{cases} \tau \max_{b\in B} \min_{a\in A} \sum_j \lambda_{ij}(a,b) v_j + \dfrac{1-\tau}{\mu}, & x_i \in \mathbb{R}^d \setminus \mathscr{T}, \\ 0, & x_i \in \mathscr{T}. \end{cases}$$

For shortness, in what follows we denote

$$d(x) = d(x, \mathcal{T}).$$

In order to obtain the convergence estimates, we need some additional control on the boundary condition. To this end we assume that there is a closed uniform neighborhood of $\mathcal{T}$, $\mathcal{T}_\delta = \{x : d(x) \le \delta\}$ and positive constants C_1, C_2 and $\alpha, \beta, \gamma \in (0,1]$ such that

$$\begin{aligned} &v(x) \le C_1 d(x)^\gamma, && x \in \mathcal{T}_\delta \backslash \mathcal{T}, \\ &v_i \le C_2 \left(d(x_i)^\alpha + \Delta t^\beta \left(1 + \left(\frac{\Delta x}{\Delta t} \right)^2 \right) \right), && x_i \in \mathcal{T}_\delta \backslash \mathcal{T}. \end{aligned} \tag{8.106}$$

Remark 8.23. *It is easy to prove that the first inequality in* (8.106) *is equivalent to the local γ-Hölder continuity of the capture-time function T. When $\mu \ge L_f$, using* (8.105), *this also implies that v is γ-Hölder continuous on the whole of $\mathbb{R}^d$. This will therefore be implicitly assumed in what follows.*

Both assumptions (8.106) are naturally related to the local controllability properties of the vector field f on $\partial\mathcal{T}$. In the control literature, estimates of this kind are typically reduced to suitable assumptions on the Lie brackets of f. To be more concrete, we give a basic example in which they are fulfilled in the case of PE games.

It is known that if $\mathcal{T}$ is the closure of an open bounded set with C^2 boundary, then a necessary and sufficient condition for the first inequality in (8.106) to hold with $\gamma = 1$ is that, for any $x \in \partial\mathcal{T}$,

$$\max_{b \in B} \min_{a \in A} \{ f(x,a,b) \cdot \nu(x) \} < 0, \tag{8.107}$$

in which $\nu(x)$ indicates as usual the exterior unit normal to $\partial\mathcal{T}$ at x (this condition may be extended to the nonsmooth and unbounded case).

The next proposition, Proposition 8.25, shows that (8.107) is also sufficient for the second condition in (8.106). Before proving this result, we give a preliminary lemma.

Lemma 8.24. *Assume that the vector W is such that $w_i \in [0, 1/\mu]$ for all i and $w_i = 0$ if $x_i \in \mathcal{T}$ and satisfies $W \ge S(W)$. Then, for all indices i we have $w_i \ge v_i$.*

Proof. Define the vector $W^{(k)}$ by the iteration

$$\begin{cases} W^{(k+1)} = S(W^{(k)}), \\ W^{(0)} = W. \end{cases}$$

By assumption, $W^{(0)} \ge W^{(1)}$, and by monotonicity of S,

$$S(W^{(0)}) \ge S(W^{(1)}),$$

so that we also obtain $W^{(1)} \ge W^{(2)}$ and, by induction, $W^{(k)} \ge W^{(k+1)}$. Therefore, the vector $W^{(k)}$ is nonincreasing, and the lemma is proved once we recall that its limit is precisely V. □

Proposition 8.25. *Assume* (8.96), (8.105), *and* (8.107). *Assume, moreover, that $\mathcal{T}$ is the closure of an open bounded set with a C^2 boundary. Then, there exist constants $C, \delta > 0$ such* *that, for all Δt and Δx and any node $x_i \in \mathcal{T}_\delta$,*

$$v_i \le C\left(d(x_i) + \Delta t\left(1 + \left(\frac{\Delta x}{\Delta t}\right)^2\right)\right).$$

Proof. We start by applying the previous lemma with a suitable vector W, which must satisfy, for any given $x_i \in \mathbb{R}^d \backslash \mathcal{T}$, the condition

$$w_i \ge S_i(W) = \tau \max_{b\in B} \min_{a\in A} \sum_j \lambda_{ij}(a,b) w_j + \frac{1-\tau}{\mu}. \tag{8.108}$$

Observe that, in view of the definition of the λ_{ij}, (8.108) is equivalent to

$$0 \ge \tau \max_{b\in B} \min_{a\in A} \sum_j \lambda_{ij}(a,b)(w_j - w_i) + \frac{1-\tau}{\mu} - (1-\tau) w_i$$

so that, dividing by τ, we obtain

$$0 \ge \max_{b\in B} \min_{a\in A} \sum_j \lambda_{ij}(a,b)(w_j - w_i) + \frac{1-\tau}{\mu\tau} - \frac{1-\tau}{\tau} w_i. \tag{8.109}$$

Since w_i is nonnegative, the last term in the right-hand side of (8.109) satisfies

$$\frac{1-\tau}{\tau} w_i \ge 0.$$

On the other hand, using a Taylor expansion for the exponential and the nonrestrictive condition $\Delta t \le 1$, we also get

$$\frac{1-\tau}{\mu\tau} \le \Delta t\, e^{\mu \Delta t} \le \Delta t\, e^{\mu},$$

and using these bounds into (8.109), we obtain that (8.108) is a fortiori satisfied if

$$0 \ge \max_{b\in B} \min_{a\in A} \sum_j \lambda_{ij}(a,b)(w_j - w_i) + \Delta t\, e^{\mu}. \tag{8.110}$$

Define now the signed distance from the target as

$$\tilde{d}(x) = \begin{cases} d(x, \mathcal{T}), & x \in \mathbb{R}^d \backslash \mathcal{T}, \\ -d(x, \partial \mathcal{T}), & x \in \mathcal{T}. \end{cases}$$

Note now that condition (8.107), along with the basic assumptions on f and $\mathcal{T}$, implies that

$$\max_{b\in B} \min_{a\in A} \{ f(x,a,b) \cdot Dd(x) \} \le -\sigma, \qquad x \in \{x : |\tilde{d}(x)| \le \rho\}, \tag{8.111}$$

for suitable constants $\rho, \sigma > 0$. Then, define the constant D_2 as

$$D_2 = \sup_{|\tilde{d}(x)| \le 2\rho} |D^2 \tilde{d}(x)|.$$

Step 1. We start by considering the case in which

$$\Delta t\left(M_f+\frac{\Delta x}{\Delta t}\right)^2\le\frac{\sigma}{D_2+1}$$

and checking (8.110) for the vector defined as follows:

$$w_i=\begin{cases}C\left(d(x_i)+\Delta t\left(M_f+\dfrac{\Delta x}{\Delta t}\right)\right)\wedge\dfrac{1}{\mu}, & x_i\in\mathbb{R}^d\setminus\mathcal{T},\\ 0, & x_i\in\mathcal{T},\end{cases}$$

where we choose C sufficiently large so that $C\rho\ge 1/\mu$ and $C\sigma/2\ge e^{\mu}$. Since the condition $w_i\ge S_i(W)$ is obvious if either $x_i\in\mathcal{T}$ or $w_i=1/\mu$, we need only prove (8.110) at all x_i such that $0<w_i<1/\mu$. Then, applying Lemma 8.24 we get

$$v_i\le C\left(d(x_i)+\Delta t\left(M_f+\frac{\Delta x}{\Delta t}\right)\right)$$

and the conclusion.

Let therefore x_i be such that $w_i\in(0,1/\mu)$. Then, by the choice of C, we obtain $x_i\in\mathcal{T}_\rho\setminus\mathcal{T}$ and by the definition of W we have, for all x_j,

$$w_j-w_i\le C(d(x_j)-d(x_i)). \tag{8.112}$$

We now use the regularity of the boundary $\partial\mathcal{T}$ to compute

$$d(x_j)-d(x_i)\le Dd(x_i)\cdot(x_j-x_i)+\frac{D_2|x_j-x_i|^2}{2}. \tag{8.113}$$

Observe now that if $\lambda_{ij}(a,b)\ne 0$, then necessarily

$$|x_j-x_i|\le\Delta x+M_f\Delta t.$$

Note that, for Δt and Δx small enough, both x_i and x_j are in $\mathcal{T}_\rho$. Now, using (8.112) and (8.113) and the definition of $\lambda_{ij}(a,b)$, we obtain

$$\begin{aligned}\max_{b\in B}\min_{a\in A}\sum_j\lambda_{ij}(a,b)(w_j-w_i)&\le C\max_{b\in B}\min_{a\in A}\sum_j\lambda_{ij}(a,b)(d(x_j)-d(x_i))\\&\le C\left(\frac{D_2(\Delta x+M_f\Delta t)^2}{2}\right.\\&\qquad\left.+\max_{b\in B}\min_{a\in A}\sum_j\lambda_{ij}(a,b)(x_j-x_i)\cdot Dd(x_i)\right)\\&\le C\left(\frac{\Delta t\sigma}{2}+\Delta t\max_{b\in B}\min_{a\in A}Dd(x_i)\cdot f(x_i,a,b)\right)\\&\le-\Delta t\,e^{\mu},\end{aligned} \tag{8.114}$$

where in the last inequality we have applied the condition on C and (8.111). Therefore, (8.110) (and hence (8.108)) is satisfied for the index i.

Step 2. Consider now the opposite case, i.e.,

$$\Delta t\left(M_f+\frac{\Delta x}{\Delta t}\right)^2\geq\frac{\sigma}{D_2+1}.$$

Then, we can choose the constant C as

$$C=\frac{D_2+1}{\sigma\mu},$$

which implies that

$$\begin{aligned}v_i\leq\frac{1}{\mu}&\leq C\Delta t\left(M_f+\frac{\Delta x}{\Delta t}\right)^2\\&=C\Delta t\left(M_f^2+2M_f\frac{\Delta x}{\Delta t}+\left(\frac{\Delta x}{\Delta t}\right)^2\right)\end{aligned}$$

and again the conclusion since $\Delta x/\Delta t\leq 1+(\Delta x/\Delta t)^2$ (applying (8.110) is unnecessary here). □

We can now prove the a priori estimate.

Theorem 8.26. *Assume* (8.96) *and* (8.105), *and let* $\mu\geq L_f$. *Then, there exists a constant* $C=C(L,\mu)$ *such that, for all* Δt, Δx, *we have*

$$||I[V]-v||_\infty\leq C\Delta t^\rho\left(1+\left(\frac{\Delta x}{\Delta t}\right)^2\right),\tag{8.115}$$

where $\rho=\min\{\beta,\alpha/2,\gamma/2\}$, α, β, *and* γ *being defined in* (8.106).

Proof. In what follows, we define $\varepsilon=\Delta t^{(2-\gamma)/2}$. We start by proving that

$$\sup_{\mathbb{R}^d}(I[V]-v)\leq C\Delta t^\rho\left(1+\left(\frac{\Delta x}{\Delta t}\right)^2\right).$$

By contradiction, we assume that for any fixed constant $C>0$, we can find $\Delta t,\Delta x$ small enough and $\overline{\sigma}(\Delta t,\Delta x)\in(0,1]$ such that, for any $0<\sigma\leq\overline{\sigma}(\Delta t,\Delta x)$,

$$\sup_{\mathbb{R}^d}(I[V]-v)\geq C\Delta t^\rho\left(1+\left(\frac{\Delta x}{\Delta t}\right)^2\right)+2\sigma\tag{8.116}$$

and divide the proof into a number of steps.

Step 1. We proceed by variable doubling and for $x,y\in\mathbb{R}^d$ introduce the function

$$\varphi(x,y)=I[V](x)-v(y)-\frac{|x-y|^2}{\varepsilon}.$$

By (8.116) it follows that

$$+\infty>\sup_{\mathbb{R}^{2d}}\varphi\geq C\Delta t^\rho\left(1+\left(\frac{\Delta x}{\Delta t}\right)^2\right)+2\sigma.$$

Choose now a point (x_1,y_1) such that $\varphi(x_1,y_1)>\sup\varphi-\sigma$, and select a function $\xi\in C_0^1(\mathbb{R}^{2d})$ satisfying $0\leq\xi\leq 1$, $\xi(x_1,y_1)=1$, $|D\xi|\leq 1$. If we denote $\psi=\varphi+\sigma\xi$, by construction the maximum point of ψ is attained at a point $(x_0,y_0)\in\operatorname{supp}\xi$.

Step 2. Observe that $I[V](x_0) > 0$. This immediately follows from the fact that v is nonnegative and that

$$\begin{aligned} I[V](x_0) - v(y_0) + \sigma &\geq \psi(x_0, y_0) \\ &\geq \psi(x_1, y_1) \\ &\geq \sup_{\mathbb{R}^{2d}} \varphi \\ &\geq C\Delta t^{\rho}\left(1 + \left(\frac{\Delta x}{\Delta t}\right)^2\right) + 2\sigma. \end{aligned} \tag{8.117}$$

We write the point x_0 as a convex combination of grid nodes, i.e., $x_0 = \sum_j \lambda_j x_j$, where $\sum_j \lambda_j = 1$, $\lambda_j \in [0,1]$, and, moreover, $|x_0 - x_j| \leq \Delta x$ if $\lambda_j \neq 0$. We prove that if $\Delta x \leq \delta$, with δ as in assumption (8.106), none of the points x_j such that $\lambda_j \neq 0$ can be in $\mathcal{T}$. In fact, if for some i, $\lambda_i \neq 0$ and $x_i \in \mathcal{T}$, by (8.106) we would obtain the bound

$$\begin{aligned} I[V](x_0) = \sum_j \lambda_j v_j &\leq C_2\left(\sum_j \lambda_j d^{\alpha}(x_j) + \Delta t^{\beta}\left(1 + \left(\frac{\Delta x}{\Delta t}\right)^2\right)\right) \\ &\leq C_2\left(\sum_j \lambda_j |x_j - x_i|^{\alpha} + \Delta t^{\beta}\left(1 + \left(\frac{\Delta x}{\Delta t}\right)^2\right)\right) \\ &\leq C_2\left(\Delta x^{\alpha} + \Delta t^{\beta}\left(1 + \left(\frac{\Delta x}{\Delta t}\right)^2\right)\right), \end{aligned}$$

which gives a contradiction with (8.116) for C sufficiently large, since

$$\Delta x^{\alpha} = \left(\frac{\Delta x}{\Delta t}\right)^{\alpha} \Delta t^{\alpha} \leq \left(1 + \left(\frac{\Delta x}{\Delta t}\right)^2\right) \Delta t^{\alpha}$$

and v is nonnegative.

Step 3. We now show that $y_0 \notin \mathcal{T}$. We use the inequality $\psi(x_0, y_0) \geq \psi(x_0, x_0)$ and get

$$\frac{|x_0 - y_0|^2}{\varepsilon} \leq \sigma(\xi(x_0, y_0) - \xi(x_0, x_0)) + v(x_0) - v(y_0).$$

Therefore, by the regularity of v as in Remark 8.23, if we indicate with L_v the γ-Hölder constant of v, we have

$$\frac{|x_0 - y_0|^2}{\varepsilon} \leq \sigma|x_0 - y_0| + L_v|x_0 - y_0|^{\gamma},$$

which implies that

$$\begin{aligned} |x_0 - y_0|^2 &\leq \varepsilon\sigma|x_0 - y_0| + \varepsilon L_v|x_0 - y_0|^{\gamma} \\ &\leq \bar{\varepsilon}|x_0 - y_0| + \bar{\varepsilon} L_v|x_0 - y_0|^{\gamma} \end{aligned}$$

(where we have used the boundedness of both ε and σ), and therefore that $|x_0 - y_0| \leq K$, with K independent of all small ε and σ. Then, choosing σ sufficiently small and bounding the linear term with a term of exponent γ, we obtain

$$|x_0 - y_0| \leq (1 + L_v)^{1/(2-\gamma)} \varepsilon^{1/(2-\gamma)}. \tag{8.118}$$

We can now conclude that $y_0 \notin \mathcal{T}$. Otherwise, from (8.106) and (8.118), for $K+(1+L_v)^{1/(2-\gamma)}\,\varepsilon^{1/(2-\gamma)} \le \delta$, we obtain as above that

$$\begin{aligned} I[V](x_0) &\le C_2\left(\sum \lambda_j d^\alpha(x_j) + \Delta t^\beta\left(1+\left(\frac{\Delta x}{\Delta t}\right)^2\right)\right) \\ &\le C_2\left(\sum \lambda_j |x_j - y_0|^\alpha + \Delta t^\beta\left(1+\left(\frac{\Delta x}{\Delta t}\right)^2\right)\right) \\ &\le C_2\left((\Delta x + |x_0 - y_0|)^\alpha + \Delta t^\beta\left(1+\left(\frac{\Delta x}{\Delta t}\right)^2\right)\right) \\ &\le C_2\left(\Delta x^\alpha + (1+L_v)^{\alpha/(2-\gamma)}\varepsilon^{\alpha/(2-\gamma)} + \Delta t^\beta\left(1+\left(\frac{\Delta x}{\Delta t}\right)^2\right)\right) \\ &\le 2C_2 \Delta t^\rho\left(1+(1+L_v)^{\alpha/(2-\gamma)} + \left(\frac{\Delta x}{\Delta t}\right)^2\right), \end{aligned} \tag{8.119}$$

in which the last inequality follows by the definition of ε. This gives again a contradiction with (8.116) when C is sufficiently large.

Step 4. We use now the equations for v at y_0 and for v_j at all vertices x_j adjacent to x_0. By the definition of viscosity solution, (8.100) and (8.104), and the maximality of (x_0, y_0), we obtain

$$\begin{aligned} \mu v(y_0) + \min_{a\in A}\max_{b\in B}\left\{-f(y_0,a,b)\cdot \sigma D_y\xi(x_0,y_0) + \frac{2}{\varepsilon}f(y_0,a,b)\cdot(y_0-x_0)\right\} - 1 \ge 0, \\ v_j + \min_{a\in A}\max_{b\in B}\{-\tau I[V](x_j + \Delta t f(x_j,a,b))\} - \frac{1-\tau}{\mu} = 0. \end{aligned}$$

We conveniently choose $b_j \in B$ and $a_j \in A$, so as to obtain

$$\mu v(y_0) - f(y_0,a_j,b_j)\cdot\sigma D_y\xi(x_0,y_0) + \frac{2}{\varepsilon}f(y_0,a_j,b_j)\cdot(y_0-x_0) - 1 \ge 0, \tag{8.120}$$

$$v_j - \tau I[V](x_j + \Delta t f(x_j,a_j,b_j)) - \frac{1-\tau}{\mu} \le 0. \tag{8.121}$$

Define $\overline{x}_j = x_j + \Delta t f(x_j,a_j,b_j)$. The optimality of (x_0,y_0) implies that $\psi(x_0,y_0) \ge \psi(\overline{x}_j, y_0)$, and therefore

$$\begin{aligned} I[V](\overline{x}_j) &\le I[V](x_0) + \frac{|\overline{x}_j - x_0|^2}{\varepsilon} + \frac{2(\overline{x}_j - x_0, x_0 - y_0)}{\varepsilon} + \sigma|\overline{x}_j - x_0| \\ &= I[V](x_0) + \frac{|\overline{x}_j - x_0|^2}{\varepsilon} + \frac{2(x_j - x_0, x_0 - y_0)}{\varepsilon} \\ &\quad - \frac{2\Delta t}{\varepsilon}f(x_j,a_j,b_j)\cdot(y_0-x_0) + \sigma|\overline{x}_j - x_0|. \end{aligned}$$

Then, the last inequality and (8.121) give

$$\begin{aligned} v_j \le \tau\Bigg[& I[V](x_0) + \frac{|\overline{x}_j - x_0|^2}{\varepsilon} + \frac{2(x_j - x_0, x_0 - y_0)}{\varepsilon} \\ & - \frac{2\Delta t}{\varepsilon}f(x_j,a_j,b_j)\cdot(y_0-x_0) + \sigma|x_0 - \overline{x}_j|\Bigg] + \frac{1-\tau}{\mu}. \end{aligned} \tag{8.122}$$

Multiplying by λ_j (as defined in Step 2), summing on the index j, and recalling that $\sum_j \lambda_j x_j = x_0$, we get

$$\mu I[V](x_0) \leq \frac{\mu\tau}{1-\tau}\sum_j \lambda_j \left[\frac{|\overline{x}_j - x_0|^2}{\varepsilon} + \frac{2(x_j - x_0, x_0 - y_0)}{\varepsilon} - \frac{2\Delta t}{\varepsilon} f(x_j, a_j, b_j)\cdot(y_0 - x_0) + \sigma|\overline{x}_j - x_0| \right] + 1 \tag{8.123}$$
$$= \frac{\mu\tau}{1-\tau}\sum_j \lambda_j \left[\frac{|\overline{x}_j - x_0|^2}{\varepsilon} - \frac{2\Delta t}{\varepsilon} f(x_j, a_j, b_j)\cdot(y_0 - x_0) + \sigma|\overline{x}_j - x_0| \right] + 1.$$

We now multiply (8.120) by λ_j and sum on the index j; then we add the result to (8.123) and obtain, also by the definition of τ,

$$\begin{aligned}
\mu(I[V](x_0) - v(y_0)) \leq{}& \frac{\mu\tau}{1-\tau}\sum_j \lambda_j \left[\frac{|\overline{x}_j - x_0|^2}{\varepsilon} + \sigma|\overline{x}_j - x_0| \right] \\
&+ \sum_j \lambda_j \left[-\frac{2\Delta t}{\varepsilon}\frac{\mu\tau}{1-\tau} f(x_j, a_j, b_j)\cdot(y_0 - x_0) \right. \\
&\left. - f(y_0, a_j, b_j)\cdot \sigma D_y \xi(x_0, y_0) + \frac{2}{\varepsilon} f(y_0, a_j, b_j)\cdot(y_0 - x_0) \right] \\
\leq{}& \Delta t^{-1} \sum_j \lambda_j \left[\frac{|\overline{x}_j - x_0|^2}{\varepsilon} + \sigma|\overline{x}_j - x_0| \right] \\
&+ \frac{2L_f}{\varepsilon}\sum_j \lambda_j \left[|x_j - y_0||x_0 - y_0| + \mu\Delta t |x_0 - y_0| \right] + \sigma L_f.
\end{aligned} \tag{8.124}$$

To obtain the second inequality, we have added and subtracted in each bracket of the second sum the terms

$$\frac{2}{\varepsilon} f(x_j, a_j, b_j)\cdot(y_0 - x_0)$$

and used the Lipschitz continuity of f and the bound

$$0 \leq 1 - \frac{\mu\Delta t \tau}{1-\tau} \leq \mu\Delta t.$$

We now proceed with the estimate. Note that if $\lambda_j \neq 0$, then

$$|\overline{x}_j - x_0| \leq \Delta x + M_f \Delta t,$$

and we define the constant

$$P = (1 + L_v)^{1/(2-\gamma)}.$$

Using now (8.118), the definition of ε, and the triangular inequality

$$|x_j - y_0| \leq |x_j - x_0| + |x_0 - y_0|$$

we obtain from (8.124)

$$
\begin{aligned}
\mu(I[V](x_0)-v(y_0)) &\le (\varepsilon\Delta t)^{-1}(\Delta x+M_f\Delta t)^2+\sigma\left(M_f+L_f+\frac{\Delta x}{\Delta t}\right)\\
&\quad +2L_fP\left(\Delta x+P\varepsilon^{1/(2-\gamma)}+\mu\Delta t\right)\varepsilon^{(\gamma-1)/(2-\gamma)}\\
&\le \Delta t^{\gamma/2}\left(\frac{\Delta x}{\Delta t}+M_f\right)^2+\sigma\left(M_f+L_f+\frac{\Delta x}{\Delta t}\right)\\
&\quad +2L_fP\left(\mu+\frac{\Delta x}{\Delta t}\right)\Delta t^{(\gamma+1)/2}+2L_fP^2\Delta t^{\gamma/2}\\
&\le \Delta t^{\gamma/2}\Bigg[\left(\frac{\Delta x}{\Delta t}\right)^2+2(M_f+L_fP)\frac{\Delta x}{\Delta t}+M_f^2\\
&\quad +2\mu L_fP+2L_fP^2\Bigg]+\sigma\left(M_f+L_f+\frac{\Delta x}{\Delta t}\right).
\end{aligned}
\tag{8.125}
$$

Using (8.117) and the fact that σ can be chosen arbitrarily small, we obtain

$$
\begin{aligned}
C\Delta t^{\rho}\left(1+\left(\frac{\Delta x}{\Delta t}\right)^2\right) &\le \mu^{-1}\Delta t^{\gamma/2}\Bigg[\left(\frac{\Delta x}{\Delta t}\right)^2+2(M_f+L_fP)\frac{\Delta x}{\Delta t}\\
&\quad +M_f^2+2\mu L_fP+2L_fP^2\Bigg].
\end{aligned}
$$

Finally, since $\Delta x/\Delta t \le 1+(\Delta x/\Delta t)^2$, this inequality can be rewritten, for a suitable constant Q, as

$$
C\Delta t^{\rho}\left(1+\left(\frac{\Delta x}{\Delta t}\right)^2\right)\le Q\Delta t^{\gamma/2}\left(1+\left(\frac{\Delta x}{\Delta t}\right)^2\right),
$$

which provides a contradiction for C sufficiently large, by the definition of ρ.

Step 5. To prove the reverse estimate,

$$
\sup_{\mathbb{R}^d}(v-I[V])\le C\Delta t^{\rho}\left(1+\left(\frac{\Delta x}{\Delta t}\right)^2\right),
$$

we argue again by contradiction and assume that for any fixed $C>0$ there are Δt, Δx arbitrarily small and $\overline{\sigma}(\Delta t,\Delta x)\in(0,1]$ such that, for all $0<\sigma\le\overline{\sigma}(\Delta t,\Delta x)$, we have

$$
\sup_{\mathbb{R}^d}(v-I[V])\ge C\Delta t^{\rho}\left(1+\left(\frac{\Delta x}{\Delta t}\right)^2\right)+2\sigma. \tag{8.126}
$$

We follow along the lines above, choosing the function

$$
\varphi(x,y)=v(x)-I[V](y)-\frac{|x-y|^2}{\varepsilon}
$$

and constructing a maximum point (x_0,y_0) for ψ. As at the beginning of Step 2, we prove that $v(x_0)>0$, so $x_0\notin\mathcal{T}$. Arguing as in Step 3, using now the inequality $\psi(x_0,y_0)\ge\psi(y_0,y_0)$, we show that

$$
|x_0-y_0|\le(1+L_v)^{1/(2-\gamma)}\varepsilon^{1/(2-\gamma)}.
$$

Moreover, if we write y_0 as a convex combination of nodes in the form $y_0 = \sum_j \lambda_j y_j$, we can prove that none of the points y_j for which $\lambda_j \neq 0$ is in $\mathscr{T}$. In fact, assume by contradiction that $y_i \in \mathscr{T}$. Then, we can estimate

$$\begin{aligned}
v(x_0) &\le L_v d^\gamma(x_0) \\
&\le L_v |x_0 - y_i|^\gamma \\
&\le L_v(|x_0 - y_0| + \Delta x)^\gamma \\
&\le L_v\left((1+L_v)^{\gamma/(2-\gamma)}\varepsilon^{\gamma/(2-\gamma)} + \Delta x^\gamma\right) \\
&\le L_v\left((1+L_v)^{1/(2-\gamma)}\Delta t^{\gamma/2} + \Delta t^\gamma\left(1+\left(\frac{\Delta x}{\Delta t}\right)^2\right)\right),
\end{aligned}$$

which provides a contradiction with (8.126) if C is sufficiently large.

Then, we apply Step 4 of the proof with obvious modifications and the final result follows. □

Approximation of optimal feedback and trajectories for games Extending the ideas used for control problems to the case of PE games, an approximate optimal feedback can be computed at every point x as

$$(a^*(x), b^*(x)) = \arg\max_b \min_a \left\{e^{-\mu\delta} I[V](x + \delta f(x,a,b))\right\} + 1 - e^{-\mu\delta}.$$

If this procedure provides multiple solutions, then a unique couple should be suitably selected, e.g., by minimizing two convex functionals. A typical choice is to introduce an inertial criterion to avoid jumps in the controls as much as possible.

8.5.3 ▪ Numerical tests for PE games

We examine here the numerical solutions to some classical games, focusing on the approximation of value functions, optimal feedbacks, and trajectories. Note that all the value functions represented in the figures have values in $[0,1]$, being computed after the Kružkov change of variables.

The Tag-Chase game Two boys P and E are running one after the other in the plane $\mathbb{R}^2$. P wants to catch E in minimal time, whereas E wants to avoid the capture. Both of them are running with constant velocity and can change their direction instantaneously. This means that the dynamics of the system is given by

$$f_P(y,a,b) = v_P a, \qquad f_E(y,a,b) = v_E b$$

(where v_P and v_E are two scalars representing the maximum speed for P and E), and the admissible controls are taken in the sets

$$A = B = B(0,1).$$

Let us give a more explicit version of the dynamics, which is useful for the discretization. Denote by (x_P, y_P) the position of P and by (x_E, y_E) the position of E, and by θ_P and θ_E the angles between the y-axis and the velocities of, respectively, P and E (see Figure 8.10).

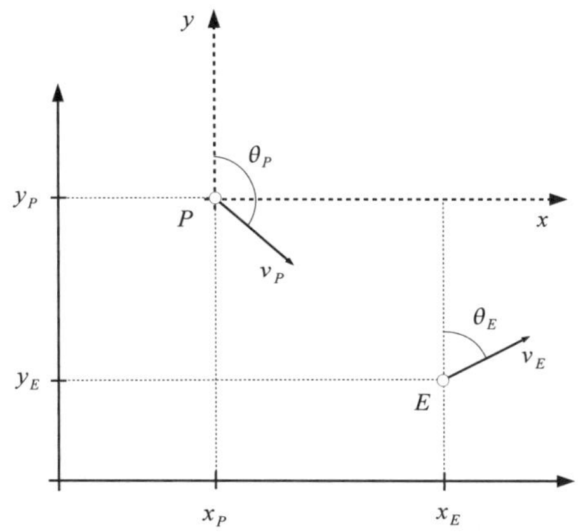

Figure 8.10. *The Tag-Chase game* [F06].

Then, we can write the dynamics as

$$\begin{cases} \dot{x}_P = v_P \sin\theta_P, \\ \dot{y}_P = v_P \cos\theta_P, \\ \dot{x}_E = v_E \sin\theta_E, \\ \dot{y}_E = v_E \cos\theta_E, \end{cases} \tag{8.127}$$

where $\theta_P \in [a_1, a_2] \subseteq [-\pi, \pi]$ and $\theta_E \in [b_1, b_2] \subseteq [-\pi, \pi]$ now play the role of one-dimensional controls for, respectively, P and E.

We say that E has been captured by P if their distance in the plane is lower than a given threshold $\epsilon > 0$. Introducing the complete state space $z = (x_P, y_P, x_E, y_E)$, we say that the capture occurs whenever $z \in \tilde{\mathcal{T}}$, where

$$\tilde{\mathcal{T}} = \left\{ z \in \mathbb{R}^4 : \sqrt{(x_P - x_E)^2 + (y_P - y_E)^2} < \epsilon \right\}. \tag{8.128}$$

The Isaacs equation is set in $\mathbb{R}^4$, since every player is represented by a point of $\mathbb{R}^2$. However, the result of the game depends only on the relative positions of P and E, since their dynamics are translation invariant. Therefore, the game can also by rephrased in the new variables

$$x = (x_E - x_P)\cos\theta - (y_E - y_P)\sin\theta,$$
$$y = (x_E - x_P)\sin\theta - (y_E - y_P)\cos\theta.$$

The new system (termed *relative coordinate* system) sets the origin on the position of P and moves with this player (see Figure 8.10). In the new coordinates, the dynamics (8.127) is rewritten as

$$\begin{cases} \dot{x} = v_E \sin\theta_E - v_P \sin\theta_P, \\ \dot{y} = v_E \cos\theta_E - v_P \cos\theta_P \end{cases} \tag{8.129}$$

and the target (8.128) as

$$\mathcal{T} = \left\{ (x, y) : \sqrt{x^2 + y^2} < \epsilon \right\}.$$

This change of variables represents a great simplification for at least three different reasons. The first is that the Isaacs equation is now solved in $\mathbb{R}^2$ instead of $\mathbb{R}^4$. The second is that

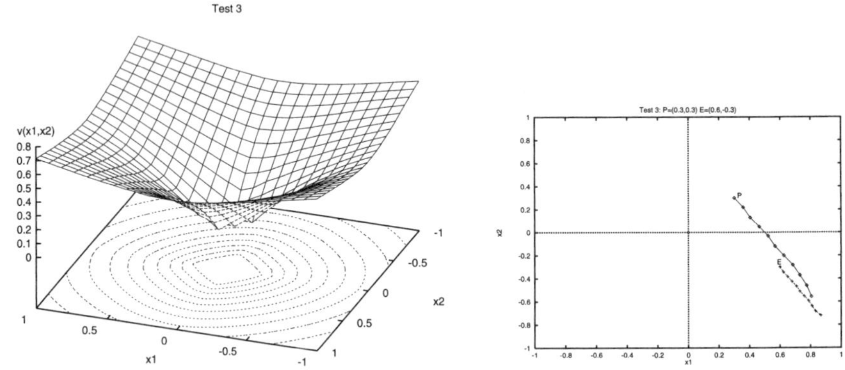

Figure 8.11. *Tag-Chase game: value function (relative coordinates) and optimal trajectories* [F06].

we now have a compact target in $\mathbb{R}^2$, whereas the original target (8.128) is unbounded, and this allows us to work on a bounded computational domain $Q \subset \mathbb{R}^2$. Finally, we get rid of the boundary conditions on ∂Q. It is easily seen that the game always has a value and that the only interesting case is $v_P > v_E$ (if the opposite inequality holds true, capture is impossible if E plays optimally). In this situation, the best strategy for E is to run at maximal velocity in the direction opposite to P along the direction passing through the initial positions of P and E. The optimal strategy for P is to run after E at maximal velocity. The corresponding minimal time of capture is

$$T(x_P, y_P, x_E, y_E) = \frac{\sqrt{(x_E - x_P)^2 + (y_E - y_P)^2}}{v_P - v_E}$$

or, in relative coordinates,

$$T(x,y) = \frac{\sqrt{x^2 + y^2}}{v_P - v_E}.$$

Figure 8.11 shows a numerical example. Here, we have chosen

$$Q = [-1,1]^2, \quad v_P = 2, \quad v_E = 1, \quad A = B = [-\pi, \pi].$$

And the discretization has been performed with a 23×23 space grid, $\Delta t = 0.05$, and $\epsilon = 0.2$. The value function is represented in the relative coordinate system, so P is fixed at the origin and the value at every point is the minimal time of capture (after the Kružkov transform), which is expected to behave as a (rescaled) distance function. Figure 8.11 shows value function (left) and optimal trajectories for the initial positions $P = (0.3, 0.3)$, $E = (0.6, -0.3)$ (right). The qualitative behavior of the solution correctly matches theoretical analysis.

The Tag-Chase game with constraints on the directions In this version of the Tag-Chase game, the dynamics is still given by (8.127), but the pursuer P has a constraint on the direction of motion θ_P, which must now be chosen in a smaller set $[a_1, a_2]$. The evader can still choose its control as $\theta_E \in [b_1, b_2] = [-\pi, \pi]$. In the numerical experiment below we have chosen

$$v_P = 2, \quad v_E = 1, \quad A = \left[\frac{3}{4}\pi, \frac{3}{4}\pi\right], \quad B = [-\pi, \pi].$$

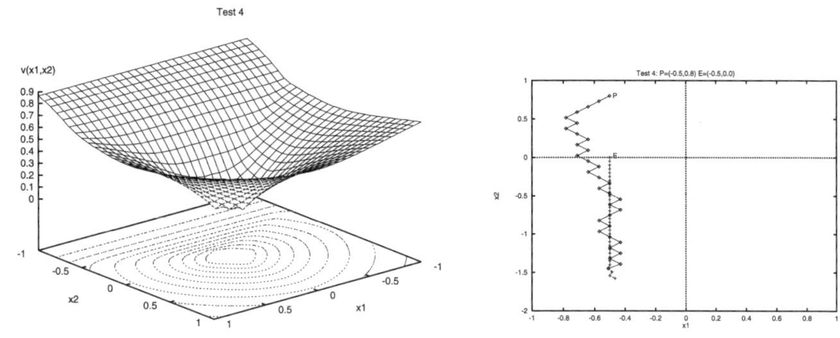

Figure 8.12. *Sector Tag-Chase game: value function and optimal trajectories* [F06].

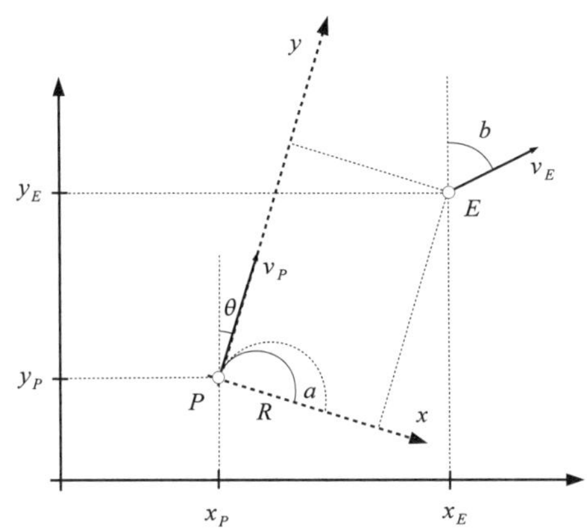

Figure 8.13. *The Homicidal Chauffeur problem* [F06].

As the left plot of Figure 8.12 shows, the capture time at points below the origin has a value bigger than at symmetric points above the origin. This happens because P has to zigzag to such points, which are not directly reachable due to directional constraints. The right plot of Figure 8.12 shows a sample trajectory.

The Homicidal Chauffeur The presentation of this game follows [FS05]. Consider two players (P and E) and the following dynamics:

$$\begin{cases} \dot{x}_P = v_P \sin\theta, \\ \dot{y}_P = v_P \cos\theta, \\ \dot{x}_E = v_E \sin b, \\ \dot{y}_E = v_E \cos b, \\ \dot{\theta} = \dfrac{R}{v_P} a, \end{cases} \tag{8.130}$$

where $a \in A = [-1, 1]$ and $b \in B = [-\pi, \pi]$ are the controls of the two players (see the sketch in Figure 8.13). While the evader is still free in its movements, the pursuer P is constrained to move with a minimum curvature radius R. The target is defined as in the Tag-Chase game, and we have used again the reduced coordinate system (8.129).

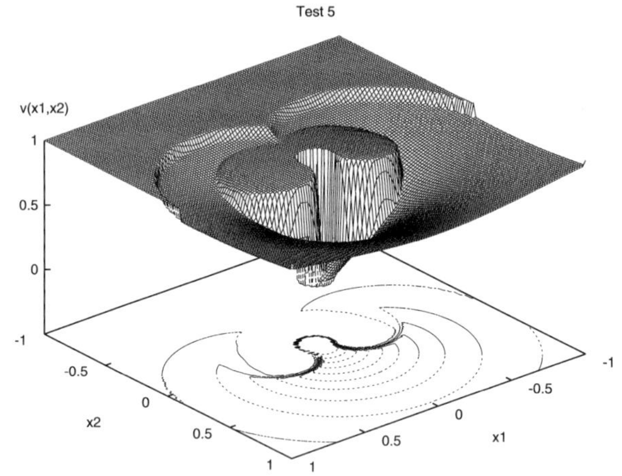

Figure 8.14. *Homicidal Chauffeur value function* [F06].

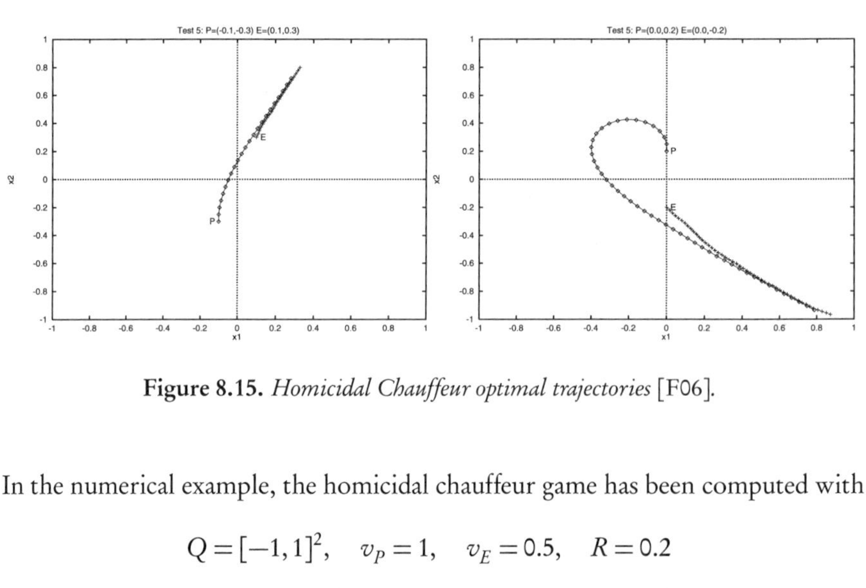

Figure 8.15. *Homicidal Chauffeur optimal trajectories* [F06].

In the numerical example, the homicidal chauffeur game has been computed with

$$Q = [-1,1]^2, \quad v_P = 1, \quad v_E = 0.5, \quad R = 0.2$$

and discretized with a space grid of 120×120 nodes, $\Delta t = 0.05$, and $\epsilon = 0.1$. Figure 8.14 shows the value function of the game. Note that when E is in front of P the two players essentially behave as in the Tag-Chase game: in this case, indeed, the radius constraint for P hardly come into play, as shown in Figure 8.15(left). However, if E is on the side or on the back of P, then the pursuer must first turn himself around to be able to catch E following a straight line (see Figure 8.15, right). This situation corresponds to the larger values of the capture time on the side lobes. Finally, there exists a region where capture is impossible ($v \equiv 1$) because the evader has the possibility of exiting Q before the pursuer can catch him, either because it is close enough to the boundary, or because it is too close to the pursuer to be reached under the radius constraint (in this situation, the optimal strategy for the pursuer requires performing a wider turn to regain a favorable position with respect to the evader). A discontinuity (barrier) appears on the boundary of this region, and it cannot be crossed if both players apply an optimal strategy.

Figure 8.16(left) is taken from [M71] and shows the optimal trajectories obtained by Merz via analytical methods. The approximate optimal trajectories (Figure 8.16, right)

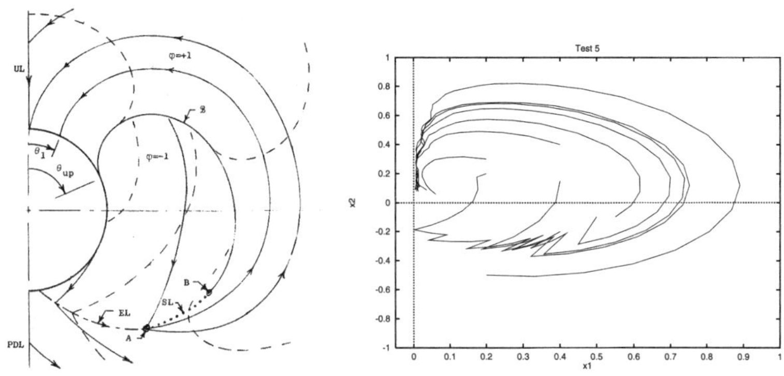

Figure 8.16. *Homicidal Chauffeur optimal trajectories: Merz description, computed trajectories. Left image reprinted from* [M71], *right image from* [F06].

show a very similar behavior. Moreover, the numerical approximation clearly locates the barrier.

8.6 ▪ Commented references

Several books review the classical theory of deterministic and stochastic control, so it is difficult to choose a specific general reference. However, we refer to classical monographs by Fleming and Rishel [FR75], and Macki and Strauss [MS82] and to the more recent books by Sontag [S98], and Bardi and Capuzzo Dolcetta [BCD97]. For a PDE-oriented theory of stochastic control problems, we quote the book by Fleming and Soner [FS93], while a more numerical approach to these problems (mainly focused on Markov chain schemes) is presented by Kushner and Dupuis [KuD01]. A classical reference for the theory of differential games is [BO82].

As we already mentioned, the application to deterministic control problems and games has been one of the main motivations for the development of the theory of viscosity solutions. Dynamic Programming (DP) singles out the *value function* $v(x)$ as a key tool in the analysis of optimal control problems and zero-sum differential games. This technique has been introduced in the classical works on calculus of variations and extensively applied to optimal control problems by Bellman [B57] (note that in the same years the alternative approach of the PMP was developed by the Russian school; see [PBGM62]). In the 1960s, a number of discrete-time control problems were investigated via the DP method, showing that v satisfies a functional (difference) equation and that its knowledge allows one to derive optimal feedbacks. We refer to the classical books [Be87] and [BeS78] for an extended presentation of these results. In continuous-time problems the DP equation is a nonlinear PDE, and in principle it would be necessary to show first that the equation has a unique solution in order to identify it with the value function. By the time DP techniques were introduced, this was a major analytical difficulty, since, in general, the value functions of deterministic optimal control problems and games may fail to be differentiable and even continuous, so the classical concept of solution could not be applied. Nevertheless, Isaacs used extensively the DP principle and the first-order PDE which is now associated with his name in his book [I65] on two-person zero-sum differential games, working mainly on the explicit solution of several examples in which the value function is regular everywhere except on some smooth surfaces.

Only at the beginning of the 1980s did Crandall and Lions [CL83] introduce the notion of weak solution in the viscosity sense for a class of first-order PDEs including Isaacs equations and prove their existence, uniqueness, and stability for the main boundary value problems. The theory was reformulated by Crandall, Evans, and Lions in [CEL84], and Lions proved in [Li82] that continuous value functions of optimal control problems are viscosity solutions of the corresponding DP equations. Similar results were obtained for the value function of some zero-sum differential games by Barron, Evans, Jensen, and Souganidis in [BEJ84, BJ89, Bar90] and [ES84, Sor93a] for various definitions of upper and lower values. Independently, Subbotin [Su80, Su84] found that the value functions in the Krasovskii–Subbotin sense [KS88] of some differential games satisfy certain inequalities for the directional derivatives which reduce to the Isaacs equation at points of differentiability. Moreover, he introduced a different notion of weak solution for first-order nonlinear PDEs, the minmax solution (where the name is related to the minmax operator appearing in the DP equation of games). The book [Su80] presents this theory with a special emphasis on the solution of differential games which motivated his study and the name for these solutions. The equivalence of this notion of weak solution with viscosity solutions has been addressed by several authors (see [EI84, LS85, LS86, SuT86]), and currently the two theories are essentially unified [Su95].

Concerning control problems with state constraints, the first characterization of the value function is due to Soner [So86a, So86b] under controllability assumptions which ensure continuity. In [CDL90] these results were extended in a more general setting, and in [Lo87] the properties of constrained viscosity solutions have been analyzed. Numerical results based on this approach can be found in [CF96]. Later, Ishii and Koike [IK96] introduced a different formulation of the constrained problem. Controllability assumptions have been later dropped in a different formulation due to Bokanowski, Forcadel, and Zidani [BFZ10a], which still obtains a characterization in terms of a DP equation.

At the present state of development, the theory of viscosity solutions applies to a wide range of boundary and Cauchy problems for first- and second-order nonlinear PDEs, including discontinuous solutions (at least for a subclass which covers DP equations for control problems and games [B89, BS89, BJ89, Bar90, BS91a, Sor93a]).

We should also mention an alternative characterization of the value function, which is based on viability theory (see, e.g., the book by Aubin and Cellina [AC84]). This approach is reviewed in the books by Aubin [A91] and Aubin and Frankowska [AF90]. In many classical problems, the graph of the value function is the boundary of the viability kernel for a new dynamics associated with the original control problem. This relationship has produced interesting results, mainly for state constrained problems where it is natural to consider (lower) semicontinuous solutions. We refer the reader to [Fr93] and [BJ90, BJ91] for two different characterizations of semicontinuous solutions of HJB equations, and to [FP00] for an application to a state constrained control problem. The numerical counterpart of the viability approach is based on the approximation of the viability kernel. Details on the related techniques can be found in [CQS99] and [CQS00]. More recent results in this framework are presented in [BFZ10a] making a connection between the viability and the DP approach. This has also created approximation schemes (see, e.g., [BFZ10b] for an a priori estimate in L^1 for a one-dimensional problem).

The approximation of the value function in the framework of viscosity solutions has been investigated by several authors starting from Capuzzo Dolcetta [CD83, CDI84, F87], while parallel works on the approximation of the value function were based on its characterization as the maximal subsolution of the Bellman equation (see, e.g., [GRo85a, GRo85b]). Abstract convergence results, as well as a priori error estimates, have been proved for the approximation of classical control problems and games [BF90a, BF90b,

CDF89, Al91a, Al91b, AL94, BS91b, BFS94], and in particular the error estimate for PE games is due to Soravia [Sor98]. Most of the material in this chapter originates from the survey papers [F97] and [BFS00], devoted, respectively, to control problems and games. Later, the same approach was applied to the approximation of impulsive control problems [CF97, CF99] and to the stabilization of dynamical systems (see [FGW99, KG11] and the references therein).

The approximation of feedback controls is still a wide open problem, especially in the case of games. A first convergence result requiring strong regularity assumptions on the value function can be found in [DS01]. The synthesis of feedback controls has also been addressed in [F01], obtaining L^1 error estimates for the approximation of optimal trajectories. The general framework presented here has been proposed in [FZ13] in the more general case of hybrid control systems. We also mention that numerical methods based on the approximation of open-loop controls have also been proposed, although this approach has not been pursued here. The advantage is to replace the approximation of the DP equation (which can be difficult or impossible in high-dimensional problems) by a large system of ODEs exploiting the necessary conditions for the optimal policy and trajectory. A general review on the related methods can be found in [P94]. Finally, the policy iteration method has been investigated in [K59, H60], proving that it corresponds to the Newton method applied to the functional equation of DP. Later, Puterman and Brumelle [PB79] gave sufficient conditions for the rate of convergence to be either superlinear or quadratic. More recent contributions on the policy iteration method and some extensions to games can be found in [SR04] and [BMZ09]. The acceleration method based on the set of subsolutions has been studied in [F87] (see also [TG92]).

Concerning differential games, a general introduction can be found in the books by Friedman [Fr71] and Basar and Olsder [BO82]. We have presented here only noncooperative zero-sum games, mainly because at present this is the only class of games for which a PDE-based theory can be developed. In particular, there exist two competing concepts of weak solutions in this area: the minmax solution and the e-solution, which coincide in many cases (as in the PE game studied in this chapter). For the characterization of the value function for games in the framework of viscosity solutions we refer the reader to the papers [Sou85b, FSo89, Sor92, GS06, DZS10] and to the book [BCD97]. The theory for the minmax solution introduced by Subbotin is reviewed in the book [Su91] and in [Su93].

Techniques for the numerical approximation of games (mainly of PE type) have originated from the Krasovskii–Subbotin theory. The numerical schemes are based on a suitable construction of generalized gradients in the finite difference operators used to approximate the value function. This approach has been developed by the Russian school (see [TUU95, PT00]) and allows one to compute an approximation of optimal feedback controls [T99]. A different approximation strategy for value and optimal policies of a dynamic zero-sum stochastic game has been proposed in [TA96, TPA97] and is based on replacing the game by a finite state approximation.

More recently, some results concerning PE games with state constraints have appeared. We refer the reader to the paper by Bardi, Koike, and Soravia [BKS00] for the theoretical setting and to [CF08] for a numerical technique which extends to games previous results for constrained control problems [CF96].

The mathematical results presented in this chapter are valid in any dimension. However, the DP approach suffers from the so-called *curse of dimensionality*. In order to overcome this problem, various techniques have been proposed in the literature. The first relies on domain decomposition methods, and consists in splitting the computational domain into several subdomains of manageable size, with suitable boundary conditions at the interfaces between subdomains. The method is well known for elliptic and parabolic problems [QV99a] and has been studied for HJB equations in [FLS94, CFLS94] and extended to games in [FLM01, FS05]. More recently, a domain decomposition technique which exploits the feature of the dynamical system constructing almost invariant "patches" has been introduced by Navasca and Krener [NK07] and improved in Cacace et al. [CCFP12]. A second technique which has been proposed to reduce the cost of the DP approach is Model Predictive Control. The goal of this technique is to obtain feedback controls at a lower cost. The theory is still under development, but the interested reader could be referred to the papers [GR08, GNP07] and to the recent book by Grüne and Pannek [GP11].

Chapter 9
Front propagation

As we have seen in the introduction of this monograph, one of the typical applications of HJ equations is the study and approximation of front propagation problems via the level set method. This technique became very popular in the 1990s due to its ability to follow front evolution after the onset of singularities and topological changes.

There exist a number of physical situations which lead to models of this class, e.g., the description of industrial etching processes, of bubbles moving in a fluid, of crystal growth, and so forth. More recently, these techniques have also been applied to image processing, one of the areas in which the introduction of nonlinear PDEs has had the strongest impact in the last decades.

In this chapter we will present some examples of models, techniques, and applications in this area, the main focus being of course on SL methods.

9.1 ▪ Front propagation via the level set method

The level set method is a simple and clever way to describe an interface separating two or more regions with different physical phases. The method describes the evolution of the front by a continuous representation function $u(x,t)$, i.e., a continuous function which is negative in the domain Ω_t corresponding to one of the phases and positive outside this domain. The interface is then identified with the level set

$$\Gamma_t = \partial\Omega_t = \{x \in \mathbb{R}^d : u(x,t) = 0\}$$

(see Chapter 1 for more details). This approach allows for changes in the topology of the front, since it does not make use of any parametrization for Γ_t, although it also entails the serious drawback of adding an extra dimension to the original problem, because a $(d-1)$-dimensional surface is represented as the level set of a function defined on $\mathbb{R}^d$.

The level set method leads to a nonlinear first-order PDE whenever the interface evolution is driven by terms of fixed normal velocity and (possibly) passive advection. More complex evolution laws consider normal velocity as a function of curvature and/or other geometric parameters of the interface, and this results in nonlinear, second-order models (or integro-differential equations).

To be more precise, we associate a scalar velocity c with the normal direction $\nu(x)$ to the front. Depending on the choice of the normal (exterior/interior), we can describe both the growth of a phase (deposition) and its reduction (erosion). Three major cases can be singled out:

- $c(x,t)$, *isotropic* velocity depending on (x,t);
- $c(x,k(x))$, velocity *depending on both x and the front curvature at x*; and
- $c(x,\nu(x))$, *anisotropic* velocity depending on both x and the normal at x.

In practical applications, these propagation laws may appear either separated or combined. In what follows, we will work mainly on the isotropic case and sketch some basic ideas on the curvature driven case.

As has been shown in Chapter 1, the standard model problem for an interface evolving in the normal direction with a given scalar velocity $c(x):\mathbb{R}^d \to \mathbb{R}$ is the first-order HJ equation

$$\begin{cases} u_t(x,t)+c(x)|Du(x,t)|=0, & (x,t)\in\mathbb{R}^d\times(0,+\infty),\\ u(x,0)=u_0(x), & x\in\mathbb{R}^d, \end{cases} \tag{9.1}$$

where the initial condition u_0 must be a representation function for the initial configuration of the front $\Gamma_0=\partial\Omega_0$. However, u_0 is itself unknown, since Γ_0 is in general the only available data of the problem. A typical choice is to use the signed distance to the front

$$\tilde{d}(x,\Gamma_0)=\begin{cases} d(x,\Gamma_0), & x\notin\Omega_0,\\ -d(x,\Gamma_0), & x\in\Omega_0. \end{cases} \tag{9.2}$$

In principle, this choice would require solving the eikonal equation

$$\begin{cases} |Du_0(x)|=1, & x\in\mathbb{R}^d\setminus\Gamma_0,\\ u_0=0, & x\in\Gamma_0. \end{cases}$$

Again, the sign of the normal depends on the choice $u_0(x)=\tilde{d}(x,\Gamma_0)$ or $u_0(x)=-\tilde{d}(x,\Gamma_0)$. Problem (9.1) can be simplified when the evolution is monotone (increasing or decreasing), a case which corresponds to a velocity c of constant sign. In this situation, the set Ω_t will have either a constant growth or a constant reduction; i.e., either $\Omega_t\subset\Omega_{t+s}$ or the reverse inclusion is satisfied for any $t,s>0$. For monotone evolutions, it can be proved that the evolutive equation (9.1) can be replaced by the stationary problem

$$\begin{cases} c(x)|DT(x)|=1, & x\in\mathbb{R}^d\backslash\Omega_0,\\ T(x)=0, & x\in\Gamma_0, \end{cases} \tag{9.3}$$

and, accordingly, the position of the front at time t is given by

$$\Gamma_t=\partial\Omega_t=\{x\in\mathbb{R}^d : T(x)=t\}.$$

The stationary approach links the propagations of fronts with the minimum time problem of control theory (see subsection 8.2.3). In fact, (9.3) corresponds to the Bellman equation (8.28) of a minimum time problem for the dynamics

$$\begin{cases} \dot{y}(t)=-c(y)\alpha(t), & t\in(0,+\infty),\\ y(0)=x \end{cases} \tag{9.4}$$

with a measurable control $\alpha:[0,+\infty)\to B(0,1)$, and with $\mathcal{T}=\Omega_0$. Using the Kružkov transform (8.32) and following the same ideas discussed for the minimum time problem, (9.3) can also be rewritten in the form

$$\begin{cases} v(x)+\max\limits_{a\in B(0,1)}\{c(x)a\cdot Dv(x)\}=1, & x\in\mathbb{R}^d\backslash\Omega_0,\\ v(x)=0, & x\in\Gamma_0. \end{cases} \tag{9.5}$$

We will strengthen the Caratheodory condition of existence and uniqueness for the trajectories of (9.4) by assuming a finite speed of propagation, that is, by requiring the following:

$$c(x) \text{ is bounded and Lipschitz continuous.} \tag{9.6}$$

The connection between front propagation and the minimum time problem can also be exploited for numerical purposes, once characteristics of the front problem (also termed as *rays*) are identified with optimal trajectories. Moreover, it is possible to treat the propagation in the presence of obstacles using a natural interpretation in terms of state constraints. Let $\mathcal{O}$ denote the set of obstacles, and define

$$D = \mathbb{R}^d \setminus (\mathcal{O} \cup \Omega_0). \tag{9.7}$$

In the presence of obstacles, trajectories are forced to avoid $\mathcal{O}$ and must remain in D. Then, the problem with obstacles corresponds to a minimum time problem in which the state is restricted to remain in D for any $t \geq 0$, that is, in which the admissible controls (or, in other words, the directions of propagation) can point only inside $\overline{D}$. This suggests using the notion of *state constrained viscosity solution*, introduced in Chapter 2.

A numerical scheme for this problem is constructed by blending the arguments of subsections 8.2.3 and 8.4.5. In particular, given a time step Δt, the discrete set of admissible controls at every point $x \in \overline{D}$ is defined by

$$A_x^{\Delta t} = \{a \in B(0,1) : x + \Delta t\, c(x)a \in \overline{D}\}. \tag{9.8}$$

At any node $x_i \in \overline{D}$, the corresponding fully discrete scheme is

$$v_i = \min_{a \in A_{x_i}^{\Delta t}} \{\tau I[V](x_i + \Delta t\, c(x_i)a)\} + 1 - \tau, \tag{9.9}$$

where $\tau = e^{-\Delta t}$, having chosen for simplicity $\mu = 1$ in the Kružkov transform (8.32). Following standard arguments, it can be proved that if I is a monotone ($\mathbb{P}_1$, $\mathbb{Q}_1$) interpolation, then there exists a unique numerical solution of (9.9) and that it converges locally uniformly to the "constrained" viscosity solution v of the problem in $\overline{D}$ for $\Delta t, \Delta x \to 0$.

Note that the direct application of this formulation would require a preliminary computation of the admissible control set $A_x^{\Delta t}$ at every node of the grid. This additional complexity could be avoided by penalizing the values inside $\mathcal{O}$, e.g., by setting in (9.9)

$$I[V](x_i + \Delta t\, c(x_i)a) = w_{\max} \quad \text{for } x_i + \Delta t\, c(x_i)a \in \mathcal{O}, \tag{9.10}$$

where $w_{\max}$ is a positive constant such that

$$w_{\max} \geq \|V\|_\infty \tag{9.11}$$

(in particular, if $\mu = 1$, then it suffices to set $w_{\max} = 1$). This penalization comes into play only at nodes close to the obstacle and is consistent with the physical model. In fact, setting the value $w_{\max} = 1$ at all the nodes $x_i \in \mathcal{O}$ corresponds to setting $c(x) \equiv 0$ in $\mathcal{O}$, so that the front cannot cross the obstacle.

9.1.1 ▪ Numerical tests

We show here the results obtained by the first-order SL scheme for the approximation of some front propagation problem, considering a variety of cases: variable normal velocity, discontinuous velocity, obstacles, and a coupling of advection with growth in the normal direction.

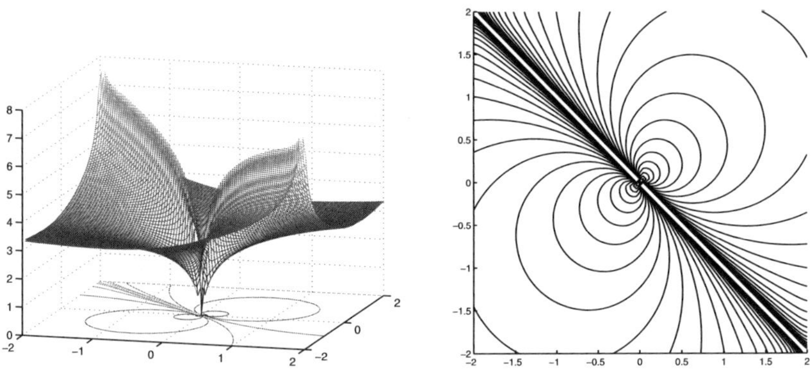

Figure 9.1. *Minimum time function T (left) and its level sets (right) for the velocity field* (9.12) [CF07].

Propagation with variable isotropic velocity We consider in this test the point source $\Gamma_0 = (0,0)$ and the nonconstant velocity field

$$c(x) = |x_1 + x_2|. \tag{9.12}$$

The speed of propagation $c(x)$ vanishes if $x_2 = -x_1$ and increases when moving away from this line. Figure 9.1 shows the minimum time function $T(x)$ and the level sets of T. On the line $x_1 + x_2 = 0$ we have $c(x) = 0$, and therefore $T = +\infty$. The solution has been computed with two fast solvers, the FM-SL and FS-SL methods (see subsections 9.4.2 and 9.4.4) on a 101×101 grid.

Propagation with a discontinuous velocity We set in this case $\Gamma_0 = (-1,0)$, and

$$c(x,y) = \begin{cases} 0.4, & (x_1,x_2) \in [0.5,1] \times [0,0.5], \\ 1 & \text{elsewhere}. \end{cases} \tag{9.13}$$

In this case the velocity field is discontinuous, and despite the basic assumption (9.6) not being satisfied, we show that both the model and the scheme are robust enough to handle this case. Figure 9.2 shows the minimum time function $T(x)$ and the level sets of T. The right plot of Figure 9.2 also shows some optimal trajectories which start from four different points and reach the target Γ_0 in minimum time with speed $c(x)$ (of course, if these trajectories are interpreted as propagation rays, they should be oriented in the opposite direction). This test uses a fast solver, the FS-SL method.

Front propagation around obstacles We set here $\Gamma_0 = (-1,-1)$, and

$$c(x,y) = \begin{cases} 0, & (x_1,x_2) \in ([0,0.5] \times [-2,1.5]) \cup ([1,1.5] \times [-1.5,2]), \\ 1 & \text{elsewhere}. \end{cases} \tag{9.14}$$

Following the ideas of the previous example, in this test the presence of obstacles has been treated by a velocity field which vanishes in two different regions. Figure 9.3 shows the computational domain and the minimum time function $T(x)$. We remark that the definition (9.14) allows us to avoid the implementation of state constraints in the numerical scheme. This example also uses the FS-SL fast solver.

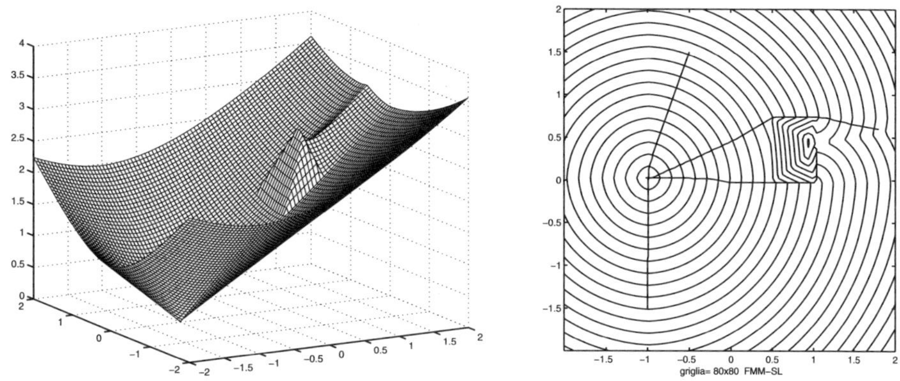

Figure 9.2. *Minimum time function T (left) and its level sets with some optimal trajectories (right) for the velocity field* (9.13) [CF07].

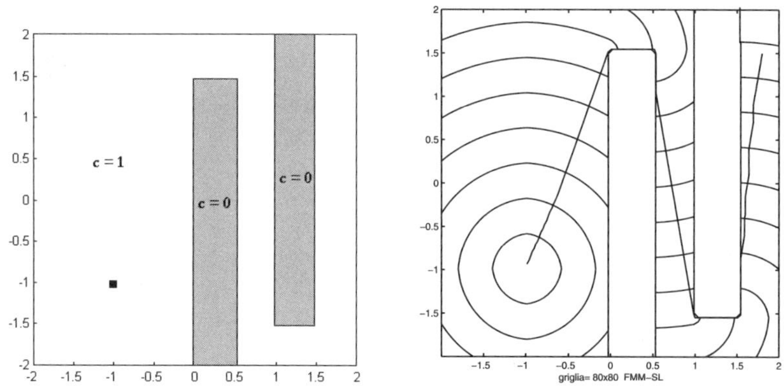

Figure 9.3. *Domain of the equation (left) and level sets of the minimum time function T with a propagation ray (right) for the velocity field* (9.14) [CF07].

A front evolving on a manifold In this test, we consider the HJ equation:

$$\begin{cases} u_t(x,t) + H(Du(x,t)) = 0, & (x,t) \in M \times [0,T], \\ u(x,0) = u_0(x), & x \in M, \end{cases} \tag{9.15}$$

posed on a Riemannian manifold M, so that the symbol Du should now be understood as the intrinsic gradient. In particular, the manifold is a torus and we set $H(p) = |p|$, so that level curves of u propagate at constant speed along geodesics.

The manifold is mapped on $Q = [-\pi, \pi]^2$ with doubly periodic boundary conditions, and (9.15) is rewritten in the planar coordinates $\xi \in Q$ as

$$u(\xi,t) + \sup_{\beta} \{-\beta \cdot Du(\xi,t) - H^*(J^t(\xi)\beta)\} = 0, \tag{9.16}$$

where $\beta \in \mathbb{R}^2$, H^* is the Legendre transform of H, and J is the Jacobian matrix of the transformation mapping Q into the torus of $\mathbb{R}^3$. The form (9.16) allows us to treat the Riemannian case as a straightforward adaptation of the euclidean case whenever the

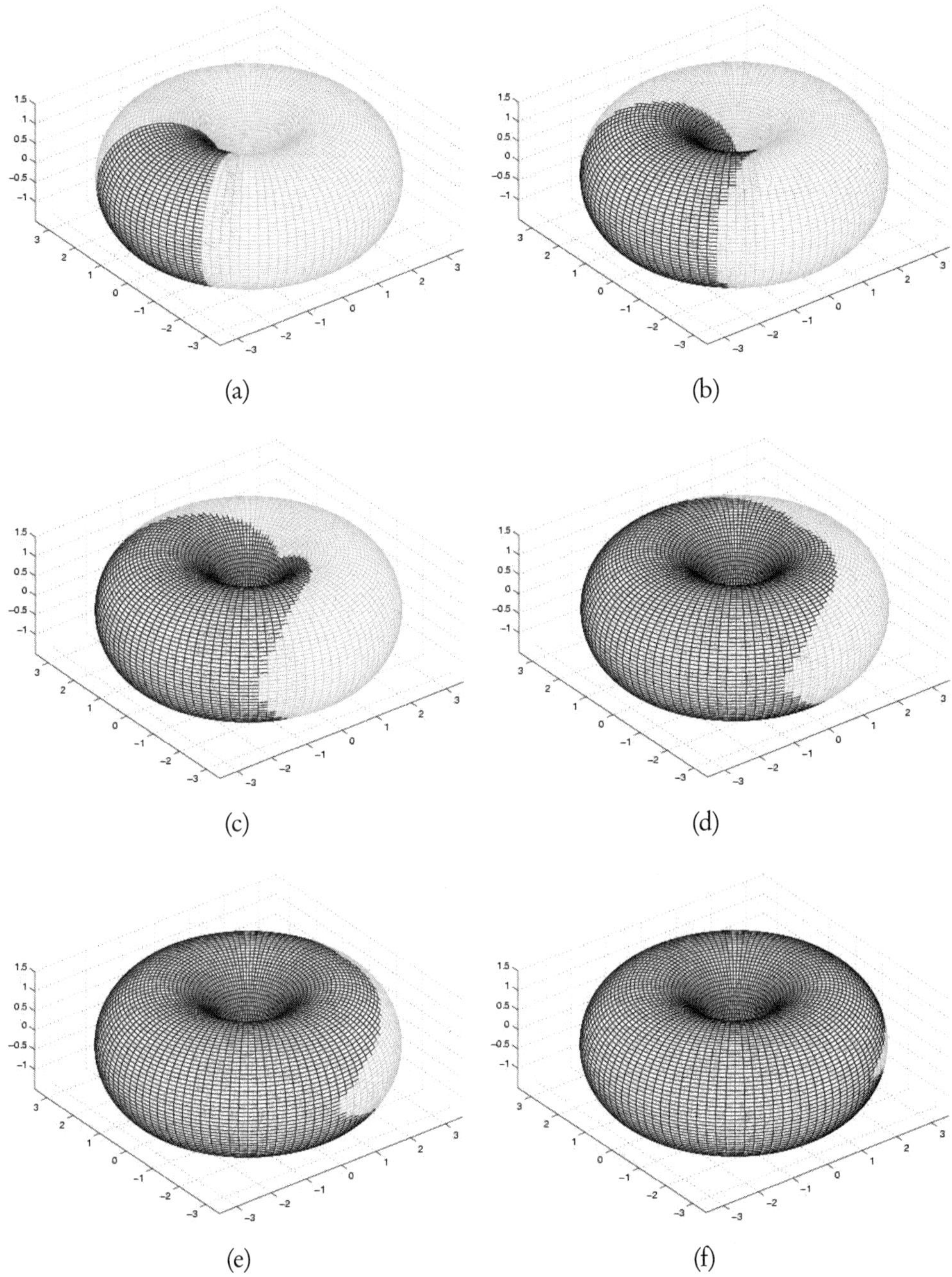

Figure 9.4. *Numerical solutions for* (9.15) [FF02].

parametrization of the manifold is known. The propagation of a level set of the numerical solution is shown in Figure 9.4.

Front propagation plus advection in $\mathbb{R}^3$ Here, we consider an initial configuration of the front corresponding to a couple of spheres in $\mathbb{R}^3$, i.e., an initial condition given by

$$v_0(x) = \min(|x - \xi_1|, |x - \xi_2|),$$

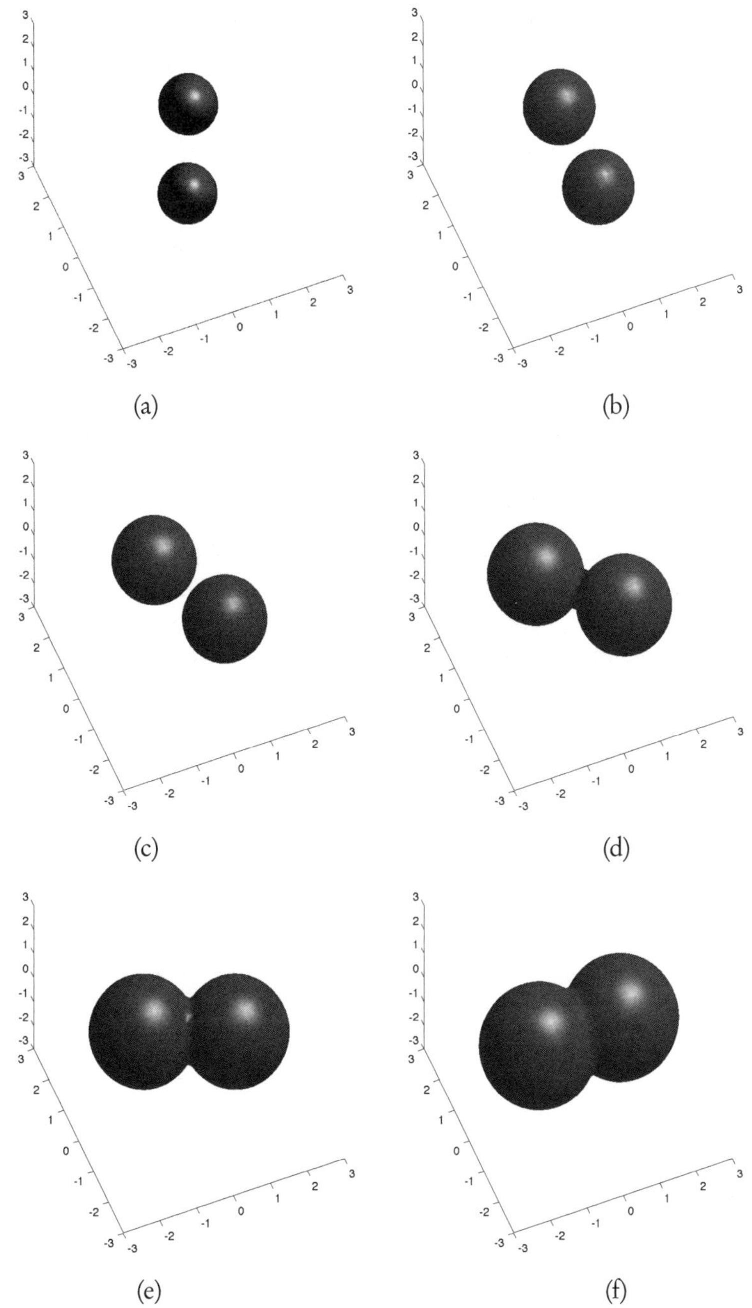

Figure 9.5. *Numerical solutions for the advection plus isotropic propagation test.*

where $\xi_1 = (1,1,0)$, $\xi_2 = (-1,-1,0)$. The front evolves by superposition of a motion at constant normal speed and a rotation around the origin on the x_1-x_2 plane.

The numerical solution has been computed with 50 nodes per side of the domain $\Omega = [-3,3]^3$ and the level surface $u(x,t) = 0.3$ plotted in Figure 9.5 from $t = 0.1$ (a) to $t = 1.1$ (f).

9.2 ▪ Fronts evolving by Mean Curvature

In this section we briefly present an SL scheme for the approximation of the level set model for the Mean Curvature Flow:

$$\begin{cases} u_t = \mathrm{tr}(A(Du)D^2u), & (x,t)\in\mathbb{R}^d\times[0,+\infty), \\ u(x,0)=u_0(x), & x\in\mathbb{R}^d. \end{cases} \tag{9.17}$$

Here, $A(Du)$ is a $d\times d$ projection matrix defined by

$$A(p)=I-\frac{pp^t}{|p|^2} \tag{9.18}$$

whose role is to project the diffusion on the space orthogonal to Du.

Equation (9.17) is *degenerate parabolic* (i.e., the diffusion operator does not have full rank) and *singular* (i.e., undefined at points where $Du=0$); moreover, there is a clear interest in treating nonsmooth initial data. An extended definition of viscosity solution has been proposed for a class of second-order equations including (9.17). We will not give details on this theoretical framework, but recall that the endpoint of this analysis is the existence and uniqueness of the viscosity solution for (9.17), as well as a result of boundedness in the Lipschitz norm.

Note that the matrix A has rank $d-1$, and there exists a $d\times(d-1)$ matrix σ, such that

$$A=\sigma\sigma^t. \tag{9.19}$$

In the case $d=2$, the matrix σ has the form

$$\sigma(Du)=\frac{1}{|Du|}\begin{pmatrix}-u_{x_2}\\ u_{x_1}\end{pmatrix}. \tag{9.20}$$

For $d=3$, we first define the vectors $d_1=(1,0,0)^t$, $d_2=(0,0,1)^t$, and

$$\nu_1=\left(\frac{-u_{x_3}}{\sqrt{u_{x_1}^2+u_{x_3}^2}},0,\frac{u_{x_1}}{\sqrt{u_{x_1}^2+u_{x_3}^2}}\right)^t, \tag{9.21}$$

$$\nu_2=\frac{1}{|Du|}\left(\frac{-u_{x_1}u_{x_2}}{\sqrt{u_{x_1}^2+u_{x_3}^2}},\sqrt{u_{x_1}^2+u_{x_3}^2},\frac{-u_{x_2}u_{x_3}}{\sqrt{u_{x_1}^2+u_{x_3}^2}}\right)^t. \tag{9.22}$$

Then, the matrix σ has the form

$$\sigma=\begin{cases}(\nu_1,\nu_2) & \text{if } \sqrt{u_{x_1}^2+u_{x_3}^2}\neq 0,\\ (d_1,d_2) & \text{otherwise.}\end{cases}$$

More details and references on this construction can be found in [CFF10].

9.2.1 ▪ Construction of the scheme

The construction of the scheme follows the guidelines already discussed for degenerate diffusion operators (see subsection 7.3.2) and for stochastic control problems (see subsection 8.4.8). In particular, the nonlinear diffusion operator is treated (by linearization) as in (7.30)–(7.31), freezing the gradients (and ultimately the diffusion matrix A) at the nth time step. In the case $d=2$, then the diffusion is one-dimensional. The approximation of the second-order term is obtained by an average in which the displacements δ_k in (7.31) are defined as

$$\delta_k = \pm\sigma\sqrt{2\Delta t}, \tag{9.23}$$

whereas, for $d = 3$, the diffusion is two-dimensional and the four displacements δ_k are defined by

$$\delta_k = \sigma \begin{pmatrix} \pm\sqrt{2\Delta t} \\ \pm\sqrt{2\Delta t} \end{pmatrix}, \tag{9.24}$$

where, in both cases, the matrix σ is computed at the node (x_j, t_n).

For simplicity, we write the approximation scheme in $\mathbb{R}^2$. First, $\sigma(Du(x_j, t_n))$ is approximated as

$$\sigma_j^n = \frac{1}{|D_j[V^n]|} \begin{pmatrix} D_{2,j}[V^n] \\ -D_{1,j}[V^n] \end{pmatrix}. \tag{9.25}$$

Here, $D_j[W]$ stands for an approximate gradient at the node $x_j = x_{j_1,j_2}$, obtained, for example, by centered differences:

$$D_j[W] = \frac{1}{2\Delta x} \begin{pmatrix} w_{j_1+1,j_2} - w_{j_1-1,j_2} \\ \\ w_{j_1,j_2+1} - w_{j_1,j_2-1} \end{pmatrix}. \tag{9.26}$$

Then the scheme reads (compare with (7.28)) as

$$v_j^{n+1} = \frac{1}{2} I[V^n](x_j + \sigma_j^n \sqrt{2\Delta t}) + \frac{1}{2} I[V^n](x_j - \sigma_j^n \sqrt{2\Delta t}). \tag{9.27}$$

Note that this construction has been carried out in the case $Du \neq 0$. Singularities can be treated either by a minmax version of the scheme, or by defining a threshold for the gradient, under which the approximation switches to isotropic diffusion. For consistency reasons, this threshold should depend on the discretization steps—for example, isotropic diffusion could be activated at points such that $|D_j^n| \le C\Delta x^s$ for proper choices of C and s. This choice is suitable to satisfy the generalized consistency condition (4.57) of the Barles–Souganidis theorem.

9.2.2 ▪ Numerical tests

We show here the scheme described above at work on two classical examples, the torus and the dumbbell evolution.

Three-dimensional torus evolution Consider the problem (9.17) in $[-2,2]^3$, with an initial condition describing a torus in $\mathbb{R}^3$:

$$u_0(x,y,z) = \left(\sqrt{x^2+y^2} - R^2\right) + z^2 - r^2.$$

Simulations are performed using a cubic interpolation with $\Delta x = 0.04$ and a time step of $\Delta t = 0.01$. Note that the collapse of a torus may have two different behaviors, depending on the ratio between the main and the secondary radii of the torus.

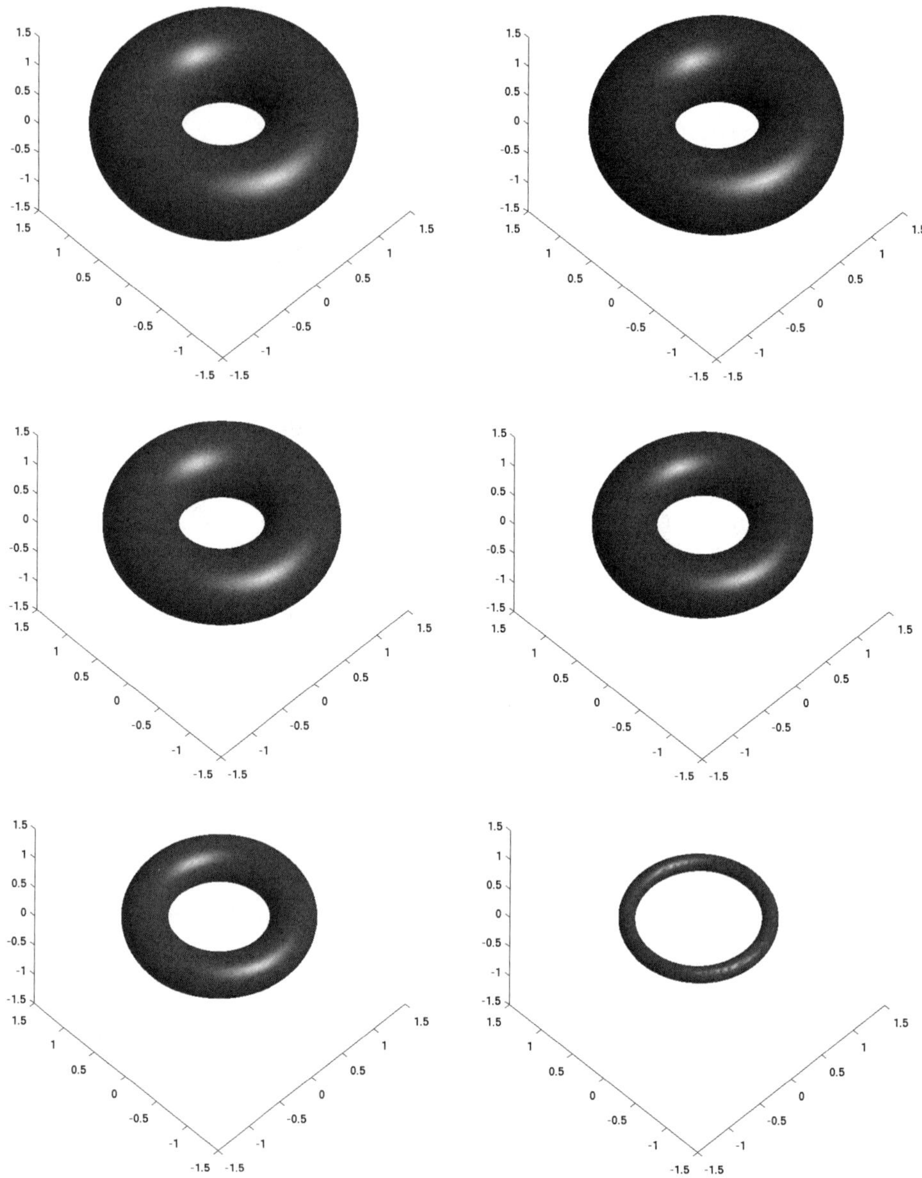

Figure 9.6. *Evolution of a torus collapsing to a circle* [CFF10].

Figure 9.6 represents the evolution starting from a torus with main radius $R = 1$ and secondary radius $r = 0.5$. The pictures are plotted for $t \in [0, 0.15]$. In this case the relationship between the main curvatures causes the front to collapse to a circle, without any topology change.

Three-dimensional dumbbell evolution In the second test, problem (9.17) is posed in $[-2, 2]^3$ with an initial condition describing a dumbbell. The test is performed using a cubic interpolation for $\Delta t = 0.001$ and $\Delta x = 0.08$. Figure 9.7 shows the evolution of the surface for $t \in [0, 0.1]$.

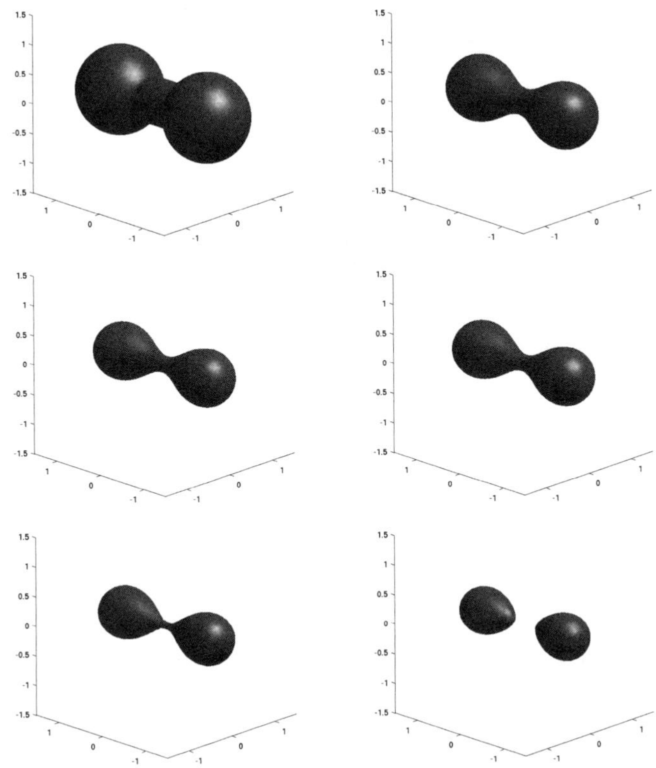

Figure 9.7. *Topology change in the evolution of a dumbbell in* $\mathbb{R}^3$ [CFF10].

This is a typical situation in which a change in topology occurs. We see that the scheme follows the breaking of the dumbbell accurately and without spurious oscillations.

9.3 ▪ Applications to image processing

We discuss in this section two classical applications taken from the area of image processing: Shape-from-Shading and segmentation with variable velocity via the level set method. We should also mention that second-order, curvature related equations appear in several models of nonlinear image filtering, but a detailed treatment of these applications goes beyond the scope of this book.

Shape-from-Shading In this inverse problem, we want to reconstruct a surface $u(x)$ from a single image. This means that, given a greyscale image $I(x)$, we are looking for the surface $u(x)$ corresponding to the object represented. The irradiance function I is given at each pixel of the image in terms of a grey level (in digital images, typically from 0 to 255).

To construct a continuous model, we will assume that I takes real values in the interval $[0,1]$, and that the height function u, which is the unknown of the problem, is defined on the "reconstruction domain" Ω.

The characterization of the surface via a PDE relies on various assumptions. First, we assume that there is a unique light source at infinity whose direction is given by the unit vector $\omega = (\omega_1, \omega_2, \omega_3) \in \mathbb{R}^3$ and, for simplicity, that ω is known. Assuming in addition

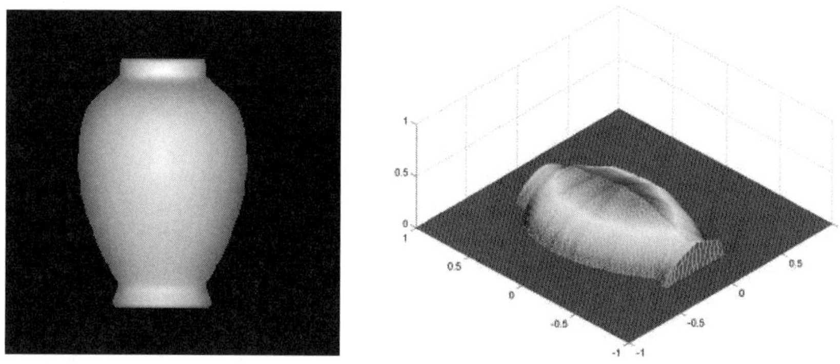

Figure 9.8. *Vase: initial image (left) and reconstructed surface (right)* [DFS08].

that the surface is Lambertian of uniform unit albedo, the reflectance function giving the brightness at a point on the surface is

$$R(\nu(x)) = \omega \cdot \nu(x),$$

where $\nu(x)$ is as usual the unit normal to the surface at $(x_1, x_2, u(x))$. This equation is called the *irradiance equation*. Recalling that, for the graph of a function u, the normal is given by

$$\nu(x) = \frac{1}{\sqrt{1+|Du(x)|^2}}(-u_{x_1}, -u_{x_2}, 1)^t,$$

we obtain the equation

$$I(x) = \frac{(\omega_1, \omega_2) \cdot Du(x) - \omega_3}{\sqrt{1+|Du(x)|^2}}, \tag{9.28}$$

which is a first-order nonlinear PDE of HJ type. We have seen that in the case of a vertical light source (i.e., $\omega = (0,0,1)$), this reduces to the eikonal equation (1.33).

As we said, this equation must be complemented with boundary conditions. A natural choice is to consider homogeneous Dirichlet boundary conditions corresponding to a flat background, that is,

$$u(x) = 0 \qquad \text{for } x \in \partial\Omega, \tag{9.29}$$

or, more in general, a Dirichlet boundary condition of the form

$$u(x) = g(x) \qquad \text{for } x \in \partial\Omega. \tag{9.30}$$

The solution of the Dirichlet problem (9.28)–(9.30) provides a surface $u(x)$ (not necessarily unique) corresponding to the image $I(x)$ measured in Ω. However, in order to set $u(x) = g(x)$ on the boundary, additional information about the surface is required. This is the case, e.g., for surfaces obtained by revolution around an axis (as for the vase in Figure 9.8).

Segmentation via the level set method In this problem, we want to detect the boundaries of objects represented in a picture. A very popular method for segmentation (based on level set techniques) is the so-called *method of active contours*, in which the segmentation is obtained following the evolution of a simple curve (a circle, for example) in its normal direction. The circle is initially set inside the object, and its evolution is tracked until the curve touches the boundaries of the object to be extracted.

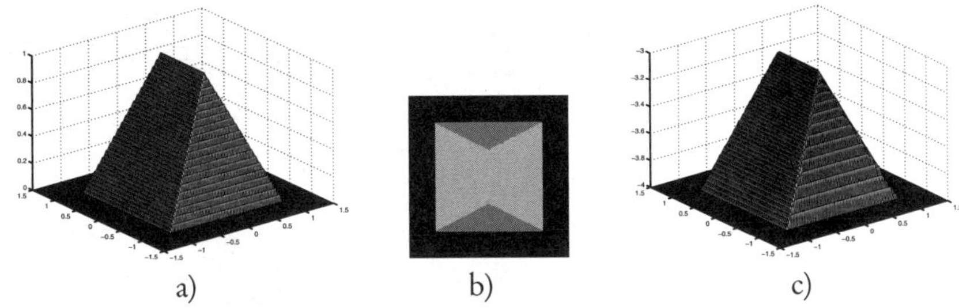

Figure 9.9. *Synthetic surface:* a) *the original surface,* b) *its brightness function in the plane,* c) *the reconstructed surface* [CFFe13].

The basic idea behind this technique is that the boundaries of a specific object inside a given image I are characterized by a rapid change in the values of I, so that the magnitude of the gradient $|DI|$ could be used as an indicator of edges. For this reason, a popular choice of the velocity when applying level set methods to the segmentation problem is

$$c(x) = (1 + |DI(x)|^p)^{-1}$$

for some $p \geq 1$. With this choice, we obtain (almost) unit velocity inside the region where the gray level is (almost) constant and a rapid decrease when x is close to an edge (which corresponds to a jump in I).

9.3.1 ▪ Numerical tests

Shape-from-Shading In the first test, we solve the Shape-from-Shading problem (1.33) in the simple case of a vase, setting the computational domain as $Q = [-1,1]^2$ and Ω as the silhouette of the vase. Figure 9.8 shows (left) the initial image and (right) the reconstructed surface. By the symmetry of the problem we expect that all characteristic curves start from the right and left sides of the image, so we can impose Dirichlet boundary conditions on both sides of the domain, while state constraints are imposed on the top and bottom sides as in [CFS05] (see also [RT92, KS01], where different boundary conditions are applied).

In the second test, consider the synthetic image corresponding to the surface

$$u(x_1,x_2) = \begin{cases} 2(1-|x_1|), & x_1 \in [-1,1], \quad x_2 \in [-1, -\frac{1}{2}|x_1| - \frac{1}{2}], \\ 2(1-|x_2|), & x_1 \in [-1,1], \quad x_2 \in [\frac{1}{2}|x_1| + \frac{1}{2}, 1], \\ 1-|x_1| & \text{otherwise.} \end{cases}$$

The surface is not smooth (see Figure 9.9), but simple homogeneous Dirichlet boundary conditions can be applied on each side of the rectangle. Numerical solution is computed by the SL scheme for the eikonal equation (1.33). Looking at the right picture of Figure 9.9, we see that the algorithm is accurate around the kinks and that there are no spurious oscillations where the gradient is discontinuous. Similar results have been obtained for other synthetic and real images. A detailed presentation of these results, as well as a comparison with other numerical methods, can be found in the survey paper [DFS08].

Segmentation of an image The first example shows the standard segmentation of an image containing several grains (Figure 9.10). The grains are separated, and the contrast

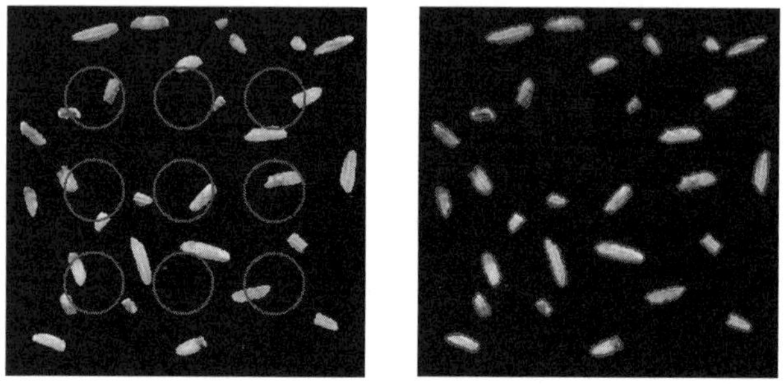

Figure 9.10. *Segmentation: initial configuration of the front (left) and segmented image (right)* [CFFe13].

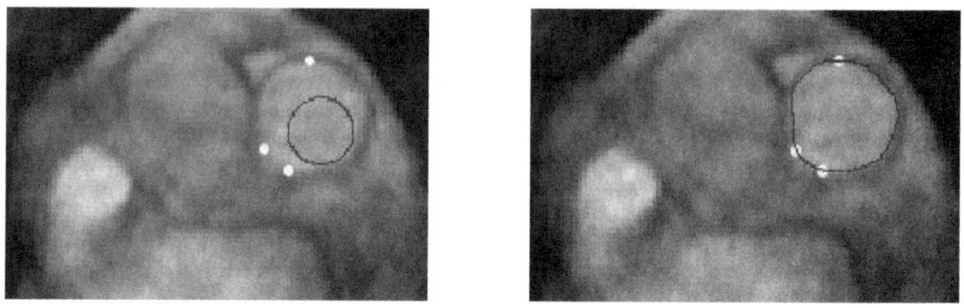

Figure 9.11. *Medical image: initial configuration of the front (left) and segmented image (right)* [CFFe13].

of the picture is very high. Starting from a regular initial configuration of the front composed of nine circles, the method is able to detect all the grains. The topology change is clearly visible.

The second example deals with a medical image with a quite high noise level. Starting from a circle inside the region of interest the method is able to detect its boundary in a rather accurate way (Figure 9.11).

9.4 ▪ Fast Marching Methods

One of the main drawbacks of the level set method is having to add an extra dimension to the original problem. In fact, to deal with the evolution of planar curves we embed the problem in $\mathbb{R}^3$ and track the evolution of a representation function $u(x,t)$ whose zero-level set gives the position of the front at time t. While the model makes *all level sets* move with the same law, we are interested in only one of them—this means computing more information than what is needed. Then, it is natural to look for models and numerical methods which concentrate their computational effort in a neighborhood of the front. The main difficulty is the following: how can we perform a local computation without losing the global information contained in (9.3)? One possible answer comes from Fast Marching Methods (FMM), which will be presented in this section.

9.4.1 ▪ FMM based on finite differences

The basic concepts and definition will be presented on the classical finite difference method for the eikonal equation, which has been the starting point of FMM in the literature.

The basic level set algorithm is based on a finite difference discretization in the iterative form $T^{(k+1)} = S(T^{(k)})$ to compute the approximate solution T in $\mathbb{R}^d \setminus \Omega_0$ at every iteration. Opposite to the iterative scheme, the FM-FD method concentrates the computation in a small neighborhood of the front, updating this neighborhood at every iteration to avoid useless computations.

This is done by dividing the grid nodes into three subsets:

- The *accepted* nodes are nodes where the solution has already been computed and where the value can no longer change in following iterations.
- The *narrow band* nodes are where the computation actually takes place and where the value can still change at following iterations.
- The *far* nodes are the remaining nodes where the approximate solution has not yet been computed.

The algorithm starts labeling as *accepted* only the nodes belonging to the initial front $\Gamma_0 = \partial\Omega_0$ and ends when all the nodes have been *accepted*. In physical terms, the *far* nodes are those in the space region which has never been touched by the front, the *accepted* nodes are those where the front has already passed through, and the *narrow band* nodes are, iteration by iteration, those lying in a neighborhood of the front.

In this section, we will briefly sketch the FM-FD scheme for (9.3). For simplicity, the presentation will be restricted to the case $n = 2$ and to a positive normal velocity; i.e., we assume that $c(x) > 0$ to ensure a monotone (increasing) evolution of the front. The results in this section can be easily generalized to the d-dimensional case and to the case of a velocity changing sign.

Assume to be working on a structured grid of $N_x \times N_y$ nodes (x_i, y_j), $i = 1, \ldots, N_x$, and $j = 1, \ldots N_y$, with Δx and Δy denoting the (uniform) discretization steps, respectively, on the x- and y-axes (for simplicity, we set here $N_x = N_y$ and $\Delta x = \Delta y$). Denote also by $T_{i,j}$ and $c_{i,j}$ the values of, respectively, T and c at (x_i, y_j). Define the computational domain where T is computed as a square Q large enough to contain Ω_0. Boundary conditions can be given on ∂Q and Γ_0, but, as a first step, we will consider the algorithm without boundary conditions on ∂Q.

First, we rewrite (9.3) as

$$T_x^2 + T_y^2 = \frac{1}{c^2(x,y)}. \tag{9.31}$$

Then, by applying a two-dimensional upwind scheme (in the spirit of (5.77)), we obtain the discrete version

$$\begin{aligned}&\left(\max\left\{\max\left\{\frac{T_{i,j}-T_{i-1,j}}{\Delta x},0\right\},-\min\left\{\frac{T_{i+1,j}-T_{i,j}}{\Delta x},0\right\}\right\}\right)^2 \\ &+\left(\max\left\{\max\left\{\frac{T_{i,j}-T_{i,j-1}}{\Delta x},0\right\},-\min\left\{\frac{T_{i,j+1}-T_{i,j}}{\Delta x},0\right\}\right\}\right)^2 = \frac{1}{c_{i,j}^2}.\end{aligned} \tag{9.32}$$

We recall a preliminary definition useful for sketching the FM-FD method.

Definition 9.1 (Neighboring nodes for the FD scheme). *Let $X = (x_i, y_j)$ be a node. We define the set of neighboring nodes to X as*

$$N_{FD}(X) = \Big\{(x_{i+1}, y_j), (x_{i-1}, y_j), (x_i, y_{j+1}), (x_i, y_{j-1})\Big\}.$$

This is the set of nodes appearing in the stencil of the finite difference discretization (9.32). The definition can be easily extended to higher dimensions.

Sketch of the FM-FD algorithm

Initialization

1. The nodes belonging to the initial front Γ_0 are located and labeled as *accepted*. Their value is set to $T = 0$ (they form the set $\widetilde{\Gamma}_0$).
2. The initial *narrow band* is defined by taking the nodes belonging to $N_{FD}(\widetilde{\Gamma}_0)$ external to Γ_0. These nodes are labeled as *narrow band*, setting the value to $T = \frac{\Delta x}{c}$.
3. The remaining nodes are labeled as *far*, and their value is set to $T = +\infty$ (in practice, the maximum floating point number).

Main Cycle

1. Among all the nodes in the *narrow band* the minimum value of T is located. The corresponding node is denoted by A.
2. A is labeled as *accepted* and removed from the *narrow band*.
3. The nodes in $N_{FD}(A)$ which are not *accepted* are labeled as *active*. If among these nodes there exist nodes labeled as *far*, they are moved into the *narrow band*.
4. The value of T in the nodes *active* is computed (or recomputed) by solving (9.32) and taking the largest root.
5. If the *narrow band* is not empty, go back to step 1; else stop.

The main point of interest in the FM-FD method is that its computational cost is bounded. In fact, every node has just four neighbors and cannot be accepted more than one time, so the maximum number of times a single node can be recomputed is four. This corresponds to a computational cost of $O(N)$ if N is the total number of nodes. We should add the search for the minimum value of T in step 1, which has a cost of $O(\ln N_{nb})$ if N_{nb} is the number of nodes in the *narrow band*. In conclusion, the algorithm has a global cost of $O(N \ln N_{nb})$ operations, a very convenient complexity when compared to the usual iterative/fixed-point algorithm, which can require in practice a high number of iterations to converge.

Observe that it is necessary to introduce some conditions or to modify the scheme in order to avoid inconsistencies due to the appearance of imaginary solutions. Assuming the CFL-type condition

$$\Delta x \leq (\sqrt{2}-1)\frac{c_{\min}}{L_c} \tag{9.33}$$

(here, $c_{\min}$ stands for the minimum value of c and L_c for its Lipschitz constant) the algorithm always computes real solutions at every node. Condition (9.33) has a clear meaning and allows one to prove convergence to the viscosity solution of (9.3) via the following result.

Proposition 9.2. *Let $X = (x_i, y_j) \in N_{FD}(A)$ be the node where the FM-FD method computes a solution. Assume that*

$$c_{min} = \min_{Q\setminus\Omega_0} c(x) > 0 \tag{9.34}$$

and that condition (9.33) *holds true. Then, we have*

$$T(A) \le T(X) \le T(A) + \frac{\Delta x}{c(X)}. \tag{9.35}$$

Proposition 9.2 is crucial to have convergence in a finite number of steps. In fact, it shows that the minimum value among the nodes in the *narrow band* (which is actually the only value *accepted* at every iteration) is exact within the consistency error of the scheme.

As a final remark on the FM-FD method, we point out that a central role is played by the choice of an upwind discretization, which makes numerical solutions propagate only on one side of each node. Central discretizations propagate information on both sides, and this feature makes them unsuitable for a Fast Marching implementation.

9.4.2 ▪ FMM based on the SL scheme

We present now a Fast Marching version of the SL scheme assuming (9.6) and, for simplicity, that the space grid is structured and orthogonal. The same scheme could be implemented on an unstructured grid as in [SS01].

As we have seen in Chapter 8, under the Kružkov transform (8.32) the numerical scheme has the form

$$\begin{cases} v_i = \min\limits_{a\in B(0,1)} \{\tau I[V](x_i - \Delta t\, c(x_i)a)\} + 1 - \tau, & x_i \in \mathbb{R}^d \backslash \Omega_0, \\ v_i = 0, & x_i \in \mathbb{R}^d \cap \Omega_0. \end{cases} \tag{9.36}$$

Note that applying (8.32) allows us to include the case $c = 0$ (e.g., when treating obstacles) in which the minimum time function could have an infinite value at some points, while v will always remain bounded. Note also that the Kružkov transform is monotone and this allows us to use the same rules for updating the *narrow band* since the ordering of the values remains the same for T and v.

For various reasons, the scheme will use a linear interpolation. Even in this simple case, the numerical domain of dependence is enlarged with respect to the finite difference scheme (9.32), and this is taken into account by the following definition.

Definition 9.3 (Neighboring nodes for the SL scheme). *Let $X = (x_i, y_j)$ be a node of the grid. We define*

$$N_{FD}(X) = \left\{(x_i, y_{j+1}), (x_i, y_{j-1}), (x_{i-1}, y_j), (x_{i+1}, y_j)\right\},$$

$$D(X) = \left\{(x_{i+1}, y_{j+1}), (x_{i+1}, y_{j-1}), (x_{i-1}, y_{j+1}), (x_{i-1}, y_{j-1})\right\},$$

$$N_{SL}(X) = N_{FD}(X) \cup D(X).$$

The FM-SL method follows the initialization and all the steps of the classical FM-FD method but the step in which the value at the node x_i is actually computed. In the original formulation, this step would require iterating until convergence the scheme

$$v_i = \min_{a\in B(0,1)} \{\tau I[V](x_i - \Delta t\, c(x_i)a)\} + 1 - \tau, \tag{9.37}$$

so that the fixed-point iteration is applied "locally" at every single node, in the order indicated by the FM-FD method. However, for an SL scheme based on a piecewise linear space reconstruction, the exact value for (9.37) can be explicitly computed, so that the computational effort becomes of the same order of the FM-FD method.

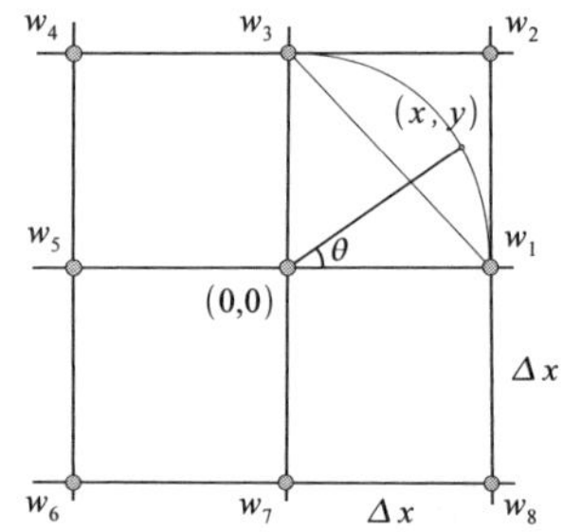

Figure 9.12. *Search for the optimal direction.*

Fast minimum search As a building block of the complete scheme, we start by improving the minimum search phase, typical of the SL scheme—in fact, in the case of a linear interpolation the minimum search in (9.37) can be solved algebraically using the known values at the nodes. To obtain an explicit minimization formula, the time step Δt in (9.37) should vary at every node to ensure that $\Delta t\, c(x_j) \leq \Delta x$. Denote by h_i the local time step for the node x_i, by $c_i = c(x_i)$ the velocity at x_i and by $\tau_i = e^{-h_i}$. When $c_i > 0$, it is always possible to choose

$$h_i = \frac{\Delta x}{c_i}. \tag{9.38}$$

With these definitions, (9.37) can be rewritten as

$$v_i = \min_{a \in B(0,1)} \{\tau_i I[V](x_i - \Delta x\, a)\} + 1 - \tau_i. \tag{9.39}$$

In this situation, nodes where $c_i = 0$ are actually treated separately from other nodes, by assigning them the value $v = 1$ (which corresponds to $T = +\infty$) without any additional computation. The minimization in (9.39) is now constrained on the boundary of the unit ball. This is not a real restriction since, for our applications, the minimum in the unit ball is actually attained at the boundary.

The typical geometry in $\mathbb{R}^2$ is a set of four cells of side Δx centered at the origin (see Figure 9.12). We want to compute the minimum of a function $w(-\Delta x\, a)$ for $a = (\cos\theta, \sin\theta)$ and $\theta \in [0, 2\pi)$. To this end, we introduce a vector $\mathbf{m} = (m_1, m_2, \ldots, m_8)$ whose components will contain the candidate values. The minimum will then be given by $p = \min\{m_1, m_2, \ldots, m_8\}$.

The first four components of $\mathbf{m}$ are defined as

$$m_1 = w(\Delta x, 0), \quad m_2 = w(0, \Delta x), \quad m_3 = w(-\Delta x, 0), \quad m_4 = w(0, -\Delta x),$$

which correspond to choosing θ as a multiple of $\pi/2$. Then, we search for constrained stationary points in every orthant.

I orthant Let w_1, w_2, and w_3 be the values of w corresponding, respectively, to the nodes $(\Delta x, 0)$, $(\Delta x, \Delta x)$, and $(0, \Delta x)$. These values are interpolated by a linear function $f(x, y)$ in the form

$$f(x,y) = \alpha x + \beta y + \gamma \tag{9.40}$$

satisfying the conditions

$$f(\Delta x, 0) = w_1, \quad f(\Delta x, \Delta x) = w_2, \quad f(0, \Delta x) = w_3,$$

which imply that

$$\alpha = \left(\frac{w_2 - w_3}{\Delta x}\right), \quad \beta = \left(\frac{w_2 - w_1}{\Delta x}\right), \quad \gamma = w_1 - w_2 + w_3.$$

We define now, for $\theta \in [0, 2\pi)$, the function

$$F(\theta) = f(\Delta x \cos\theta, \Delta x \sin\theta) = \alpha \Delta x \cos\theta + \beta \Delta x \sin\theta + \gamma \tag{9.41}$$

and look for the minimum of $F(\theta)$ in the interval $[0, \pi/2]$. Since the values at the extreme points $\theta = 0$ and $\theta = \pi/2$ are already included in $\mathbf{m}$, we look for internal stationary points. The condition $F'(\theta) = 0$ is satisfied for

$$\theta = \theta_1^* = \arctan(\beta/\alpha),$$

which provides a minimum in $(0, \pi/2)$ if and only if $w_2 < w_1$ and $w_2 < w_3$. In this case, we set $m_5 = F(\theta_1^*)$, while if $w_2 \geq w_1$ or $w_2 \geq w_3$, we set $m_5 = +\infty$ (or the highest machine number).

II orthant Let w_3, w_4, and w_5 be the values of w, respectively, at the nodes $(0, \Delta x)$, $(-\Delta x, \Delta x)$, and $(-\Delta x, 0)$. The linear function $f(x, y)$ defined by (9.40) satisfies the conditions

$$f(0, \Delta x) = w_3, \quad f(-\Delta x, \Delta x) = w_4, \quad f(-\Delta x, 0) = w_5$$

for

$$\alpha = \left(\frac{w_3 - w_4}{\Delta x}\right), \quad \beta = \left(\frac{w_4 - w_5}{\Delta x}\right), \quad \gamma = w_3 - w_4 + w_5.$$

Considering again the composite function $F(\theta)$ defined in (9.41), we observe that it has a local minimum in $(\pi/2, \pi)$ if and only if $w_4 < w_3$ and $w_4 < w_5$. In this case, we have

$$\theta = \theta_2^* = \arctan(\beta/\alpha),$$

and operating as in the first orthant, we set $m_6 = F(\theta_2^*)$, while if $w_4 \geq w_3$ or $w_4 \geq w_5$, then we set $m_6 = +\infty$.

The analysis of the third and fourth orthants follows the same arguments and will be skipped.

Once all the components of $\mathbf{m}$ have been assigned, the value $p = \min\{m_1, m_2, \ldots, m_8\}$ is substituted in the expression

$$w(0) = \tau p + 1 - \tau \tag{9.42}$$

to compute the numerical solution at the central node. Note that the use of a linear interpolation allows us to avoid an iterative procedure for computing the correct value of $w(0)$ given the values at the neighboring nodes, since $F(\theta)$ does not depend on $w(0)$. Another advantage of this technique is that it gives the exact value of the optimal direction without actually performing any interpolation. Both properties would *not* hold for interpolations of higher orders.

We are now ready to outline the construction of the Fast Marching version of the SL algorithm (for simplicity, still in $\mathbb{R}^2$). In adapting the FM-FD algorithm to the SL case, we should consider that, according to Definition 9.3, the nodes in the *narrow band* should include also the diagonal directions and not only the four main directions North, South, East, and West as in the FM-FD method.

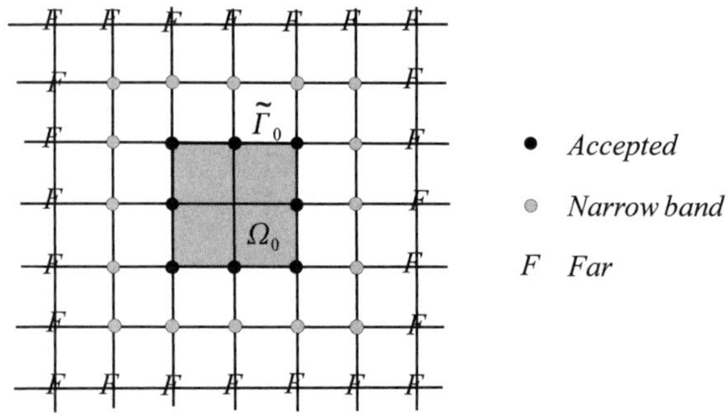

Figure 9.13. *Initialization for the FM-SL method, case $c > 0$* [CF07].

Sketch of the FM-SL algorithm

Initialization

1. The nodes belonging to the initial front Γ_0 are located and labeled as *accepted*. Their value is set to $v = 0$ (they form the set $\widetilde{\Gamma}_0$).

2. The initial *narrow band* is defined, according to Definition 9.3, taking the nodes belonging to $N_{SL}(\widetilde{\Gamma}_0)$ external to Γ_0. These nodes are labeled as *narrow band*. Their value is set to
$$v_i = \begin{cases} 1 - e^{-\Delta x/c_i}, & x_i \in N_{FD}(\widetilde{\Gamma}_0), \\ 1 - e^{-\sqrt{2}\Delta x/c_i}, & x_i \in D(\widetilde{\Gamma}_0). \end{cases}$$

3. The remaining nodes of the grid are labeled as *far*, and their value is set to $v = 1$ (see Figure 9.13).

Main Cycle

1. Among all the nodes in the *narrow band* the minimum value of v is located. The corresponding node is denoted by A.

2. A is labeled as *accepted* and removed from the *narrow band*.

3. The nodes in $N_{SL}(A)$ which are not *accepted* are labeled as *active* . If among these nodes there exist nodes labeled as *far*, they are moved into the *narrow band*.

4. The value v at the nodes belonging to $N_{FD}(A)$ which are *active* is computed (or recomputed) as
$$v_i = \min_{a \in B(0,1)} \{\tau_i I[V](x_i - h_i c_i a)\} + 1 - \tau_i. \tag{9.43}$$
Then, by the same formula the value at the remaining *active* nodes in $N_{SL}(A) \setminus N_{FD}(A)$ is also computed.

5. If the *narrow band* is not empty go back to Step 1; else stop.

Although the algorithm, according to the new definition, advances the *narrow band* also in the diagonal directions, it computes first the values at the neighboring nodes in the

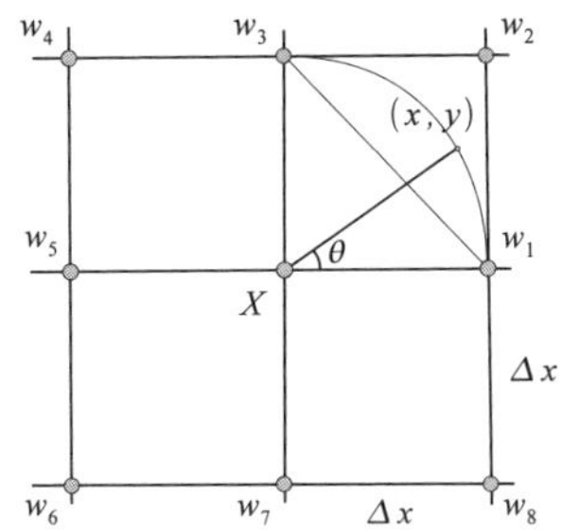

Figure 9.14. *Analysis of the minimum in the* I *orthant* [CF07].

directions North, South, East, and West (i.e., the FD directions) and then passes to the diagonal directions.

Properties of the FM-SL scheme In order to prove convergence of the FM-SL scheme, we start by some technical results. In what follows, for the sake of a simpler notation, we will denote a generic node x_i by X and the reconstructed numerical solution $I[V](x)$ by $w(x)$. Moreover, we will possibly denote by w_k some samples v_{i_k} of the numerical solution, according to the geometry and notation of Figure 9.14.

Proposition 9.4. *Let X be a node, and assume that $w(X)$, defined by* (9.43), *is computed by interpolation using the three values w_{k_1}, w_{k_2}, w_{k_3}. Then,*

$$w(X) \geq \min\left\{w_{k_1}, w_{k_2}, w_{k_3}\right\}. \tag{9.44}$$

Proof. Let a^* be the optimal direction/control at X. The inequality

$$\tau_i w(X - h_i c_i a^*) + 1 - \tau_i \geq w(X - h_i c_i a^*)$$

is satisfied if and only if $w(X - h_i c_i a^*) \leq 1$. Since (due to the Kružkov transform) $w \leq 1$, we have proved that

$$w(X) \geq w(X - h_i c_i a^*). \tag{9.45}$$

On the other hand, by monotonicity of linear interpolation we have

$$\max\left\{w_{k_1}, w_{k_2}, w_{k_3}\right\} \geq w(X - h_i c_i a^*) \geq \min\left\{w_{k_1}, w_{k_2}, w_{k_3}\right\}, \tag{9.46}$$

and therefore, by (9.45) and (9.46), (9.44) is proved. □

As a next step, we have to prove that the minimization method constructed in subsection 9.4.2 is compatible with the Fast Marching technique. This is necessary since the *narrow band* is larger for the FM-SL method than for the FM-FD method, due to the different definition of neighboring nodes. In particular, we will show that the algorithm automatically rejects *far* nodes from the computation as in the standard upwind finite difference discretization. We assume for simplicity that $c > 0$, in which case a node is labeled as *far* if and only if its value is $w = 1$.

Proposition 9.5. *Let X be a node, and let $w(X)$ be defined by* (9.43). *Then, the value $w(X)$ will* not *be computed by interpolation using nodes labeled as* far.

Proof. We give the proof for a minimum in the I orthant. The analysis for the other orthants is similar and can be easily obtained by symmetry arguments.

1. $w_1 = w_2 = w_3 = 1$.
   This configuration cannot occur. In fact, since the minimum is attained in the I orthant, we should have $w_4 = w_5 = w_6 = w_7 = w_8 = 1$. But this is not possible, since we compute the value at X only when at least one of the nodes belonging to $N_{SL}(X)$ has been labeled as *accepted* in one of the previous iterations, and an *accepted* node must have a value lower than 1.

2. Among w_1, w_2, and w_3, there are two values equal to 1.

   (a) $w_1 = w_3 = 1$: this case cannot occur. In fact, since the minimum is attained in the I orthant, we must have $w_2 \leq w_1, w_3, w_4, \ldots, w_8$. The node to be labeled as *accepted* is the one corresponding to the value w_2. This implies that the values w_1 and w_3 must be computed before X (see the sketch of the algorithm).

   (b) $w_1 = w_2 = 1$: the minimum value is w_3. A new iteration for computing $w(X)$ would not give a lower value, so the optimal value is obtained at the first iteration.

   (c) $w_2 = w_3 = 1$: the minimum value is w_1. Again, we would not obtain a lower value by performing further iterations.

3. Among w_1, w_2, and w_3, only one value is equal to 1.

   (a) $w_2 = 1$: since f is linear, the minimum will be attained by w_1 or w_3. The optimal value is obtained at the first iteration.

   (b) $w_1 = 1$, $w_3 \leq w_2$: the minimum is w_3.

   (c) $w_1 = 1$, $w_3 > w_2$: this is the most delicate case, since $w_2 < w_1, w_3$. The minimum for F will be attained at some $\theta^* \in (0, \pi/2)$. The value $w(X)$, obtained by linear interpolation, will not be correct, since it depends on $w_1 = 1$, which is a conventional value. Moreover, note that a new iteration of the fixed-point map at X will not make $w(X)$ decrease, since w_1 is frozen and so is $w(X)$. If this case could occur, we would not get convergence to the correct value even at successive iterations. Note that this is not the case in both the global SL scheme, where *all* the nodes are computed at the same iteration, and in the FM-FD method, where the values corresponding to *far* nodes are not used in the stencil.

   The following argument shows that this case cannot occur also for the FM-SL scheme. Since $w_1 = 1$, the corresponding node is labeled as *far* at the current iteration. This implies that the nodes labeled as *accepted* at the previous iteration do not belong to $N_{SL}(w_1)$. As a consequence, w_2 belongs to the *narrow band*. By Proposition 9.4, we have $w(X) > w_2$. This implies that X cannot be labeled as *accepted* before the node corresponding to w_2. Once w_2 becomes *accepted*, the algorithm computes w_1 and w_3 before computing $w(X)$ so that the values at nodes labeled as *far* will not contribute.

   (d) $w_3 = 1$, $w_1 \leq w_2$: the minimum is w_1. The optimal value is obtained at the first iteration.

   (e) $w_3 = 1$, $w_1 > w_2$: analogous to case (3c). □

9.4.3 ▪ Convergence of the FM-SL scheme

As we have seen, not all the nodes belonging to the *narrow band* are accepted together at every iteration, but only the node corresponding to the minimal value (this is a cautious strategy which simplifies the theoretical results). The central point in convergence analysis is to prove that the minimal value of the nodes of the *narrow band* cannot decrease if we iterate the fixed-point operator, or in other terms that it coincides with the value obtained by the discrete operator working on all the nodes.

The following proposition gives a bound on the number of times one node can be recomputed and is a building block for the convergence of the scheme.

Proposition 9.6. *Let X be a node in the* narrow band *such that $w(X) = w_{old}(X)$, for which at the current iteration the algorithm needs to compute a new value $w_{new}(X)$. Assume that at the current iteration the following property holds true:*

$$\text{If } A \text{ belongs to the narrow band and } B \text{ is accepted, then } w(A) \geq w(B). \tag{9.47}$$

Then, the following properties hold:

1. *If the value $w_{old}(X)$ was computed at an iteration in which a grid point $A_1 \in N_{FD}(X)$ was labeled as* accepted, *then it is impossible that $w_{new}(X) < w_{old}(X)$.*

2. *If the value $w_{old}(X)$ was computed at an iteration in which a grid point $A_2 \in D(X)$ was labeled as* accepted, *then a new value $w_{new}(X) < w_{old}(X)$ can be assigned to the node X, but it will always satisfy the inequality $w_{new}(X) \geq w(A_2)$.*

Proof. Let us start from the first statement.

1. Assume that when the value w_{old} was assigned to X, the node A_1 was the (unique) node belonging to $N_{FD}(X)$ which was labeled as *accepted*. When the algorithm computed $w(X) = w_{old}(X)$ we certainly had

$$\min_{a \in \partial B(0,1)} w(X - \Delta x\, a) = w^* \leq w(A_1),$$

since there exists a direction/control $\overline{a}$ such that $w(X - \Delta x\, \overline{a}) = w(A_1)$. The only possibility of having at X a value lower than $w_{old}(X)$ in the next iterations of the algorithm is that a value assigned to a node belonging to $N_{SL}(X)$ would be lower than w^*. However, by Proposition 9.4, we know that this value cannot be computed using in the stencil the values at the nodes of the current *narrow band* because they are all greater than $w(A_1) \geq w^*$, which has been *accepted* (as (9.47) ensures). A lower value could only be computed using a stencil which contains nodes already *accepted* in one of the previous iterations, since they all have values lower than $w(A_1)$. This is not possible because all the neighbors of those *accepted* nodes have already been computed and have a value greater than or equal to $w(A_1)$, since they have not yet been labeled as *accepted*.

2. Assume, for simplicity, that the node A_2 is the unique node in $D(X)$ labeled as *accepted*, and let $w_{old}(X)$ be the value assigned at X at the same iteration. When a node $A_1 \in N_{FD}(X)$ has been labeled as *accepted* before A_2, the result holds true by the arguments of the above case 1.

Assume that A_2 is the unique neighbor of X which has been labeled as *accepted*. Then, we have

$$\min_{a\in\partial B(0,1)} w(X-\Delta x\, a)=w^*\geq w(A_2).$$

Using $w(A_2)$, it is always possible to obtain a new value $w_{new}(X)$ lower than $w_{old}(X)$. However, by (9.47) and Proposition 9.4, all the new values will be greater than or equal to $w(A_2)$, and therefore $w_{new}(X)\geq w(A_2)$. □

Remark 9.7. *Note that the previous proposition allows one to accelerate the algorithm by avoiding recomputing the values at the nodes corresponding to case* 1. *However, they cannot be labeled as* accepted *before their value is the minimum in the* narrow band. *An important consequence of this observation and Proposition* 9.6 *is that every node can be computed at most* 5 *times, compared to the* 4 *times of the FM-FD method. This is one of the reasons for FM-SL to perform slightly worse than FM-FD in terms of complexity, although (as we will see in the last section) this small increment of CPU time is compensated by a higher accuracy.*

The following result is an analogue of Proposition 9.2.

Proposition 9.8. *Let w be defined by* (9.43), *and let $w(X)$ be the value assigned at X at the same iteration when a node $Z\in N_{SL}(X)$ is labeled as* accepted. *Assume that $c(x)\geq 0$. Then,*

$$w(X)\geq w(Z). \tag{9.48}$$

Proof. We examine all the cases corresponding to a minimum in the I orthant (see Figure 9.15). The proof will be obtained by induction on the number of iterations of the algorithm.

At the first step the result holds true as a result of the initialization procedure.

Consider now the nth step of the algorithm. The inductive hypothesis implies that at the current iteration all nodes in the *narrow band* have values greater than the values of nodes labeled as *accepted*. Therefore, (9.47) holds true and we can apply Proposition 9.6. We single out three cases.

Case 1. $w_1,\dots,w_8$ are *narrow band* or *far* (before Z is labeled as *accepted*).
If Z belongs to the I orthant, then by Proposition 9.4

$$w(X)\geq \min\{w_1,w_2,w_3\}=w(Z).$$

If Z does not belong to the I orthant, then we have

$$w(X)\geq \min\{w_1,w_2,w_3\}\geq w(Z)$$

since Z has been labeled as *accepted*.

Case 2. One node among $w_1,\dots,w_8$ is *accepted* (before Z is labeled as *accepted*).
Denote by P this node. When P was accepted the value at X was $w_{old}(X)$. Now, the value at X has to be recomputed. We can have only one of the following situations:

1. P belongs to the I orthant.

   (a) Z belongs to the I orthant

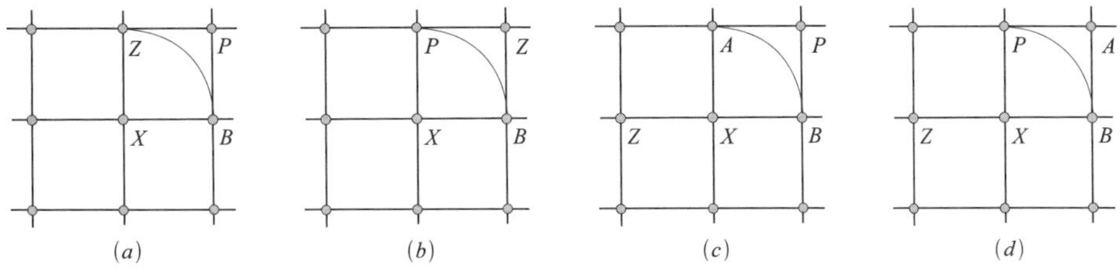

Figure 9.15. *Four different configurations for Case* 2 [CF07].

i. See Figure 9.15(a). By Proposition 9.6, Z and B cannot be assigned to a lower value after P is *accepted*, so $w_{new}(X) = w_{old}(X)$ and $w_{old}(X) \geq w(Z)$, since Z is the node chosen to be labeled as *accepted*.

ii. See Figure 9.15(b). When Z is *accepted*, the minimum is attained at P and this implies again $w_{new}(X) = w_{old}(X)$.

(b) Z does not belong to the I orthant

i. See Figure 9.15(c). In the iterations between the acceptance of P and that of Z, the values $w(A)$ and $w(B)$ cannot be changed. Moreover, the minimum is attained in the I orthant, so we have $w_{new}(X) = w_{old}(X)$.

ii. See Figure 9.15(d). We know that the value $w(A)$ has not been replaced, that $w(B) \geq w(P)$, and that the minimum is attained in the I orthant. Then, the minimum is attained at P and $w_{new}(X) = w_{old}(X)$.

2. P does not belong to the I orthant.
Since the minimum is attained in the I orthant this means that P does not have an effect on the computation at X and we are back to Case 1.

Case 3. More than one value among $w_1, \ldots, w_8$ has been labeled as *accepted* (before Z is labeled as *accepted*).

This case can be solved by the same arguments of Case 2. □

As for the FM-FD method (see [Se99]) we can now conclude that the value of the node which is labeled as *accepted* at every iteration cannot be decreased by iterating the fixed-point operator. In fact, denoting this value $w_{\min}$, since all the nodes in the *narrow band* have values greater than $w_{\min}$, the previous result implies that using such nodes we cannot assign to a node a value lower than $w_{\min}$. In conclusion, the correct upwinding is respected and the value $w_{\min}$ can be considered exact, since it cannot be improved on the same grid (of course, it does depend on the discretization steps).

Convergence to the viscosity solution The SL scheme is consistent and monotone, as has already been proved. We will now prove that the FM-SL method computes the same solution of the standard SL in iterative form. Then, convergence of the FM-SL method follows from convergence of the standard scheme.

Theorem 9.9. *Let the scheme* (9.43) *be rewritten in compact form as*

$$V = S(V). \tag{9.49}$$

Let $\widehat{V}$ be the solution of the fixed-point iteration (9.49), *and let $\overline{V}$ be the solution provided by the FM-SL algorithm (i.e., the result obtained when the* narrow band *is empty). Then, $\overline{V} = \widehat{V}$.*

Proof. The two solutions coincide if and only if

$$\overline{V} = S(\overline{V}). \tag{9.50}$$

Assume now that $\overline{V}$ is the final solution of the FM-SL algorithm (obtained when the *narrow band* is empty), and take $\overline{V}$ as initial guess for the fixed-point technique. Since the value is computed by the same scheme, when all the nodes are *accepted* the equality (9.50) must hold. In fact, if the equality is not true at some node, then it would be possible for the associated value to be improved, and this implies that the list of *narrow band* or *far* nodes is not empty, which gives a contradiction. □

The above results allow us to draw some conclusions about the order of complexity of the FM-SL scheme. The value $w(X)$ computed by (9.43) is an approximation of $v(X)$ which has been (re)computed at most 5 times for every node. This means that the computational cost can be estimated as in the FM-FD scheme. One component is given by the cost of the heap-sort method for selecting the minimum value in the *narrow band* and the other by the computational cost at every node. This gives a global cost of $O(N \log(N_{nb}))$, where N is the total number of nodes and N_{nb} the number of nodes in the *narrow band* (see [Se99]).

Last, since the iterative scheme and the Fast Marching scheme provide the same solution, the same error estimates also hold for both schemes. In the last section we will present several tests which confirm these theoretical results.

9.4.4 ▪ Fast Sweeping Method

For completeness, we briefly mention the basic ideas behind an alternative fast solver for front problems. The *Fast Sweeping Method* (FSM) is more similar to the classical iterative method, because no attempt is made to find the exact sequence of the nodes to be computed. Hence, there is no need to divide the nodes into the *accepted*, *narrow band*, and *far* regions and to update such information at every iteration.

The underlying idea of the FSM is rather to apply the scheme in a multiple-direction mode using a *predefined ordering of the nodes*. Usually, a two-dimensional rectangular grid is iteratively swept along four directions: North → South, East → West, South → North, and West → East.

This method has been shown to be much faster than the classical iterative method which works on the whole grid and searches in every direction, but it is also not *local*, in the sense that it requires storing all the values on the grid nodes. In general, depending on the geometry, the method could need several sweeps to converge (and convergence is in general not guaranteed). A well-known exception is given by the eikonal equation, for which it is proved that a single sweep (i.e., four visits from the whole grid) is enough to reach convergence. The FSM computes the same solution of the global iterative method, provided the same scheme and the same stopping rule are used.

An SL version of the FSM will be tested and compared with the other methods in the next section.

9.4.5 ▪ Numerical tests

This section presents some numerical experiments performed with MATLAB 7 on a Pentium IV 2.80 GHz processor, with the main goal of comparing the FM-FD method and the FM-SL method described in previous sections, as well as the SL iterative method and

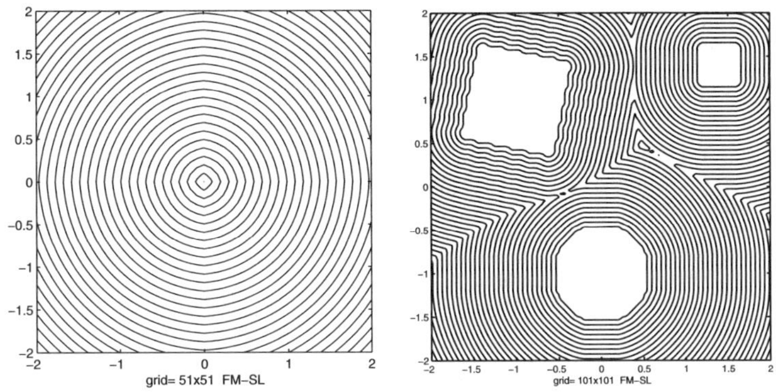

Figure 9.16. *Evolution at constant velocity: level sets of $T(x)$ computed by FM-SL method, with a* 51×51 *grid* [CCF].

the SL-based Fast Sweeping (FS-SL) method (in the implementation of the FM-SL algorithm we have used the observation in Remark 9.7 to speed up the computation). While some more complicated problems have been presented earlier in this chapter, the following tests are devoted to model problems in which we know the exact solution, so that the (normalized) ∞- and 1-norms of the error will be computed and compared, along with the convergence rate.

If not specified otherwise, we choose a computational domain $Q = [-2,2]^2$. We consider 51×51, 101×101, and 201×201 grids, corresponding, respectively, to $\Delta x = 0.08$, $\Delta x = 0.04$, and $\Delta x = 0.02$. In the iterative SL method, we choose the stopping criterion

$$||V^{(k)} - V^{(k-1)}||_\infty < 10^{-7},$$

whereas the FS-SL method is implemented with four iterations in different orders.

Evolution of a circular front at constant velocity We set here $\Gamma_0 = (0,0)$, and $c(x,y) \equiv 1$, while the exact solution is $T(x,y) = \sqrt{(x^2 + y^2)}$ (see Figure 9.16, left).

Results are summarized in Table 9.1. In all cases, errors converge with a similar rate (in the interval $[0.5, 1]$ for both methods) as Δx decreases, although errors are nearly halved in the SL schemes. The FM-SL method and SL give exactly the same errors in agreement with Theorem 9.9, and they match the errors of FS-SL as expected (since FS-SL converges in four iterations in the case where c is constant).

Both methods based on Fast Marching techniques are dramatically faster than the iterative SL method, even though it should be noted that a single iteration of the iterative scheme is less expensive than an iteration of Fast Marching–based algorithms. This happens because the *narrow band* technique requires both computing a minimum over the nodes in the *narrow band* and accessing the data in an almost random manner, rather than a systematic way along the loop indices (see [K01]).

Finally, we note that CPU time is slightly larger for the FM-SL method than for the FM-FD method, due to the larger size of the *narrow band* and to the additional complexity of the minimum search.

Evolution of a complex set at constant velocity Here, Γ_0 is the union of a unit square centered in $(-1,1)$ and rotated by $11.25°$, a circle with radius $R = 0.5$ centered in $(0,-1)$, and a square with side 0.4 centered in $(1.4,1.4)$, and $c(x,y) \equiv 1$. The exact solution $T(x,y)$ is the minimum among the distance functions of the three sets (see Figure 9.16, right).

Table 9.1. *Normalized errors and CPU times (seconds), and convergence rates for the evolution of a circular front.*

Method	Δx	L^∞ error	L^1 error	CPU
FM-FD	0.08	0.0875	0.7807	0.5
FM-SL	0.08	0.0329	0.3757	0.7
SL (46 it)	0.08	0.0329	0.3757	8.4
FS-SL	0.08	0.0329	0.3757	0.8
FM-FD	0.04	0.0526	0.4762	2.1
FM-SL	0.04	0.0204	0.2340	3.1
SL (86 it)	0.04	0.0204	0.2340	60
FS-SL	0.04	0.0204	0.2340	3.2
FM-FD	0.02	0.0309	0.2834	9.4
FM-SL	0.02	0.0122	0.1406	14
SL (162 it)	0.02	0.0122	0.1406	443.7
FS-SL	0.02	0.0122	0.1406	12.5

Method	L^∞ (0.08 → 0.04)	L^∞ (0.04 → 0.02)	L^1 (0.08 → 0.04)	L^1 (0.04 → 0.02)
FM-FD	0.7342	0.7675	0.7132	0.7487
FM-SL	0.6895	0.7417	0.6831	0.7349

Results are summarized in Table 9.2. In this test the shape of the initial front is more complicated, but errors have about the same behavior as in the previous test, although with a smaller variation. This is probably due to the more complicated structure of the *narrow band*.

Rates of convergence are again in the interval $[0.5, 1]$.

Comparison between the FMM and FSM In this test, we compare the two fast solvers on the equation

$$\max_{a\in B(0,1)} \{f(x,a)\cdot DT(x)\} = 1 \tag{9.51}$$

in which front propagation is changed from isotropic to anisotropic by setting, respectively,

$$f(x,a) = \begin{cases} a & \text{(isotropic)}, \\ \left(1+(\lambda a_1 + \mu a_2)^2\right)^{-1/2} & \text{(anisotropic)}, \end{cases}$$

where it is clear that in the second case the speed of propagation depends on the direction a. The target set is a neighborhood $B(0,\varepsilon)$ of the origin; i.e., the solution is set to zero on $B(0,\varepsilon)$.

In the isotropic case, FSM and FMM compute exactly the same solution and lead to the same error (see Table 9.3). In this case, the solution is regular since the velocity is constant at every point and does not depend on the normal direction to the front.

In the anisotropic case, the velocity depends on the components of a, and by varying the positive parameters λ and μ the weight of a component can be increased. Table 9.4 shows a difference between the FSM and FMM. Actually, the FSM computes the same solution as the iterative SL scheme, but this is not the case for the FMM. This shows a well-known feature of Fast Marching schemes, their unsatisfactory behavior in problems with substantial anisotropies.

Table 9.2. *Normalized errors and CPU times (seconds), and convergence rates for the evolution of a complex set.*

Method	Δx	L^∞ error	L^1 error	CPU
FM-FD	0.08	0.0625	0.2154	0.5
FM-SL	0.08	0.0440	0.1849	0.7
SL (30 it)	0.08	0.0440	0.1849	4.9
FS-SL	0.08	0.0440	0.1849	0.7
FM-FD	0.04	0.0393	0.1120	2.2
FM-SL	0.04	0.0215	0.1044	3.1
SL (55 it)	0.04	0.0215	0.1044	34.1
FS-SL	0.04	0.0215	0.1044	2.9
FM-FD	0.02	0.0248	0.0669	10.2
FM-SL	0.02	0.0135	0.0633	14.5
SL (102 it)	0.02	0.0135	0.0633	246.6
FS-SL	0.02	0.0135	0.0633	11.4

Method	L^∞ (0.08 → 0.04)	L^∞ (0.04 → 0.02)	L^1 (0.08 → 0.04)	L^1 (0.04 → 0.02)
FM-FD	0.6693	0.6642	0.9435	0.7434
FM-SL	1.0332	0.6714	0.8246	0.7218

Table 9.3. *Comparison between the FSM and FMM, isotropic case.*

Method	Grid	L^∞ error	L^1 error
FSM	101^2	0.079	0.011
FMM	101^2	0.079	0.011
FSM	201^2	0.057	0.006
FMM	201^2	0.057	0.006
FSM	401^2	0.029	0.002
FMM	401^2	0.029	0.002

Table 9.4. *Test: Comparison between the FSM and FMM, anisotropic case.*

Method	Grid	L^∞ error	L^1 error
FSM	101^2	0.635	0.029
FMM	101^2	0.635	0.058
FSM	201^2	0.404	0.014
FMM	201^2	0.408	0.049
FSM	401^2	0.189	0.005
FMM	401^2	0.290	0.044

9.5 ▪ Commented references

General, comprehensive references on the level set method are provided by [OF03, Se99]. Both books present a lot of information about models and schemes and a wide range of applications. In fact, since its first appearance [OS88], the level set method has been successfully applied to problems in combustion, image processing, Fluid Dynamics, and more, and we will not even try to present a satisfactory overview of the related literature here.

The SL method for the minimum time problem has been proposed by Bardi and Falcone in [BF90a, BF90b], and its adaptation to the front propagation problem, as well as some a priori estimates of the error on the front position, can be found in [F94]. A game theoretical interpretation of anisotropic motion has been proposed by Soravia in [Sor94]. This interpretation allows one to deal with a stationary version of the problem, and is also useful for constructing Fast Marching schemes (see also [O93] for the level set interpretation of a Dirichlet problem). The numerical approximation of a front propagation problem in the presence of obstacles is treated in [FT09], while an investigation on more efficient data structures for this class of problems is contained in [BCZ10].

The Mean Curvature Motion (MCM) has been studied by a number of authors. For a review of the analytical results, we refer the reader to the *User's Guide* by Crandall, Ishii, and Lions [CIL92] and to the references therein. The SL scheme presented in this chapter has been introduced by the authors in [FF03] and analyzed in [CFF10]—this latter paper also presents a long list of references on the approximation of the MCM equation. The original idea stems from a stochastic representation formula [ST02a, ST02b, ST03] and uses tools from the numerical analysis of SDEs, for which [KP92] represents a classical reference. We also quote two extensions of the scheme, to the case of Neumann boundary condition [AF13] and to the MCM in codimension 2 [CFF07]. Finally, we mention various contributions by Strain [St99, St99a, St99b, St00, St01], who have proposed efficient methods based on a tree structure and SL/finite difference approximations.

Image processing is one of the most popular applications of the level set method, which has been very successful in segmentation problems (see [OF03, Se99]). For the Shape-from-Shading problem, we refer the reader to the classical book by Horn and Brooks [HB89] and the survey [ZTCS99], while more recent references can be found in the survey [DFS08]. A level set approach has been proposed in [KSKB95], and the role of maximal solutions of the eikonal equation is used as a theoretical background in [CG00]. Recently, the classical orthographic model has been modified in order to deal with perspective deformations (see, e.g., [PF03, CCDG04, TSY03] for the models and [CFS05, BCDFV10, BCDFV12, PCF06] for the numerical approximation). We refer the reader to the books by Sapiro [Sa01] and Aubert and Kornprobst [AK06] for a general introduction to mathematical methods in image processing and to [Ch05] for the applications to biomedical images.

The introduction of Fast Marching methods is due to Tsitsiklis [Ts95], who presented the method for solving the eikonal equation related to the minimum time problem. The application of this technique to the front propagation problem, as well as a clear description of the algorithm, is due to Sethian [Se96b]. The analysis of a first-order SL method for the eikonal equation has been presented in [CF07] (where Proposition 9.2 is also proved), while a more general version of the SL scheme has been proposed by Sethian and Vladimirsky in [SV01, SV03] in a framework which includes anisotropic front propagation on unstructured grids. Finally, we refer the reader to [TSY04] for the application of a Fast Marching method to the Shape-from-Shading model.

FSMs have been proposed by several authors starting with Danielsson [D80]. The method has been studied in detail by Zhao [Z05], who has proved that the method based on a finite difference solver converges in a finite number of steps for the eikonal equation (see also [QZZ07a] for an extension to triangular grids). Other sweeping methods which can be applied to more general HJ equations have been proposed in [KOQ04, KOT05, TCOZ03, QZZ07b].

Bibliography

[A96] R. Abgrall, *Numerical discretization of the first-order Hamilton-Jacobi equation on triangular meshes*, Comm. Pure Appl. Math., **49** (1996), 1339–1373. (Cited on p. 187)

[A03] R. Abgrall, *Numerical discretization of boundary conditions for first order Hamilton–Jacobi equations*, SIAM J. Numer. Anal., **41** (2003), 2233–2261. (Cited on p. 187)

[AS64] M. Abramowitz, I.A. Stegun (eds.), Handbook of Mathematical Functions with Formulas, Graphs and Mathematical Tables, National Bureau of Standards, Washington, D.C. 1964. (Cited on p. 59)

[AF13] Y. Achdou, M. Falcone, *A numerical scheme for mean curvature motion with nonlinear Neumann conditions*, Interfaces Free Bound., **14** (2012), 455–485. (Cited on p. 298)

[A67] R.G.D. Allen, Macro-economic theory: A Mathematical Treatment, MacMillan, 1967. (Cited on p. 241)

[Al91a] B. Alziary de Roquefort, *Jeux différentiels et approximation numérique de fonctions valeur, 1re partie: étude théorique*, RAIRO Modél. Math. Anal. Numér., **25** (1991), 517–533. (Cited on p. 267)

[Al91b] B. Alziary de Roquefort, *Jeux différentiels et approximation numérique de fonctions valuer, 2e partie: étude numérique*, RAIRO Modél. Math. Anal. Numér., **25** (1991), 535–560. (Cited on p. 267)

[AL94] B. Alziary de Roquefort, P.L. Lions, *A grid refinement method for deterministic control and differential games*, Math. Models Methods Appl. Sci., **4** (1994), 899–910. (Cited on p. 267)

[AK06] G. Aubert, P. Kornprobst, Mathematical Problems in Image Processing: Partial Differential Equations and the Calculus of Variations, Springer, New York, 2006 (Cited on p. 298)

[A91] J.-P. Aubin, Viability Theory, Birkhäuser, Boston-Basel-Berlin, 1991. (Cited on p. 266)

[AC84] J.-P. Aubin, A. Cellina, Differential Inclusions, Springer-Verlag, Berlin-Heidelberg-New York, 1984. (Cited on p. 266)

[AF90] J.-P. Aubin, H. Frankowska, Set-Valued Analysis, Birkhäuser, Boston-Basel-Berlin, 1990. (Cited on p. 266)

[AA00] S. Augoula, R. Abgrall, *High order numerical discretization for Hamilton–Jacobi equations on triangular meshes*, J. Sci. Comput., **15** (2000), 197–229. (Cited on p. 187)

[B89] M. Bardi, *A boundary value problem for the minimum-time function*, SIAM J. Control Optim., **27** (1989), 776–785. (Cited on p. 266)

[BCD97] M. Bardi, I. Capuzzo Dolcetta, Optimal Control and Viscosity Solutions of Hamilton–Jacobi–Bellman Equations, Berlin, Birkhäuser, 1997. (Cited on pp. x, 6, 23, 25, 31, 32, 39, 219, 265, 267, 306)

[BF90a] M. Bardi, M. Falcone, *An approximation scheme for the minimum time function*, SIAM J. Control Optim., **28** (1990), 950–965. (Cited on pp. 267, 298)

[BF90b] M. Bardi, M. Falcone, *Discrete approximation of the minimal time function for systems with regular optimal trajectories*, in A. Bensoussan, J.L. Lions (eds.), Analysis and Optimization of Systems (Antibes, 1990), Lecture Notes in Control and Inform. Sci., vol. 144, Springer, Berlin, 1990, 103–112. (Cited on pp. 267, 298)

[BFS94] M. Bardi, M. Falcone, P. Soravia, *Fully discrete schemes for the value function of pursuit-evasion games*, in Advances in Dynamic Games and Applications, T. Basar, A. Haurie (eds.), Ann. Internat. Soc. Dynam. Games, vol. 1, Birkhäuser, Boston, 1994, 89–105. (Cited on p. 267)

[BFS00] M. Bardi, M. Falcone, P. Soravia, *Numerical methods for pursuit-evasion games via viscosity solutions*, in [BPR00], 289–303. (Cited on p. 267)

[BKS00] M. Bardi, S. Koike, P. Soravia, *Pursuit-evasion games with state constraints: Dynamic programming and discrete-time approximations*, Discrete Contin. Dynam. Systems, **6** (2000), 361–380. (Cited on p. 267)

[BPR00] M. Bardi, T. Parthasarathy, T.E.S. Raghavan (eds.), Stochastic and Differential Games: Theory and Numerical Methods, Ann. Internat. Soc. Dynam. Games 4, Birkhäuser, Boston, 2000. (Cited on p. 300)

[BS89] M. Bardi, P. Soravia, *A PDE framework for differential games of pursuit-evasion type*, in T. Basar, P. Bernhard (eds.), Differential Games and Applications, Lecture Notes in Control and Inform. Sci., vol. 119, Springer-Verlag, Berlin, 1989, 62–71. (Cited on p. 266)

[BS91a] M. Bardi, P. Soravia, *Hamilton–Jacobi equations with singular boundary conditions on a free boundary and applications to differential games*, Trans. Amer. Math. Soc., **325** (1991), 205–229. (Cited on p. 266)

[BS91b] M. Bardi, P. Soravia, *Approximation of differential games of pursuit-evasion by discrete-time games*, in R.P. Hamalainen, H.K. Ethamo (eds.), Differential Games—Developments in Modelling and Computation, Lecture Notes in Control and Inform. Sci., vol. 156, Springer-Verlag, Berlin, 1991, 131–143. (Cited on p. 267)

[Ba93] G. Barles, *Discontinuous viscosity solutions of first order Hamilton–Jacobi equations: a guided visit*, Nonlinear Anal., **20** (1993), 1123–1134. (Cited on pp. 20, 38, 39)

[Ba98] G. Barles, Solutions de viscositè des equations d'Hamilton–Jacobi, Springer-Verlag, Paris, 1998. (Cited on pp. x, 25, 39)

[BaJ07] G. Barles, E.R. Jakobsen, *Error bounds for monotone approximation schemes for parabolic Hamilton–Jacobi–Bellman equations*, Math. Comp., **76** (2007), 1861–1893. (Cited on p. 100)

[BaS91] G. Barles, P.E. Souganidis, *Convergence of approximation schemes for fully nonlinear second order equations*, Asymptotic Anal., **4** (1991), 271–283. (Cited on pp. 91, 93, 94, 100)

[Bar90] E.N. Barron, *Differential games with maximum cost*, Nonlinear Anal., **14** (1990), 971–989. (Cited on p. 266)

[BEJ84] E.N. Barron, L.C. Evans, R. Jensen, *Viscosity solutions of Isaacs' equations and differential games with Lipschitz controls*, J. Differential Equations, **53** (1984), 213–233. (Cited on p. 266)

[BJ89] E.N. Barron, R. Jensen, *Total risk aversion, stochastic optimal control, and differential games*, Appl. Math. Optim., **19** (1989), 313–327. (Cited on p. 266)

[BJ90] E.N. Barron, R. Jensen, *Semicontinuous viscosity solutions for Hamilton–Jacobi equations with convex Hamiltonian*, Comm. Partial Differential Equations, **15** (1990), 1713–1742. (Cited on pp. 39, 266)

[BJ91] E.N. Barron, R. Jensen, *Optimal control and semicontinuous viscosity solutions*, Proc. Amer. Math. Soc., **113** (1991), 397–402. (Cited on pp. 39, 266)

[BO82] T. Basar, G.J. Olsder, Dynamic Non-cooperative Game Theory, Academic Press, New York, 1982. (Cited on pp. 265, 267)

[BM82] J.R. Bates, A. McDonald, *Multiply-upstream, semi-Lagrangian advective schemes: Analysis and application to a multi-level primitive equation model*, Mon. Wea. Rev., **110** (1982), 1831–1842. (Cited on p. 186)

[BGS06] F. Bauer, L. Grüne, W. Semmler, *Adaptive spline interpolation for Hamilton–Jacobi–Bellman equations*, Appl. Numer. Math., **56** (2006), 1196–1210. (Cited on p. 187)

[B57] R. Bellman, Dynamic Programming, Princeton University Press, Princeton, NJ, 1957. (Cited on p. 265)

[BM03] Y. Belopolskaya, G.N. Milstein, *An approximation method for Navier–Stokes equations based on probabilistic approach*, Statist. Probab. Lett., **64** (2003), 201–211. (Cited on p. 206)

[BS12] R. Bermejo, L. Saavedra, *Modified Lagrange–Galerkin methods of first and second order in time for convection-diffusion problems*, Numer. Math., **120** (2012), 601–638. (Cited on p. 187)

[Be87] D.P. Bertsekas, Dynamic programming: Deterministic and Stochastic Models, Prentice–Hall, Englewood Cliffs, NJ, 1987. (Cited on p. 265)

[BeS78] D. P. Bertsekas, S.E. Shreve, Stochastic Optimal Control: The Discrete Time Case, Academic Press, New York–London, 1978. (Cited on p. 265)

[BM08] N. Besse, M. Mehrenberger, *Convergence of classes of high-order semi-Lagrangian schemes for the Vlasov–Poisson system*, Math. Comp., **77** (2008), 93–123. (Cited on pp. 145, 150, 187)

[BCZ10] O. Bokanowski, E. Cristiani, H. Zidani, *An efficient data structure and accurate scheme to solve front propagation problems*, J. Sci. Comput., **42** (2010), 251–273. (Cited on p. 298)

[BFZ10a] O. Bokanowski, N. Forcadel, H. Zidani, *Reachability and minimal times for state constrained nonlinear problems without any controllability assumption*, SIAM J. Control Optim., **48** (2010), 4292–4316. (Cited on p. 266)

[BFZ10b] O. Bokanowski, N. Forcadel, H. Zidani, *L^1-error estimates for numerical approximations of Hamilton–Jacobi–Bellman equations in dimension* 1, Math. Comp., **79** (2010), 1395–1426. (Cited on p. 266)

[BGGK13] O. Bokanowski, J. Garcke, M. Griebel, I. Klompmaker, *An adaptive sparse grid semi-Lagrangian scheme for first order Hamilton–Jacobi–Bellman equations*, J. Sci. Comput. **55** (2013), 575–605. (Cited on p. 187)

[BMZ09] O. Bokanowski, S. Maroso, H. Zidani, *Some convergence results for Howard's algorithm*, SIAM J. Numer. Anal., **47** (2009), 3001–3026. (Cited on p. 267)

[BZ07] O. Bokanowski, H. Zidani, *Anti-dissipative schemes for advection and application to Hamilton–Jacobi–Bellman equations*, J. Sci. Comput., **30** (2007), 1–33. (Cited on p. 187)

[B] L. Bonaventura, Semi-Lagrangian methods for environmental flows, in preparation. (Cited on p. 206)

[Br89] Y. Brenier, *Un algorithme rapide pour le calcul de trasformées de Legendre–Fenchel discretes*, C.R. Acad., Sci. Paris Sér. I Math., **308** (1989), 587–589. (Cited on p. 73)

[BS08] S.C. Brenner, L.R. Scott, The Mathematical Theory of Finite Element Method, Springer-Verlag, New York, 2008. (Cited on p. 73)

[Bre73] R. Brent, Algorithms for Minimization without Derivatives, Prentice–Hall, Englewood Cliffs, NJ, 1973. (Cited on p. 73)

[BCDFV10] M. Breuss, E. Cristiani, J.-D. Durou, M. Falcone, O. Vogel, *Numerical algorithms for Perspective Shape from Shading*, Kybernetika, **46** (2010), 207–225. (Cited on p. 298)

[BCDFV12] M. Breuss, E. Cristiani, J.-D. Durou, M. Falcone, O. Vogel, *Perspective shape from shading: Ambiguity analysis and numerical approximations*, SIAM J. Imaging Sci., **5** (2012), 311–342. (Cited on p. 298)

[BrL03] S. Bryson, D. Levy, *Central schemes for multidimensional Hamilton–Jacobi equations*, SIAM J. Sci. Comput., **25** (2003), 767–791. (Cited on p. 187)

[BrL04] S. Bryson, D. Levy, *High-order central schemes for multidimensional Hamilton–Jacobi Equations*, SIAM J. Numer. Anal., **41** (2004), 1339–1369. (Cited on p. 187)

[Buh03] M.D. Buhmann, Radial Basis Functions, Cambridge University Press, Cambridge, UK, 2003. (Cited on p. 73)

[BG04] H.-J. Bungartz, M. Griebel, *Sparse grids*, Acta Numer., **13** (2004), 147–269. (Cited on p. 73)

[But03] J.C. Butcher, Numerical Analysis of Ordinary Differential Equations, Wiley, Chichester, 2003. (Cited on p. 72)

[CCF] S. Cacace, E. Cristiani, M. Falcone, *Can local single-pass methods solve any stationary Hamilton-Jacobi-Bellman equation?*, submitted to SIAM J. Sci. Comput. (Cited on p. 295)

[CCFP12] S. Cacace, E. Cristiani, M. Falcone, A. Picarelli, *A patchy dynamic programming scheme for a class of Hamilton–Jacobi–Bellman equations*, SIAM J. Sci. Comput., **34** (2012), A625–A649. (Cited on p. 268)

[CF95] F. Camilli, M. Falcone, *An approximation scheme for the optimal control of diffusion processes*, RAIRO Modél. Math. Anal. Numér., **29** (1995), 97–122. (Cited on p. 236)

[CF96] F. Camilli, M. Falcone, *Approximation of optimal control problems with state constraints: Estimates and applications*, in B.S. Mordukhovic, H.J. Sussman (eds.), in Nonsmooth Analysis and Geometric Methods in Deterministic Optimal Control, IMA Vol. Appl. Math., vol. 78, Springer Verlag, Berlin, 1996, 23–57. (Cited on pp. 266, 267)

[CF97] F. Camilli, M. Falcone, *Analysis and approximation of the infinite horizon problem with impulsive controls*, Automatika i Telemekanika, **7** (1997), 169–184. (Cited on p. 267)

[CF99] F. Camilli, M. Falcone, *Approximation of control problems involving ordinary and impulsive controls*, ESAIM Control Optim. Calc. Var., **4** (1999), 159–176. (Cited on p. 267)

[CFLS94] F. Camilli, M. Falcone, P. Lanucara, A. Seghini, *A domain decomposition method for Bellman equations*, in D.E. Keyes and J. Xu (eds.), Domain Decomposition Methods in Scientific and Engineering Computing, Contemp. Math., vol. 180, AMS, Providence, RI, 1994, 477–483. (Cited on p. 268)

[CG00] F. Camilli, L. Grüne, *Numerical approximation of the maximal solutions for a class of degenerate Hamilton–Jacobi equations*, SIAM J. Numer. Anal., **38** (2000), 1540–1560. (Cited on p. 298)

[CS03] F. Camilli, A. Siconolfi, *Hamilton–Jacobi equations with measurable dependence on the state variable*, Adv. Differential Equations, **8** (2003), 733–768. (Cited on p. 39)

[CS04] P. Cannarsa, C. Sinestrari, Semiconcave Functions, Hamilton–Jacobi equations, and Optimal Control, Birkhäuser, Boston, 2004. (Cited on p. 30)

[CD83] I. Capuzzo Dolcetta, *On a discrete approximation of the Hamilton–Jacobi equation of dynamic programming*, Appl. Math. Optim., **10** (1983), 367–377. (Cited on p. 266)

[CDF89] I. Capuzzo Dolcetta, M. Falcone, *Viscosity solutions and discrete dynamic programming*, Ann. Inst. H. Poincaré Anal. Non Linéaire, **6** (supplement) (1989), 161–183. (Cited on p. 267)

[CDI84] I. Capuzzo Dolcetta, H. Ishii, *Approximate solutions of the Bellman equation of deterministic control theory*, Appl. Math. Optim., **11** (1984), 161–181. (Cited on pp. 223, 266)

[CDL90] I. Capuzzo Dolcetta, P.L. Lions , *Hamilton–Jacobi equations with state constraints*, Trans. Amer. Math. Soc., **318** (1990), 643–683. (Cited on p. 266)

[CQS99] P. Cardaliaguet, M. Quincampoix, P. Saint-Pierre, *Set valued numerical analysis for optimal control and differential games*, in M. Bardi, T. Parthasarathy, T.E.S. Raghavan (eds.), Stochastic and Differential Games: Theory and Numerical Methods, Ann. Internat. Soc. Dynam. Games, vol. 4, Birkhäuser, Boston, 1999, 177–247. (Cited on p. 266)

[CQS00] P. Cardaliaguet, M. Quincampoix, P. Saint-Pierre, *Numerical schemes for discontinuous value functions of optimal control. Set-valued analysis in control theory*, Set-Valued Anal., **8** (2000), 111–126. (Cited on p. 266)

[CFF04] E. Carlini, M. Falcone, R. Ferretti, *An efficient algorithm for Hamilton–Jacobi equations in high dimension*, Comput. Vis. Sci., **7** (2004), 15–29. (Cited on pp. 73, 187, 241, 242, 243, 245)

[CFF07] E. Carlini, M. Falcone, R. Ferretti, *A semi-Lagrangian scheme for the curve shortening flow in codimension* 2, J. Comput. Phys., **225** (2007), 1388–1408. (Cited on p. 298)

[CFF10] E. Carlini, M. Falcone, R. Ferretti, *Convergence of a large time-step scheme for mean curvature motion*, Interfaces Free Bound., **12** (2010), 409–441. (Cited on pp. 100, 276, 278, 279, 298)

[CFFe13] E. Carlini, M. Falcone, A. Festa, *A brief survey on semi-Lagrangian schemes for image processing*, in M. Breuss, A. Bruckstein, P. Maragos, (eds.), Innovations for Shape Analysis: Models and Algorithms, Springer-Verlag, Berlin, 2013, 191–218. (Cited on pp. 281, 282)

[CFR05] E. Carlini, R. Ferretti, G. Russo, *A weighted essentially nonoscillatory, large time-step scheme for Hamilton–Jacobi equations*, SIAM J. Sci. Comput., **27** (2005), 1071–1091. (Cited on pp. 73, 186, 187)

[Ca90] V. Casulli, *Semi-implicit finite difference methods for the two dimensional shallow water equations*, J. Comput. Phys., **86** (1990), 56–74. (Cited on p. 195)

[CC90] V. Casulli, R.T. Cheng, *Stability analysis of Eulerian-Lagrangian methods for the one-dimensional shallow-water equations*, Appl. Math. Modelling, **14** (1990), 122–131. (Cited on p. 206)

[CDM13] F. Charles, B. Després, M. Mehrenberger, *Enhanced convergence estimates for semi-lagrangian schemes application to the Vlasov–Poisson equation*, SIAM J. Numer. Anal., **51** (2013), 840–863. (Cited on p. 145)

[CK76] C.Z. Cheng, G. Knorr, *The integration of the Vlasov equation in configuration space*, J. Comput. Phys., **22** (1976), 330–351. (Cited on p. 145)

[Ch05] D.L. Chopp, *Recent advances in the level set method*, in Handbook of Biomedical Image Analysis, vol. 1, Kluwer Academic, Dordrecht, The Netherlands, 2005. (Cited on p. 298)

[Cia02] Ph.G. Ciarlet, The Finite Element Method for Elliptic Problems, SIAM, Philadelphia, 2002. (Cited on p. 73)

[Cl83] F.H. Clarke, Optimization and Nonsmooth Analysis, Wiley, New York, 1983. (Cited on p. 31)

[CST97] A.R. Conn, K. Scheinberg, Ph.L. Toint, *Recent progress in unconstrained nonlinear optimization without derivatives*, Math. Programming, **79** (1997), 397–414. (Cited on p. 73)

[Co96] L. Corrias, *Fast Legendre–Fenchel transform and applications to Hamilton–Jacobi equations and conservation laws*, SIAM J. Numer. Anal., **33** (1996), 1534–1558. (Cited on p. 73)

[CFL67] R. Courant, K. Friedrichs, H. Lewy, *On the partial difference equations of mathematical physics*, IBM J. Res. Develop., **11** (1967), 215–234. (Cited on p. 99)

[CIR52] R. Courant, E. Isaacson, M. Rees, *On the solution of nonlinear hyperbolic differential equations by finite differences*, Comm. Pure Appl. Math., **5** (1952), 243–255. (Cited on pp. ix, 145)

[CCDG04] F. Courteille, A. Crouzil, J.-D. Durou, P. Gurdjos, *Towards shape from shading under realistic photographic conditions*, in Proc. 17th Int. Conf. Patt. Recog. (vol. II), Cambridge, UK, 2004, 277–280. (Cited on p. 298)

[CEL84] M.G. Crandall, L.C. Evans, P.L. Lions, *Some properties of viscosity solutions of Hamilton-Jacobi equations*, Trans. Amer. Math. Soc., **282** (1984), 487–502. (Cited on pp. 38, 266)

[CIL92] M.G. Crandall, I. Ishii, P.L. Lions, *User's guide to viscosity solutions of second order partial differential equations*, Bull. Amer. Math. Soc., **27** (1992), 1–67. (Cited on pp. 39, 298)

[CL83] M.G. Crandall, P.L. Lions, *Viscosity solutions of Hamilton–Jacobi equations*, Trans. Amer. Math. Soc., **277** (1983), 1–42. (Cited on p. 266)

[CL84] M.G. Crandall, P.L. Lions, *Two approximations of solutions of Hamilton-Jacobi equations*, Math. Comp., **43** (1984), 1–19. (Cited on pp. 38, 99, 100, 127, 129, 144, 145)

[CT80] M.G. Crandall, L. Tartar, *Some relations between nonexpansive and order preserving mappings*, Proc. Amer. Math. Soc., **78** (1980), 385–390. (Cited on p. 100)

[CF07] E. Cristiani, M. Falcone, *Fast semi-Lagrangian schemes for the eikonal equation and applications*, SIAM J. Numer. Anal., **45** (2007), 1979–2011. (Cited on pp. 272, 273, 288, 289, 293, 298)

[CF08] E. Cristiani, M. Falcone, *Numerical solution of the Isaacs equation for differential games with state constraints*, in Proceedings of the Conference IFAC 2008, Seoul, 2008. (Cited on p. 267)

[CFS05] E. Cristiani, M. Falcone, A. Seghini, *Numerical solution of the perspective Shape from Shading problem*, in Proceedings of Control Systems: Theory, Numerics and Applications PoS (CSTNA2005) 008, `http://pos.sissa.it/`. (Cited on pp. 281, 298)

[CMS10] N. Crouseilles, M. Mehrenberger, E. Sonnendrücker, *Conservative semi-Lagrangian schemes for Vlasov equations*, J. Comput. Phys., **229** (2010), 1927–1953. (Cited on p. 206)

[D80] P. Danielsson, *Euclidean distance mapping*, Comput. Graphics Image Process., **14** (1980), 227–248. (Cited on p. 298)

[D05] T. Davies, M.J.P. Cullen, A.J. Malcolm, M.H. Mawson, A. Staniforth, A.A. White, N. Wood, *A new dynamical core for the Met Office's global and regional modeling of the atmosphere*, Q. J. R. Meteorol. Soc., **131** (2005), 1759–1782. (Cited on p. 195)

[DZS10] C. De Zan, P. Soravia, *Cauchy problems for noncoercive Hamilton–Jacobi–Isaacs equations with discontinuous coefficients*, Interfaces Free Bound., **12** (2010), 347–368. (Cited on p. 267)

[DR82] J. Douglas Jr., T.F. Russell, *Numerical methods for convection-dominated diffusion problems based on combining the method of characteristics with finite element or finite difference procedures*, SIAM J. Numer. Anal., **19** (1982), 871–885. (Cited on p. 187)

[DS01] P. Dupuis, A. Szpiro, *Convergence of the optimal feedback policies in a numerical method for a class of deterministic optimal control problems*, SIAM J. Control Optim., **40** (2001), 393–420. (Cited on p. 267)

[DFS08] J.D. Durou, M. Falcone, M. Sagona, *Numerical methods for shape from shading: A new survey with benchmarks*, Comput. Vis. Image Understand., **109** (2008), 22–43. (Cited on pp. 280, 281, 298)

[EK72] R.J. Elliott, N.J. Kalton, The existence of value in differential games, Mem. Amer. Math. Soc., **126**, 1972. (Cited on p. 247)

[E98] L. C. Evans, *Partial Differential Equations*, Grad. Stud. Math., vol. 19, AMS, Providence, RI, 2010. (Cited on pp. 2, 23, 27, 28, 29, 214)

[EI84] L.C. Evans, H. Ishii, *Differential games and nonlinear first order PDEs in bounded domains*, Manuscripta Math., **49** (1984), 109–139. (Cited on p. 266)

[ES84] L.C. Evans, P.E. Souganidis. *Differential games and representation formulas for solutions of Hamilton–Jacobi equations*, Indiana Univ. Math. J., **33** (1984), 773–797. (Cited on pp. 248, 266)

[F87] M. Falcone, *A numerical approach to the infinite horizon problem of deterministic control theory*, Appl. Math. Optim., **15** (1987), 1–13. (Cited on pp. 266, 267)

[F94] M. Falcone, *The minimum time problem and its applications to front propagation*, in A. Visintin, G. Buttazzo (eds.), Motion by Mean Curvature and Related Topics, De Gruyter Verlag, Berlin, 1994. (Cited on p. 298)

[F97] M. Falcone, *Numerical solution of Dynamic Programming equations*, Appendix A in [BCD97]. (Cited on pp. 145, 228, 267)

[F01] M. Falcone, *Some remarks on the synthesis of feedback controls via numerical methods*, in J.L. Menaldi, E. Rofman, A. Sulem (eds.), Optimal Control and Partial Differential Equations, IOS Press, Amsterdam, 2001, 456–465. (Cited on p. 267)

[F06] M. Falcone, *Numerical methods for differential games via partial differential equations*, Int. Game Theor. Rev., **8** (2006), 231–272. (Cited on pp. 261, 262, 263, 264, 265)

[FF94] M. Falcone, R. Ferretti, *Discrete-time high-order schemes for viscosity solutions of Hamilton–Jacobi equations*, Numer. Math., **67** (1994), 315–344. (Cited on pp. 145, 220, 240)

[FF98] M. Falcone, R. Ferretti, *Convergence analysis for a class of high-order semi-Lagrangian advection schemes*, SIAM J. Numer. Anal., **35** (1998), 909–940. (Cited on pp. 117, 145, 187)

[FF02] M. Falcone, R. Ferretti, *Semi-Lagrangian schemes for Hamilton–Jacobi equations, discrete representation formulae and Godunov methods*, J. Comput. Phys., **175** (2002), 559–575. (Cited on pp. 145, 187, 238, 239, 274)

[FF03] M. Falcone, R. Ferretti, *Consistency of a large time-step scheme for mean curvature motion*, in F. Brezzi, A. Buffa, S. Corsaro, A. Murli (eds.), Numerical Mathematics and Advanced Applications- ENUMATH 2001, Springer-Verlag, Milan, 2003, 495–502. (Cited on p. 298)

[FFM01] M. Falcone, R. Ferretti, T. Manfroni, *Optimal discretization steps for a class of semi-lagrangian schemes*, in Numerical Methods for Viscosity Solutions and Applications (Heraklion, 1999), Ser. Adv. Math. Appl. Sci. vol. 59, World Scientific, River Edge, NJ, 2001. (Cited on p. 187)

[FG98] M. Falcone, T. Giorgi, *An approximation scheme for evolutive Hamilton-Jacobi equations*, in W.M. McEneaney, G. Yin, Q. Zhang (eds.), Stochastic Analysis, Control, Optimization and Applications: A Volume in Honor of W.H. Fleming, Birkhäuser, Boston, 1998, 289–303. (Cited on pp. 224, 225)

[FGW99] M. Falcone, L. Grüne, F. Wirth, *A maximum time approach to the computation of robust domains of attraction*, in International Conference on Differential Equations, Vols. 1, 2 (Berlin, 1999), World Scientific, River Edge, NJ, 2000, 844–849. (Cited on p. 267)

[FLM01] M. Falcone, P. Lanucara, M. Marinucci, *Parallel Algorithms for the Isaacs equation*, in O. Pourtallier, E. Altman (eds.), Advances in Dynamic Games and Applications, Ann. Internat. Soc. Dynam. Games 7, Boston, Birkhäuser, 2001, 203–223. (Cited on p. 268)

[FLS94] M. Falcone, P. Lanucara, A. Seghini, *A splitting algorithm for Hamilton–Jacobi–Bellman equations*, Appl. Numer. Math., **15** (1994), 207–218. (Cited on p. 268)

[FS05] M. Falcone, P. Stefani, *Advances in parallel algorithms for the Isaacs equation*, in A.S. Nowak, K. Szajowski (eds.), Advances in Dynamic Games, Ann. Internat. Soc. Dynam. Games, Birkhäuser, Boston, 2005, 515–544. (Cited on pp. 263, 268)

[FT09] M. Falcone, C. Truini, *A level-set algorithm for front propagation in the presence of obstacles*, Rend. Mat. Appl. (7), **29** (2009), 29–50. (Cited on p. 298)

[Fe03] R. Ferretti, *Convergence of semi-Lagrangian approximations to convex Hamilton–Jacobi equations under (very) large Courant numbers*, SIAM J. Numer. Anal., **40** (2003), 2240–2253. (Cited on p. 187)

[Fe10a] R. Ferretti, *Equivalence of semi-Lagrangian and Lagrange–Galerkin schemes under constant advection speed*, J. Comput. Math., **28** (2010), 461–473. (Cited on pp. 153, 187)

[Fe10b] R. Ferretti, *A technique for high-order treatment of diffusion terms in semi-Lagrangian schemes*, Commun. Comput. Phys., **8** (2010), 445–470. (Cited on pp. 199, 202, 206)

[Fe13a] R. Ferretti, *On the relationship between semi-Lagrangian and Lagrange–Galerkin schemes*, Numer. Math., **124** (2013), 31–56. (Cited on pp. 49, 153, 154, 160, 163, 165, 169, 187)

[Fe13b] R. Ferretti, *Stability of some generalized Godunov schemes with linear high-order reconstructions*, J. Sci. Comput., **57** (2013), 213–228. (Cited on p. 206)

[FP07] R. Ferretti, G. Perrone, *On the stability of semi-Lagrangian advection schemes under finite element interpolation*, in Applied and Industrial Mathematics in Italy II, Ser. Adv. Math. Appl. Sci., vol. 75, World Scientific, Singapore, 2007, 339–350. (Cited on pp. 156, 158, 187)

[FZ13] R. Ferretti, H. Zidani, *Monotone numerical schemes for hybrid systems*, in preparation. (Cited on p. 267)

[FR75] W.H. Fleming, R.W. Rishel, Deterministic and Stochastic Optimal Control, Springer-Verlag, New York, 1975. (Cited on pp. 239, 265)

[FS93] W.H. Fleming, H.M. Soner, Controlled Markov Processes and Viscosity Solutions, Springer-Verlag, New York, 1993. (Cited on pp. x, 39, 265)

[FSo89] W.H. Fleming, P.E. Souganidis, *On the existence of value function of two-players, zero-sum differential games*, Indiana Univ. Math. J., **38** (1989), 293–314. (Cited on p. 267)

[Fr93] H. Frankowska, *Lower semicontinuous solutions of Hamilton–Jacobi–Bellman equations*, SIAM J. Control Optim., **31** (1993), 257–272. (Cited on pp. 39, 266)

[FP00] H. Frankowska, S. Plaskacz, *Semicontinuous solutions of Hamilton–Jacobi–Bellman equations with degenerate state constraints*, J. Math. Anal. Appl., **251** (2000), 818–838. (Cited on p. 266)

[Fr71] A. Friedman, Differential Games, Wiley, New York, 1971. (Cited on p. 267)

[GS06] M. Garavello, P. Soravia, *Representation formulas for solutions of the HJI equations with discontinuous coefficients and existence of value in differential games*, J. Optim. Theory Appl., **130** (2006), 209–229. (Cited on p. 267)

[GBSJFF90] A. Ghizzo, P. Bertrand, M.M. Shoucri, T.W. Johnston, E. Fijalkow, M.R. Feix, *A Vlasov code for the numerical simulation of stimulated Raman scattering*, J. Comput. Phys., **90** (1990), 431–457. (Cited on p. 145)

[GR96] E. Godlewski, P.A. Raviart, Numerical Approximation of Hyperbolic Systems of Conservation Laws, Appl. Math. Sci., vol. 118, Springer-Verlag, New York, 1996. (Cited on p. 145)

[GRo85a] R. Gonzalez, E. Rofman, *On deterministic control problems: An approximation procedure for the optimal cost. I. The stationary problem*, SIAM J. Control Optim., **23** (1985), 242–266. (Cited on p. 266)

[GRo85b] R. Gonzalez, E. Rofman, *On deterministic control problems: An approximation procedure for the optimal cost. II. The nonstationary case*, SIAM J. Control Optim., **23** (1985), 267–285. (Cited on p. 266)

[G97] L. Grüne, *An adaptive grid scheme for the discrete Hamilton–Jacobi–Bellman equation*, Numer. Math., **75** (1997), 319–337. (Cited on p. 187)

[G04] L. Grüne, *Error estimation and adaptive discretization for the discrete stochastic Hamilton–Jacobi–Bellman equation*, Numer. Math., **99** (2004), 85–112. (Cited on p. 187)

[GNP07] L. Grüne, D. Nesic, J. Pannek, Model Predictive Control for Nonlinear Sampled-Data Systems. Assessment and Future Directions of Nonlinear Model Predictive Control, Lecture Notes in Control and Inform. Sci., vol. 358, Springer, Berlin, 2007, 105–113 (Cited on p. 268)

[GP11] L. Grüne, J. Pannek, Nonlinear Model Predictive Control: Theory and Algorithms, Springer-Verlag, London, 2011. (Cited on p. 268)

[GR08] L. Grüne, A. Rantzer, *On the infinite horizon performance of receding horizon controllers*, IEEE Trans. Automat. Control, **53** (2008), 2100–2111. (Cited on p. 268)

[HNW93] E. Hairer, S.P. Nørsett, G. Wanner, Solving Ordinary Differential Equations: Nonstiff Problems, Springer-Verlag, Berlin, 1993. (Cited on p. 72)

[Ho65] E. Hopf, *Generalized solutions of nonlinear equations of first order*, J. Math. Mech., **14** (1965), 951–973. (Cited on p. 27)

[HB89] B.K.P. Horn, M.J. Brooks (eds.), Shape from Shading, The MIT Press, Cambridge, MA, 1989. (Cited on pp. 9, 298)

[H60] R.A. Howard, Dynamic Programming and Markov Processes, Wiley, New York, 1960. (Cited on p. 267)

[HS99] C. Hu, C.-W. Shu, *A discontinuous Galerkin finite element method for Hamilton–Jacobi equations*, SIAM J. Sci. Comput., **21** (1999), 666–690. (Cited on p. 187)

[I65] R. Isaacs, Differential Games, Wiley, New York, 1965. (Cited on pp. 249, 265)

[Is89] H. Ishii, *A boundary value problem of the Dirichlet type for Hamilton–Jacobi equations*, Ann. Scuola Norm. Sup. Pisa 9 Cl. Sci. (4), **16** (1989), 105–135. (Cited on p. 39)

[IK96] H. Ishii, S. Koike, *A new formulation of state constraint problem for first-order PDEs*, SIAM J. Control Optim., **34** (1996), 554–571. (Cited on p. 266)

[Is03] A. Iske, *Radial basis functions: Basics, advanced topics and meshfree methods for transport problems*, Rend. Sem. Mat. Univ. Pol. Torino, **61** (2003), 247–285. (Cited on p. 73)

[JP00] G.-S. Jiang, D. Peng, *Weighted ENO schemes for Hamilton–Jacobi equations*, SIAM J. Sci. Comput., **21** (2000), 2126–2143. (Cited on p. 185)

[K59] R. Kalaba, *On nonlinear differential equations, the maximum operation and monotone convergence*, J. Math. Mech., **8** (1959), 519–574. (Cited on p. 267)

[KOQ04] C.-Y. Kao, S. Osher, J. Qian, *Lax-Friedrichs sweeping scheme for static Hamilton-Jacobi equations*, J. Comput. Phys., **196** (2004), 367–391. (Cited on p. 298)

[KOT05] C.-Y. Kao, S. Osher, Y.-H. Tsai, *Fast sweeping methods for static Hamilton-Jacobi equations*, SIAM J. Numer. Anal., **42** (2005), 2612–2632. (Cited on p. 298)

[KG11] I. Karafyllis, L. Grüne, *Feedback stabilization methods for the numerical solution of ordinary differential equations*, Discrete Contin. Dyn. Syst. Ser. B, **16** (2011), 283–317. (Cited on p. 267)

[K01] S. Kim, *An $\mathcal{O}(N)$ level set method for eikonal equations*, SIAM J. Sci. Comput., **22** (2001), 2178–2193. (Cited on p. 295)

[KS98] R. Kimmel, J.A. Sethian, *Computing geodesic paths on manifold*, Proc. Natl. Acad. Sci. USA, **95** (1998), 8431–8435. (Not cited)

[KS01] R. Kimmel, J.A. Sethian, *Optimal algorithm for shape from shading and path planning*, J. Math. Imaging Vision, **14** (2001), 237–244. (Cited on p. 281)

[KSKB95] R. Kimmel, K. Siddiqi, B.B. Kimia, A. Bruckstein, *Shape from shading: Level set propagation and viscosity solutions*, Int. J. Comput. Vis., **16** (1995), 107–133. (Cited on p. 298)

[KP92] P.E. Kloeden, E. Platen, Numerical Solution of Stochastic Differential Equations, Springer-Verlag, Berlin, 1992. (Cited on pp. 206, 236, 298)

[KLT03] T.G. Kolda, R.M. Lewis, V. Torczon, *Optimization by direct search: New perspectives on some classical and modern methods*, SIAM Rev., **45** (2003), 385–482. (Cited on p. 73)

[KS88] N.N. Krasovskii, A.I. Subbotin, Game Theoretical Control Problems, Springer, New York, 1988. (Cited on p. 266)

[K62] T.N. Krishnamurthi, *Numerical integration of primitive equation by a quasi-Lagrangian advective scheme*, J. Appl. Meteorol., **1** (1962), 508–521. (Cited on p. 205)

[Kr60] S.N. Kružkov, *The Cauchy problem in the large for certain nonlinear first order equations*, Soviet Math. Dokl., **1** (1960), 474–477. (Cited on p. 38)

[Kr64] S.N. Kružkov, *The Cauchy problem in the large for nonlinear equations and for certain quasilinear systems of the first order with several variables*, Soviet Math. Dokl., **5** (1964), 493–496. (Cited on p. 38)

[Kr66] S.N. Kružkov, *The method of finite differences for a first-order nonlinear equation with many independent variables*, U.S.S.R. Comput. Math. and Math. Phys., **6** (1966), 136–151. (Cited on p. 38)

[KuD01] H.J. Kushner, P. Dupuis, Numerical Methods for Stochastic Control Problems in Continuous Time, Springer-Verlag, Berlin, 2001. (Cited on pp. x, 145, 265)

[LR56] P.D. Lax, R.D. Richtmeyer, *Survey of the stability of linear finite difference equations*, Comm. Pure Appl. Math., **9** (1956), 267–293. (Cited on p. 99)

[L92] R.J. Leveque, Numerical Methods for Conservation Laws, Birkhäuser, Basel, 1992. (Cited on pp. 38, 145)

[LT00] C.-T. Lin, E. Tadmor, *High-resolution nonoscillatory central schemes for Hamilton-Jacobi equations*, SIAM J. Sci. Comput., **21** (2000), 2163–2186. (Cited on p. 185)

[LT01] C.-T. Lin, E. Tadmor, *L^1 stability and error estimates for Hamilton–Jacobi solutions*, Numer. Math., **87** (2001), 701–735. (Cited on pp. 98, 100, 133)

[LR96] S.-J. Lin, R.B. Rood, *Multi-dimensional flux-form semi-Lagrangian transport schemes*, Mon. Wea. Rev., **124** (1996), 2046–2070. (Cited on p. 206)

[Li82] P.L. Lions, Generalized Solutions of Hamilton–Jacobi Equations, Res. Notes in Math. 69, Pitman (Advanced Publishing Program), Boston-London, 1982. (Cited on pp. 35, 39, 266)

[LS95] P.L. Lions, P.E. Souganidis, *Convergence of MUSCL and filtered schemes for scalar conservation laws and Hamilton–Jacobi equations*, Numer. Math., **69** (1995), 441–470. (Cited on p. 187)

[LS85] P.-L. Lions, P.E. Souganidis, *Differential games, optimal control and directional derivatives of viscosity solutions of Bellman's and Isaacs' equations*, SIAM J. Control Optim., **23** (1985), 566–583. (Cited on p. 266)

[LS86] P.-L. Lions, P.E. Souganidis, *Differential games and directional derivatives of viscosity solutions of Isaacs' equations* II, SIAM J. Control Optim., **24** (1986), 1086–1089. (Cited on p. 266)

[Lo87] P. Loreti, *Some properties of constrained viscosity solutions of Hamilton–Jacobi–Bellman equations*, SIAM J. Control Optim., **25** (1987), 1244–1252. (Cited on p. 266)

[Lu97] Y. Lucet, *Faster than the fast Legendre transform, the linear-time Legendre transform*, Numer. Algorithms, **16** (1997), 171–185. (Cited on p. 73)

[MS82] J. Macki, A. Strauss, Introduction to Optimal Control Theory, Springer-Verlag, Berlin-Heidelberg-New York, 1982. (Cited on p. 265)

[MPR90] Y. Maday, A.T. Patera, E.M. Rønquist, *An operator-integration-factor splitting method for time-dependent problems: Application to incompressible fluid flow*, J. Sci. Comput., **5** (1990), 263–292. (Cited on p. 206)

[MP94] C. Marchioro, M. Pulvirenti, Mathematical Theory of Incompressible Nonviscous Fluids, Springer, New York, 1994. (Cited on p. 205)

[M70] M.B. Mathur, *A note on an improved quasi-Lagrangian advective scheme for primitive equations*, Mon. Wea. Rev., **98** (1970), 214–219. (Cited on p. 205)

[M84] A. McDonald, *Accuracy of multiply-upstream semi-Lagrangian advective schemes*, Mon. Wea. Rev., **112** (1984), 1267–1275. (Cited on p. 186)

[M87] A. McDonald, *Accuracy of multiply-upstream semi-Lagrangian advective schemes* II, Mon. Wea. Rev., **115** (1987), 1446–1450. (Cited on p. 186)

[M71] A.W. Merz, The Homicidal Chauffeur - A Differential Game, Ph.D. Dissertation, Stanford University, Stanford, CA, 1971. (Cited on pp. 264, 265)

[M02] G.N. Milstein, *The probability approach to numerical solution of nonlinear parabolic equations*, Numer. Methods Partial Differential Equations, **18** (2002), 490–522. (Cited on p. 206)

[MT00] G.N. Milstein, M.V. Tretyakov, *Numerical algorithms for semilinear parabolic equations with small parameter based on approximation of stochastic equations*, Math. Comp., **69** (2000), 237–267. (Cited on p. 206)

[MT01] G.N. Milstein, M.V. Tretyakov, *Numerical solution of the Dirichlet problem for non-linear parabolic equations by a probabilistic approach*, IMA J. Numer. Anal., **21** (2001), 887–917. (Cited on p. 206)

[M98] K.W. Morton, *On the analysis of finite volume methods for evolutionary problems*, SIAM J. Numer. Anal., **35** (1998), 2195–2222. (Cited on p. 206)

[MPS88] K.W. Morton, A. Priestley, E. Süli, *Stability of the Lagrange-Galerkin method with non-exact integration*, RAIRO Modél. Math. Anal. Numér., **22** (1988), 625–653. (Cited on pp. 161, 166, 187)

[NK07] C. Navasca, A. J. Krener, *Patchy solutions of Hamilton–Jacobi–Bellman partial differential equations*, in A. Chiuso et al. (eds.), Modeling, Estimation and Control, Lecture Notes in Control and Inform. Sci., vol. 364, Springer, Berlin, 2007, 251–270. (Cited on p. 268)

[NW06] J. Nocedal, S.J. Wright, Numerical Optimization, Springer-Verlag, New York, 2006. (Cited on p. 73)

[O93] S. Osher, *A level set formulation for the solution of the Dirichlet problem for Hamilton–Jacobi equations*, SIAM J. Math. Anal., 24 (1993), 1145–1152. (Cited on p. 298)

[OF03] S. Osher, R.P. Fedkiw, Level Set Methods and Dynamic Implicit Surfaces, Springer-Verlag, New York, 2003. (Cited on pp. x, 39, 297, 298)

[OS88] S. Osher, J.A. Sethian, *Fronts propagating with curvature-dependent speed: Algorithms based on Hamilton–Jacobi formulations*, J. Comput. Phys., **79** (1988), 12–49. (Cited on p. 298)

[OS91] S. Osher, C.-W. Shu, *High-order essentially nonoscillatory schemes for Hamilton–Jacobi equations*, SIAM J. Numer. Anal., **28** (1991), 907–922. (Cited on p. 187)

[P77] A. Papoulis, Signal Analysis, McGraw–Hill, New York, 1977. (Cited on p. 151)

[PT00] V.S. Patsko, V.L. Turova, *Numerical study of differential games with the homicidal chauffeur dynamics*, Scientific report of the Russian Academy of Science, Ural Branch, Ekaterinburg, 2000. (Cited on p. 267)

[P87] J. Pedlosky, Geophysical Fluid Dynamics, Springer, New York, 1987. (Cited on p. 205)

[PKB93] A.S. Perelson, D.E. Kirschner, R. de Boer, *Dynamics of HIV infection of CD4$^+$ cells*, Math. Biosci., **114** (1993), 81–125. (Cited on p. 244)

[PN99] A.S. Perelson, P.W. Nelson, *Mathematical analysis of HIV-1 dynamics in vivo*, SIAM Rev., **41** (1999), 3–44. (Cited on p. 242)

[P06] G. Perrone, Sulla stabilità dei metodi Semi-Lagrangiani di ordine alto, Tesi di laurea, Università di Roma Tre, 2006. (Cited on pp. 157, 187)

[P94] H.J. Pesch, *Solving optimal control and pursuit-evasion game problems of high complexity*, in Computational Optimal Control (Munich, 1992), Internat. Ser. Numer. Math., vol. 115, Birkhäuser, Basel, 1994, 43–61. (Cited on p. 267)

[Pi82] O. Pironneau, *On the transport-diffusion algorithm and its application to the Navier–Stokes equations*, Numer. Math., **38** (1982), 309–332. (Cited on p. 187)

[PBGM62] L. S. Pontryagin, V. G. Boltyanskii, R. V. Gamkrelidze, E. F. Mishchenko, Mathematical Theory of Control Processes, Interscience, New York, 1962. (Cited on p. 265)

[Pow08] M.J.D. Powell, *Developments of NEWUOA for minimization without derivatives*, IMA J. Numer. Anal., **28** (2008), 649–664. (Cited on p. 73)

[PCF06] E. Prados, F. Camilli, O. Faugeras, *A unifying and rigorous Shape from Shading method adapted to realistic data and applications*, J. Math. Imaging Vision, **25** (2006), 307–328. (Cited on p. 298)

[PF03] E. Prados, O. Faugeras, *"Perspective shape from shading" and viscosity solutions*, in Proc. 9th IEEE Int. Conf. Comp. Vis. (vol. II), Nice, France, 2003, 826–831. (Cited on p. 298)

[PB79] M.L. Puterman, S.L. Brumelle, *On the convergence of policy iteration in stationary dynamic programming*, Math. Oper. Res., **4** (1979), 60–69. (Cited on p. 267)

[QZZ07a] J. Qian, Y.-T. Zhang, H.-K. Zhao, *Fast sweeping methods for eikonal equations on triangular meshes*, SIAM J. Numer. Anal., **45** (2007), 83–107. (Cited on p. 298)

[QZZ07b] J. Qian, Y.-T. Zhang, H. Zhao, *A fast sweeping method for static convex Hamilton–Jacobi equations*, J. Sci. Comput., **31** (2007), 237–271. (Cited on p. 298)

[QS08] J.-M. Qiu, C.-W. Shu, *Convergence of Godunov-type schemes for scalar conservation laws under large time steps*, SIAM J. Numer. Anal., **46** (2008), 2211–2237. (Cited on p. 206)

[QS11] J.-M. Qiu, C.-W. Shu, *Conservative high order semi-Lagrangian finite difference WENO methods for advection in incompressible flow*, J. Comput. Phys., **230** (2011), 863–889. (Cited on pp. 202, 203, 206)

[QV99a] A. Quarteroni, A. Valli, Domain Decomposition Methods for Partial Differential Equations, Oxford University Press, New York, 1999. (Cited on p. 268)

[Ra94] D.A. Randall, *Geostrophic adjustment and the finite-difference shallow water equations*, Mon. Wea. Rev., **122** (1994), 1371–1377. (Cited on p. 195)

[RW90] P.J. Rasch, D.L. Williamson, *On shape-preserving interpolation and semi-Lagrangian transport*, SIAM J. Sci. Statist. Comput., **11** (1990), 656–687. (Cited on p. 73)

[RBS06] M. Restelli, L. Bonaventura, R. Sacco, *A semi-Lagrangian discontinuous Galerkin method for scalar advection by incompressible flows*, J. Comput. Phys., **216** (2006), 195–215. (Cited on p. 206)

[RM67] R.D. Richtmeyer, K.W. Morton, Difference Methods for Initial Value Problems, Interscience, New York, 1967. (Cited on p. 99)

[Ro81] A. Robert, *A stable numerical integration scheme for the primitive meteorological equations*, Atmos. Ocean., **19** (1981), 35–46. (Cited on pp. 145, 206)

[RBDG11] G. Rosatti, L. Bonaventura, A. Deponti, G. Garegnani, *An accurate and efficient semi-implicit method for section-averaged free-surface flow modelling*, Int. J. Numer. Methods Fluids, **65** (2011), 448–473. (Cited on pp. 203, 204)

[RT92] E. Rouy, A. Tourin, *A viscosity solutions approach to shape-from-shading*, SIAM J. Numer. Anal., **29** (1992), 867–884. (Cited on p. 281)

[SS01] M. Sagona, A. Seghini, *An adaptive scheme for the shape from shading problem*, in M. Falcone, Ch. Makridakis (eds.), Numerical Methods for Viscosity Solutions and Application, World Scientific, Singapore, 2001. (Cited on pp. 187, 285)

[SR04] M.S. Santos, J. Rust, *Convergence properties of policy iteration*, SIAM J. Control Optim., **42** (2004), 2094–2115. (Cited on p. 267)

[Sa01] G. Sapiro, Geometric Partial Differential Equations and Image Analysis, Cambridge University Press, Cambridge, UK, 2001. (Cited on p. 298)

[S63] J. Sawyer, *A semi-Lagrangian method of solving the vorticity advection equation*, Tellus, **15** (1963), 336–342. (Cited on pp. 205, 206)

[Sch99] M. Schlather, *Introduction to positive definite functions and to unconditional simulation of random fields*, Technical Report ST-99-10, Department of Mathematics and Statistics, Lancaster University, Lancaster, UK, 1999. (Cited on p. 152)

[Se99a] D. Serre, Systems of Conservation Laws. 1. Hyperbolicity, Entropies, Shock Waves, Cambridge University Press, Cambridge, UK, 1999. (Cited on p. 38)

[Se99b] D. Serre, Systems of Conservation Laws. 2. Geometric Structures, Oscillations, and Initial-Boundary Value Problems, Cambridge University Press, Cambridge, UK, 1999. (Cited on p. 38)

[Se99] J.A. Sethian, Level Set Methods and Fase Moving Methods: Evolving Interfaces in Geometry, Fluid Mechanics, Computer Vision, and Materials Science, Cambridge Monogr. Appl. Comput. Math., vol. 3, Cambridge University Press, Cambridge, UK, 1999. (Cited on pp. x, 39, 244, 293, 294, 297, 298)

[Se96b] J.A. Sethian, *A fast marching level set method for monotonically advancing fronts*, Proc. Natl. Acad. Sci. USA, **93** (1996), 1591–1595. (Cited on pp. 244, 298)

[SV01] J. A. Sethian, A. Vladimirsky, *Ordered upwind methods for static Hamilton-Jacobi equations*, Proc. Natl. Acad. Sci. USA, **98** (2001), 11069–11074. (Cited on p. 298)

[SV03] J. A. Sethian, A. Vladimirsky, *Ordered upwind methods for static Hamilton-Jacobi equations: Theory and algorithms*, SIAM J. Numer. Anal., **41** (2003), 325–363. (Cited on p. 298)

[Sh98] C.W. Shu, *Essentially non-oscillatory and weighted essentially non-oscillatory schemes for hyperbolic conservation laws*, in Advanced Numerical Approximation of Nonlinear Hyperbolic Equations (Cetraro, 1997), Lecture Notes in Math., vol. 1697, Springer-Verlag, Berlin, 1998, 325–342. (Cited on pp. 54, 73)

[So86a] H.M. Soner, *Optimal control with state-space constraint* I, SIAM J. Control Optim., **24** (1986), 552–561. (Cited on pp. 229, 266)

[So86b] H.M. Soner, *Optimal control with state-space constraint* II. SIAM J. Control Optim., **24** (1986), 1110–1122. (Cited on pp. 229, 266)

[ST02a] H.M. Soner, N. Touzi, *A stochastic representation for the level set equations*, Comm. Partial Differential Equations, **27** (2002), 2031–2053. (Cited on p. 298)

[ST02b] H.M. Soner, N. Touzi, *Dynamic programming for stochastic target problems and geometric flows*, J. Eur. Math. Soc. (JEMS), **4** (2002), 201–236. (Cited on p. 298)

[ST03] H.M. Soner, N. Touzi, *A stochastic representation for mean curvature type geometric flows*, Ann. Probab., **31** (2003), 1145–1165. (Cited on p. 298)

[SBG99] E. Sonnendrücker, J. Roche, P. Bertrand, A. Ghizzo, *The semi-Lagrangian method for the numerical resolution of the Vlasov equation*, J. Comput. Phys., **149** (1999), 201–220. (Cited on p. 145)

[S98] E.D. Sontag, Mathematical Control Theory: Deterministic Finite Dimensional Systems, Springer, New York, 1998. (Cited on p. 265)

[Sor92] P. Soravia, *The concept of value in differential games of survival and viscosity solutions of Hamilton–Jacobi equations*, Differential Integral Equations, **5** (1992), 1049–1068. (Cited on p. 267)

[Sor93a] P. Soravia, *Pursuit–evasion problems and viscosity solutions of Isaacs equations*, SIAM J. Control Optim., **31** (1993), 604–623. (Cited on p. 266)

[Sor93b] P. Soravia, *Discontinuous viscosity solutions to Dirichlet problems for Hamilton–Jacobi equations with convex Hamiltonians*, Comm. Partial Differential Equations, **18** (1993), 1493–1514. (Cited on p. 39)

[Sor94] P. Soravia, *Generalized motion of a front propagating along its normal direction: A differential games approach*, Nonlinear Anal., **22** (1994), 1247–1262. (Cited on p. 298)

[Sor98] P. Soravia, *Estimates of convergence of fully discrete schemes for the Isaacs equation of pursuit-evasion games via maximum principle*, SIAM J. Control Optim., **36** (1998), 1–11. (Cited on pp. 251, 267)

[Sor02] P. Soravia, *Boundary value problems for Hamilton–Jacobi equations with discontinuous Lagrangian*, Indiana Univ. Math. J., **51** (2002), 451–477. (Cited on p. 39)

[Sou85a] P.E. Souganidis, *Approximation schemes for viscosity solutions of Hamilton–Jacobi equations*, J. Differential Equations, **59** (1985), 1–43. (Cited on p. 100)

[Sou85b] P.E. Souganidis, *Max-min representations and product formulas for the viscosity solutions of Hamilton–Jacobi equations with applications to differential games*, Nonlinear Anal., **9** (1985), 217–257. (Cited on p. 267)

[SC91] A.N. Staniforth, J. Côtè, *Semi-Lagrangian integration schemes for atmospheric models - A review*, Mon. Wea. Rev., **119** (1991), 2206–2223. (Cited on pp. ix, 145, 206)

[St99] J. Strain, *Semi-Lagrangian methods for level set equations*, J. Comput. Phys., **151** (1999), 498–533. (Cited on p. 298)

[St99a] J. Strain, *Fast tree-based redistancing for level set computations*, J. Comput. Phys. **152** (1999), 664–686. (Cited on p. 298)

[St99b] J. Strain, *Tree methods for moving interfaces*, J. Comput. Phys., **151** (1999), 616–648. (Cited on p. 298)

[St00] J. Strain, *A fast modular semi-Lagrangian method for moving interfaces*, J. Comput. Phys., **161** (2000), 512–536. (Cited on p. 298)

[St01] J. Strain, *A fast semi-Lagrangian contouring method for moving interfaces*, J. Comput. Phys., **170** (2001), 373–394. (Cited on p. 298)

[Str89] J.C. Strickwerda, Finite Difference Schemes and Partial Differential Equations, Wadsworth and Brooks/Cole, Pacific Grove, CA, 1989. (Cited on pp. 99, 145)

[Su80] A.I. Subbotin, *A generalization of the basic equation of the theory of differential games*, Soviet Math. Dokl., **22** (1980), 358–362. (Cited on p. 266)

[Su84] A.I. Subbotin, *Generalization of the main equation of differential game theory*, J. Optim. Theory Appl., **43** (1984), 103–133. (Cited on p. 266)

[Su91] A.I. Subbotin, Minimax and Viscosity Solutions of Hamilton–Jacobi Equations, Nauka, Moscow, 1991 (in Russian). (Cited on p. 267)

[Su93] A.I. Subbotin, *Discontinuous solutions of a Dirichlet type boundary value problem for first order partial differential equations*, Russian J. Numer. Anal. Math. Modelling, **8** (1993), 145–164. (Cited on p. 267)

[Su95] A.I. Subbotin, Generalized Solutions of First-Order PDEs, Birkhäuser, Boston, 1995. (Cited on p. 266)

[SuT86] A.I. Subbotin, A.M. Tarasyev, *Stability properties of the value function of a differential game and viscosity solutions of Hamilton–Jacobi equations*, Probl. Control Inf. Theory, **15** (1986), 451–463. (Cited on p. 266)

[T91] E. Tadmor, *Local error estimates for discontinuous solutions of nonlinear hyperbolic equations*, SIAM J. Numer. Anal., **28** (1991), 891–906. (Cited on p. 100)

[TSY03] A. Tankus, N. Sochen, Y. Yeshurun, *A new perspective [on] Shape-from-Shading*, in Proc. 9th IEEE Int. Conf. Comp. Vis. (vol. II), Nice, France, 2003, 862–869. (Cited on p. 298)

[TSY04] A. Tankus, N. Sochen, Y. Yeshurun, *Perspective shape-from-shading by fast marching*, in Proc. IEEE Conf. Comp. Vis. and Patt. Recog. (vol. I), Washington, D.C., 2004, 43–49. (Cited on p. 298)

[T99] A.M. Tarasyev, *Control synthesis in grid schemes for Hamilton–Jacobi equations*, Ann. Oper. Res., **88** (1999), 337–359. (Cited on pp. 206, 267)

[TUU95] A.M. Tarasyev, A.A. Uspenskiy, V.N. Ushakov, *Approximation schemes and finite-difference operators for constructing generalized solutions of Hamilton–Jacobi equations*, J. Comput. Systems Sci. Internat., **33** (1995), 127–139; translated from Izv. Ross. Akad. Nauk Tekhn. Kibernet, 1994, no. 3, 173–185 (in Russian). (Cited on p. 267)

[TS87] C. Temperton, A.N. Staniforth, *An efficient two-time-level semi-Lagrangian semi-implicit integration scheme*, Q. J. R. Meteorol. Soc., **113** (1987), 1025–1039. (Cited on p. 121)

[TA96] M.M. Tidball, E. Altman, *Approximations in dynamic zero-sum games* I, SIAM J. Control Optim., **34** (1996), 311–328. (Cited on p. 267)

[TG92] M.M. Tidball, R.L.V. Gonzalez, *Fast solution of general nonlinear fixed point problems*, in System Modelling and Optimization (Zurich, 1991), Lecture Notes in Control and Inform. Sci., vol. 180, Springer, Berlin, 1992, 35–44. (Cited on p. 267)

[TPA97] M.M. Tidball, O. Pourtallier, E. Altman, *Approximations in dynamic zero-sum games* II, SIAM J. Control Optim., **35** (1997), 2101–2117. (Cited on p. 267)

[TCOZ03] Y.-H. R. Tsai, L.-T. Cheng, S. Osher, H.-K. Zhao, *Fast sweeping algorithms for a class of Hamilton–Jacobi equations*, SIAM J. Numer. Anal., **41** (2003), 673–694. (Cited on p. 298)

[Ts95] J. N. Tsitsiklis, *Efficient algorithms for globally optimal trajectories*, IEEE Trans. Automat. Control, **40** (1995), 1528–1538. (Cited on p. 298)

[TBR13] G. Tumolo, L. Bonaventura, M. Restelli, *A semi-implicit, semi-Lagrangian, p-adaptive discontinuous Galerkin method for the shallow water equations*, J. Comput. Phys., **232** (2013), 46–67. (Cited on pp. 195, 204, 205)

[VR50] J. Von Neumann, R.D. Richtmeyer, *A method for the numerical calculation of hydrodynamic shocks*, J. Appl. Phys., **21** (1950), 232–237. (Cited on p. 99)

[W59] A. Wiin-Nielsen, *On the application of trajectory methods in numerical forecasting*, Tellus, **11** (1959), 180–196. (Cited on pp. 145, 191, 205)

[XK01] D. Xiu, G.E. Karniadakis, *A semi-Lagrangian high-order method for Navier–Stokes equations*, J. Comput. Phys. **172** (2001), 658–684. (Cited on p. 206)

[ZTCS99] R. Zhang, P.-S. Tsai, J. E. Cryer, M. Shah, *Shape from shading: A survey*, IEEE Trans. Pattern Anal. Mach. Intell., **21** (1999), 690–706. (Cited on p. 298)

[ZS02] Y.-T. Zhang, C.-W. Shu, *High-order WENO schemes for Hamilton–Jacobi equations on triangular meshes*, SIAM J. Sci. Comput., **24** (2003), 1005–1030. (Cited on p. 187)

[Z05] H. Zhao, *A fast sweeping method for eikonal equations*, Math. Comp., **74** (2005), 603–627. (Cited on p. 298)

Index